Modern Automotive Antenna Measurements

For a listing of recent titles in the
Artech House Antennas Library,
turn to the back of this book.

Modern Automotive Antenna Measurements

Lars J. Foged
Manuel Sierra Castañer

Editors

ARTECH HOUSE
BOSTON | LONDON
artechhouse.com

Library of Congress Cataloging-in-Publication Data
A catalog record for this book is available from the U.S. Library of Congress.

British Library Cataloguing in Publication Data
A catalog record for this book is available from the British Library.

ISBN-13: 978-1-63081-849-4

Cover design by Andy Meaden Creative

685 Canton Street
Norwood, MA 02062

10 9 8 7 6 5 4 3 2

Contents

CHAPTER 1
Introduction to Automotive Antenna and Device Measurements 1
References 4

CHAPTER 2
Challenges and Figures of Merit in Automotive Measurements 7
2.1 Antenna Measurement 7
2.1.1 Antenna Gain and OTA Measurements 8
2.1.2 Calibration Antenna 12
2.1.3 Testing Parameters 14
2.2 NF versus FF Measurements 18
2.3 Introduction to Near Transformation Theory 23
2.4 Radiofrequency System 25
References 27

CHAPTER 3
NFFF Transformations 29
3.1 NF/FF Transformation based on Spherical Wave Expansion 30
3.1.1 SWE 30
3.1.2 Truncated SNF Measurements 43
3.1.3 Advanced SWE Techniques 49
3.2 Equivalent Current Technique 60
3.2.1 Formulation 60
3.2.2 Features 61
3.3 New Methods for NF-to-FF Transformations 63
3.3.1 Sparse Recovery of SWC 64
3.3.2 Optimal Sampling Interpolation 64
References 65

CHAPTER 4
Chamber Design and Analysis 71
4.1 Measurement Ranges 71

4.2 Design Criteria 73
4.2.1 Absorbers and Ferrites 73
4.2.2 Rectangular Chambers 76
4.2.3 Tapered Chambers 79
4.2.4 Chambers with Conductive Floors 80
4.3 Analysis Methods 81
4.3.1 Full-Wave Models 81
4.3.2 Ray-Tracer Solvers 81
4.3.3 Image Theory–Based Technique 83
4.3.4 Measurements with the Scaled Model Method 85
4.4 QZ Evaluation with Measurements 90
4.4.1 Field Probe Measurements Over QZ 90
4.4.2 Antenna Pattern Comparison 91
References 93

CHAPTER 5
Implementation and Calibration of Automotive Ranges 97
5.1 Absorber and Conductive Floor Systems 98
5.1.1 Gain Calibration 99
5.1.2 Gain Calibration Accuracy 101
5.1.3 Measurement Examples 103
5.2 FF Ranges 106
5.3 Single-Probe NF Systems 108
5.3.1 Use of Wideband Probes 111
5.3.2 Use of Wideband and Dual-Polarized Probes 114
5.4 Multiprobe NF Systems 120
5.4.1 Probe-Related Design Challenges 121
5.4.2 Sampling in MP Systems 122
5.4.3 Different Implementations of Automotive MP Systems 123
5.4.4 Low-Frequency Measurements 124
5.4.5 High-Frequency Measurements 125
5.5 Scaled Measurements 129
References 131

CHAPTER 6
OTA Measurements 133
6.1 OTA Test Setup 133
6.2 OTA Testing Parameters 134
6.3 OTA System Calibration 135
6.4 OTA Measurement Methods 136
6.4.1 NF to FF Transform Using Phase Recovery 136
6.4.2 Separate Measurements with Conducted RF 137
6.4.3 Two-Step Measurement or Combinational Method 138
6.4.4 Direct OTA Measurements 138
6.4.5 NF to Quasi-FF Transform Through Parallax 139
6.4.6 TRP/TIS Testing of Devices on a Limited Ground Plane 140

6.5 OTA Sampling 141
6.6 OTA Measurement Examples 142
6.7 MIMO OTA Testing 146
6.7.1 Introduction 146
6.7.2 Direct MIMO OTA Testing 147
6.7.3 Two-Stage Methods 150
References 151

CHAPTER 7
Advanced Post-Processing Techniques 155
7.1 Post-Processing by the Equivalent Current Method 156
7.1.1 Diagnostics and Filtering 156
7.1.2 Extrapolation of Truncated Areas 158
7.1.3 NF Calculation from the Equivalent Current 160
7.1.4 Link Between Measurements and Simulations 160
7.2 Investigation and Mitigation of Truncation Errors in Free-Space Automotive Systems 163
7.2.1 Case Study Description and Investigation of Truncation Errors Effect 163
7.2.2 Mitigation of Truncation Errors 165
7.3 Free-Space Response Retrieval in PEC-Based Automotive Systems 168
7.3.1 Spatial Filtering Techniques for PEC Removal 168
7.3.2 Example with Scaled Measurements 170
References 174

CHAPTER 8
Virtual Drive Testing 177
8.1 VDT 177
8.2 The Link Between Measurement and Simulation 180
8.3 Simulation of Complex Scenarios from Measurements 181
8.4 Different Ground Emulation of Measured Vehicle Antennas 183
8.5 V2V and V2X Coupling Evaluation 190
References 192

CHAPTER 9
In-Situ Acquisition Systems for Automotive Measurements: Drone and Handheld Approaches 195
9.1 Introduction 195
9.2 Airborne-Based Antenna Measurement Techniques 197
9.2.1 An Overview of Airborne-Based Antenna Measurement Techniques 197
9.2.2 Antenna Characterization by Means of Equivalent Currents 199
9.2.3 Application Example 202
9.3 Handheld Systems 208
9.3.1 An Overview of the Handheld System 208
9.3.2 Handheld System Description 208

9.3.3 Particular Characteristics of the Handheld System 210
9.3.4 Application Example 212
9.4 Final Remarks 213
References 216

CHAPTER 10
Practical Aspects of Automotive Measurements and Virtual-Drive Testing 221
10.1 Challenges for Automotive Antenna Measurements 221
10.1.1 Limitations of Automotive Antenna Measurement Range 221
10.1.2 Implications of the Car Body for Antenna Design and Measurements in the Installed State 225
10.2 Application-Oriented Post-Processing of Automotive Antenna Measurements 229
10.2.1 Phase Center Determination 229
10.2.2 Automotive Antenna Performance Indicators 232
10.3 Over-the-Air Vehicle-in-the-Loop Approach for System Validation in a Virtual Environment 234
10.3.1 Concept 234
10.3.2 LTE and V2X Emulation 236
10.3.3 Automotive Radar 240
References 246

About the Editors 253
About the Contributors 254
Index 259

CHAPTER 1

Introduction to Automotive Antenna and Device Measurements

Lars J. Foged and Manuel Sierra Castañer

Antenna and electromagnetic characterization have been challenging topics for more than 50 years. Antenna measurements are commonly an evaluation of the radiation properties in the far field (FF), assuming that the antenna or device origin is positioned at an infinite distance from the observation point. The FF condition is also often referred to as plane wave condition since radiation tends to approximate a plane wave at infinity. Far from its origin, electromagnetic radiation will have spread out enough that it will appear to have uniform amplitude and phase on a plane perpendicular to its direction of travel. Thus, at sufficient distances, an antenna will appear to radiate plane waves in all directions. Due to reciprocity, the FF radiation of an antenna can also be measured by exposing the entire antenna locally to an approximate plane wave as shown in Figure 1.1.

Thriving to perform the measurements in sufficiently good approximation to FF condition is a paradox since a major part of communications supported by antennas and devices with antennas occurs at a finite distance and often in the near

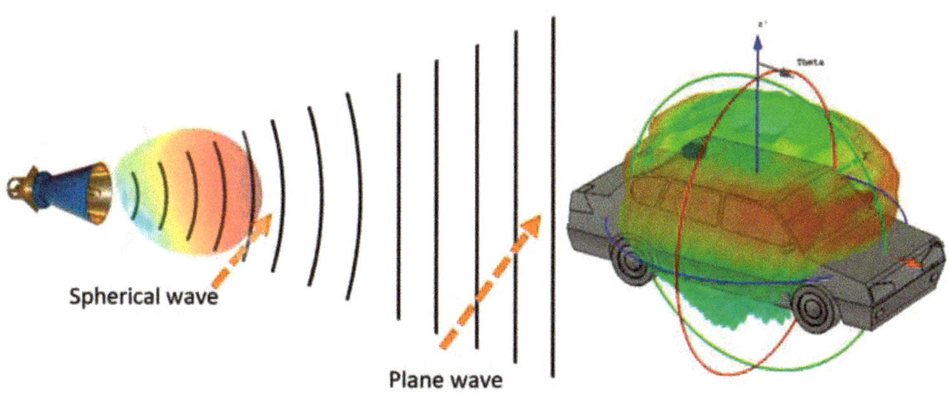

Figure 1.1 Example of automotive measurement by exposing the vehicle-mounted antenna to an approximate plane wave.

field (NF). However, the FF state is a convenient reference condition that allows for traceable and comparable results from different measurement ranges and systems.

Conventional performance parameters, such as gain, directivity, and radiation pattern, are defined in the FF state [1]. Since the main antenna electrical parameters to be characterized are defined and discussed by IEEE and ETSI standards, these are often considered the reference documents for test engineers [2–6]. With the focus on over-the-air (OTA) testing [6–8] different parameters are becoming more relevant such as radiated power and sensitivity equivalent isotropic radiated power (EIRP), total radiated power (TRP), effective isotropic sensitivity (EIS), and total isotropic sensitivity (TIS). Other dedicated organizations, such as the 5G Automotive Association (5GAA) [9], the 3G/5G Partnership Project (3GPP/5GPP) [10], and the Cellular Telecommunications and Internet Association (CTIA) [11] have developed documents with proposed standards for specific measurements.

The trend in automotive applications is for stronger systems integration of the antenna and devices with the vehicle. We thus see more and more hidden antennas in the vehicle body and antennas and devices spread around on the vehicles. Common positions for antenna/device placements and integration, as shown in Figure 1.2, include front and rear windows, side mirrors, front and rear bumpers, under the dash, and on the roof.

The full spectrum for automotive antenna systems is vast, as illustrated in Figure 1.3. The measurement frequency range covered by this book is limited to what can be measured in an anechoic chamber, which is in the range 50 MHz–100 GHz. For the purposes of this book, anything above and below this range is considered a special measurement technique.

The desired FF measurement condition is not always possible to achieve in a simple test setup due to the widely different antenna positions and wide frequency range. This book provides an overview of state-of-the art measurement techniques to determine antenna and antenna device performances. The book is divided into eleven chapters, combining theory and application. The main challenge in automotive antenna/device measurements today is to approximate the desired free-space, FF performance of the antenna under test (AUT)/device under test (DUT) and to reduce the influence of the probe/range antenna, measurements setup, and environment on the measured quantities. In Chapter 2 the pertinent test parameters

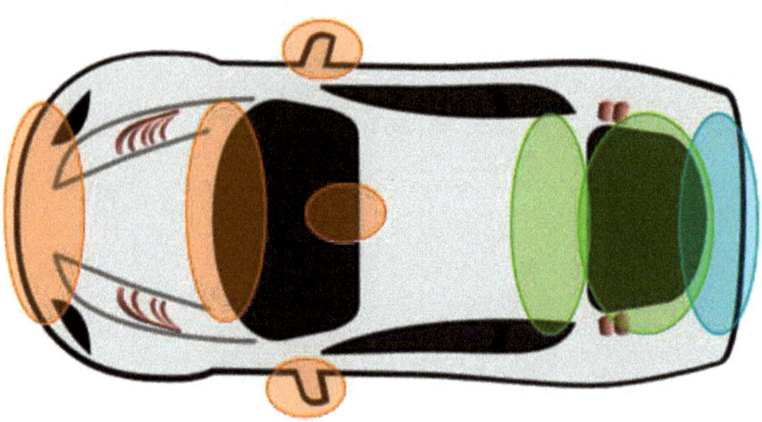

Figure 1.2 Illustration of common antenna positions on vehicles.

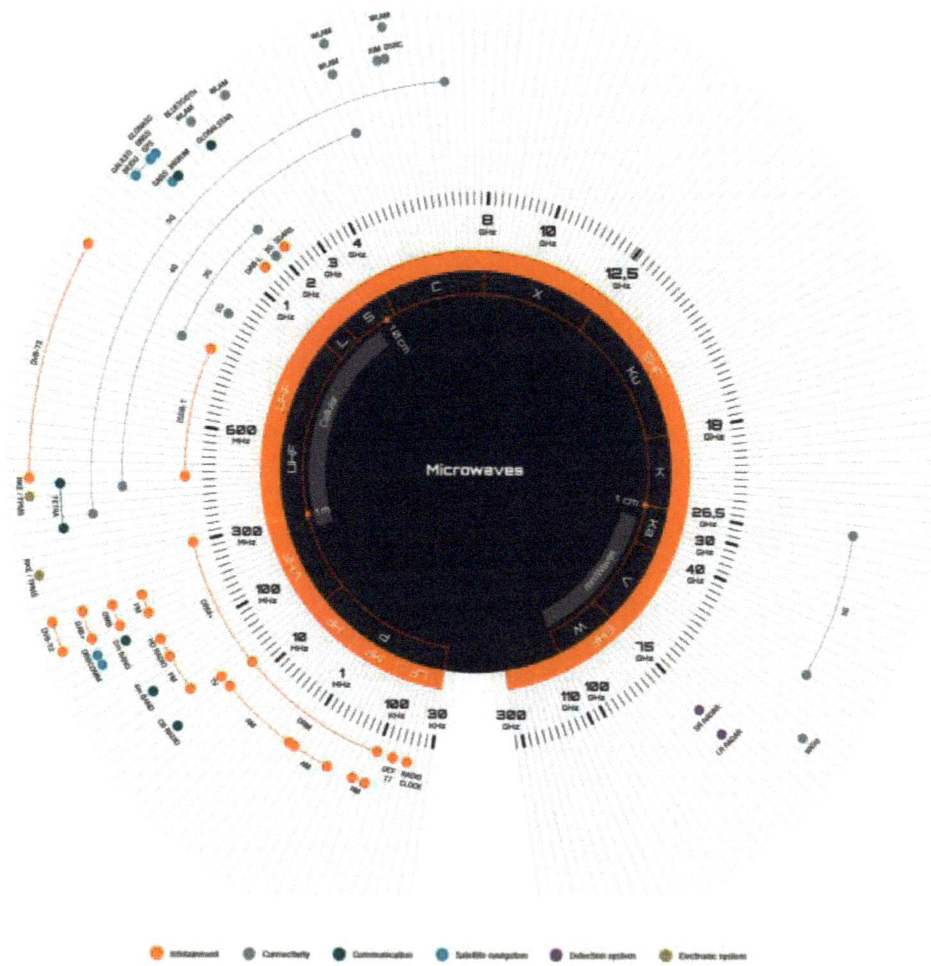

Figure 1.3 The frequency spectrum occupied by modern automotive antennas.

for both passive and OTA automotive measurements are introduced. We also discuss the need for NF to FF (NFFF) transformation to determine the desired FF performances.

The underlying theory for modern NFFF transformation is explained in Chapter 3, which discusses the classical spherical mode theory, the equivalent currents methods, and new methods such as helicoidal sampling and transformation from noncanonical sampling. Chapter 3 is also an introduction to the modern post-processing techniques that are further discussed in Chapter 7 [12].

Chapter 4 is dedicated to the measurement environment, such as chamber analysis and design, and provides a summary of the kind of anechoic chambers for automotive measurements, the specifications of the absorbing material, and the evaluation methods of the quiet zones. In practical systems, the relative movement of the probe/range antenna and the AUT/DUT can be performed in different ways. Automotive systems measure the full sphere or hemispherical space by angular scanning using a representative ground-material. Angular scanning is often achieved by a combination of two mechanical rotations, one for the probe/range antenna and one for the AUT/DUT. In many cases it can be advantageous to replace the slower mechanical movement of a single probe in a single-probe (SP) system, such as gantry arm or arch rail system, with an array of probes as in a multiprobe (MP) system. This is discussed in detail in Chapter 5.

Chapter 6 focuses on OTA measurements, introducing parameters such as TIS and TRP and new environments close to actual communication systems. Chapter 7 introduces advanced post-processing techniques in antenna measurements. The use of equivalent currents can improve the measurement results through mitigation of some errors (i.e., echoes) or extrapolation of the radiation pattern. Also, they can be used to emulate free-space conditions from perfect electric conductor (PEC) measurements. Chapter 8 introduces the concept of virtual drive testing. Exhaustive testing consumes significant amounts of time and money, while, through post-processing, different situations can be emulated. For example, the emulation of any kind of ground, but combining the measurements with simulation tools, complex scenarios for vehicle-to-vehicle (V2V) or vehicle-to-everything (V2X) can be emulated.

Chapters 9–10 deal with applications where this book benefits from the experience of two reference laboratories in automotive measurements. Chapter 9 explains two useful alternatives for automotive measurements from the University of Oviedo; the first uses drones to deal with problems of reference, localization, acquisition trajectories, and sampling. The second alternative involves the use of hand-held systems working in the extreme NF close to the vehicle. The acquisition scheme, the tracking, the probes, and the post-processing are the key elements in hand-held systems. Finally, in Chapter 10, Ilmenau University brings its practical experience in automotive measurements to the challenges of implementation, post-processing, performance indicators, and emulation of real environments.

References

[1] IEEE Standard Definitions of Terms for Antennas, IEEE Std 145-2013.

[2] IEEE Recommended Practice for Antenna Measurements, IEEE Std 149-2021. (Revision of IEEE Std 149-1977 Standard Test Procedures for Antennas.)

[3] IEEE Recommended Practice for Near-Field Antenna Measurements. IEEE Std 1720-2012.

[4] IEEE Std 802.11a-1999, Part 11: Wireless LAN Medium Access Control (MAC) and Physical Layer (PHY) Specifications, High-Speed Physical Layer in the 5GHz Band.

[5] IEEE Std 802.11pTM-2010; Part 11: Wireless LAN Medium Access Control (MAC) and Physical Layer (PHY) Specifications; Amendment 6: Wireless Access in Vehicular Environments.

[6] ETSI EN 302 571 V2.1.1 (2017-02), Intelligent Transport System (ITS), Radiocommunications equipment Operating in the 5 855 MHz to 5 925 MHz Frequency Band, Harmonised Standard Covering the essential requirements of article 3.2 of Directive 2014/53/EU.

[7] Huang, Y., *Antennas: From Theory to Practice* (Second Edition), Hoboken, NJ: John Wiley and Sons, 2021.

[8] Loh, T. H. (ed.), *Metrology for 5G and Emerging Wireless Technologies,* London: SciTech Publishing (Institution of Engineering and Technology Electromagnetic Waves series), 2021.

[9] 5GAA, TR P-180092, 5G Automotive Association, Working Group Evaluation, test beds and pilots: "V2X Functional and Performance Test Procedures—Elected Assessment of Device to Device Communication Aspects."

[10] 3GPP, TR 25.914, "Measurement of Radio Performances for UMTS Terminals in Speech Mode."

[11] CTIA, OTA Test Plan version 3.9.1, May 2020, www.ctia.org.

[12] Sierra-Castaner, M., and L. J. Foged (eds.), *Post-Processing Techniques in Antenna Measurement* , London: SciTech Publishing (Institution of Engineering and Technology Electromagnetic Waves series), 2019.

CHAPTER 2

Challenges and Figures of Merit in Automotive Measurements

Lars J. Foged, Francesco Saccardi, Alessandro Scannavini, and Manuel Sierra Castañer

This chapter introduces NF versus FF measurement techniques applied to automotive antennas. Some important concepts necessary to understand the following chapters of the book are described in this chapter. The main antenna measurement metrics in classical systems and those more specific to over-the-air (OTA) measurement setups are discussed. In addition, the chapter introduces some of the differences between NF and FF measurements, beginning with measurement distance, attenuation, sampling, and other relevant aspects of the NF to FF transformation algorithm. The chapter concludes with a discussion of basic radiofrequency systems and instrumentation.

2.1 Antenna Measurement

The first successful attempt to measure electromagnetic radiation in free space or antenna patterns as we call them today was made by Heinrich Hertz in 1886 [1]. In the Hertz experiment, he confirmed Maxwell's theories about the existence of electromagnetic radiation. In fact, the unit of frequency, one cycle per second, was named "hertz" in recognition of this discovery and his other pioneering breakthroughs. Hertz used a spark transmitter consisting of a dipole antenna with a spark gap powered by high-voltage pulses from a Ruhmkorff coil as his AUT or DUT. The receiver in the experiment was a loop antenna with a micrometer spark gap between the elements. He was able to determine the radiation intensity as a function of spatial direction by observing the intensity of the fields (sparks) in the receiver. Today, the technologies employed in antenna measurements have greatly improved. However, even today, the spatial radiation from an antenna or device is still measured as the coupling between the source of the radiation and a probe/range antenna [2–4]. An example of a typical measurement setup is shown in Figure 2.1. In most automotive measurements, the movements of the DUT are often limited to a simple azimuth rotation while the probe/range antenna is moved in elevation. A convenient representation of the 3-D pattern radiated by an antenna or device is the spherical coordinate system. The antenna or device measurement data is collected on a surface, which is often spherical; this same surface also defines the reference coordinate system of the measurement. As we will see later, the discrete positions of the moving probe can be substituted by an array of highly similar probes that can be scanned electronically to achieve fast measurements [5, 6].

Both passive and active devices can be measured in a system like the one shown in Figure 2.1. Passive devices can be connected to an external signal source or receiver (connectorized device), whereas the active devices have an internal signal generator or receiver. For example, antennas of modern wireless devices such as cell phones fall into the active device category. Antennas on active devices are integrated with the whole RF system; hence their performance is affected by the internal active components and the structure of the device. Measurements of wireless devices in active mode, typically called OTA testing, are studied and standardized by different organizations such as the 3GPP [7–9] and CTIA [10]. Currently, some wireless systems are embedded on board the vehicle with a telematic control unit (TCU). This enables the vehicle to wirelessly connect to cloud services or other vehicles via V2X standards over a cellular network. Antenna measurements, and specifically OTA measurements of systems consisting of antennas and TCUs on vehicles are gaining momentum in the automotive industry. Organizations, such as 5GAA, have published technical reports that standardize passive and OTA measurements of vehicle-mounted antennas [11]. To measure any active and/or passive device the measurement system must be calibrated using a well-charactered antenna with gain or efficiency as the reference.

2.1.1 Antenna Gain and OTA Measurements

Classical antenna gain measurements are based on the Friis transmission formula shown in (2.1). Harald Trap Friis, a Danish scientist working at the United States–based Bell Laboratories, made significant contributions to radio propagation, radio astronomy, and radar. His two Friis formulas on noise and transmission are still

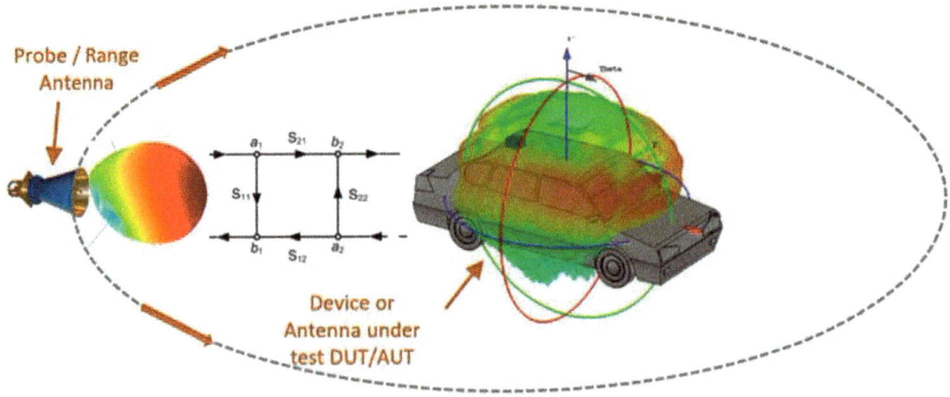

Figure 2.1 Conceptual illustration of a typical automotive measurements scenario. The coupling between the probe/range antenna and the DUT is measured in different spatial directions.

widely used today [12]. Applying the Friis transmission formula to the scenario in Figure 2.1, it links the input power of the transmitting antenna to the output power of the receiving antenna, which is proportional to the wavelength square and the gains of the two antennas and inversely proportional to the square of the separation distance between the Probe and DUT. The term $\left(\frac{4\pi R}{\lambda}\right)^2$, also called the free-space path loss or pathloss, $L_{freespace}$, results from the spherical spreading of the radiated energy.

$$S_{21}^{dB} = Power_{OUT}^{dBm} - Power_{IN}^{dBm} = G_{Probe}^{dB} + G_{DUT}^{dB} - 10log\left(\frac{4\pi R}{\lambda}\right)^2 \quad (2.1)$$

The powers on the left side of (2.1) are expressed in decibels relative to one milliwatt (dBm), a decibel-based unit of power in which 0 dBm is equivalent to 1 mW of power. The subscripts "*OUT*" and "*IN*" indicate the output/received and input/transmitted powers so that the equation is valid for both a transmitting and receiving DUT. For a transmitting DUT, $Power_{OUT}^{dBm}$ is the power received by the probe and $Power_{IN}^{dBm}$ is the power available at the DUT port. For a receiving DUT, $Power_{OUT}^{dBm}$ is the power received by the DUT, and $Power_{IN}^{dBm}$ is the power available at the probe port. The antenna gain considered in (2.1) is *realized gain* [2], meaning that it is defined considering the input power at the antenna terminals. Other versions of the Friis equation instead consider the *absolute gain,* defined considering the power accepted by the antenna. When the absolute gain is used, a perfect matching of two antennas is assumed. In practice, it is more convenient to consider the *realized gain* since it already includes the mismatch losses of the antennas. Alternatively, if the reflection coefficients of the antennas (Γ) are known, a correction factor of the form $1-|\Gamma|^2$ can be inserted in the Friis formula to correct for mismatch [4]. Throughout this book and unless implicitly stated, we will use *realized gain G* in derivations, since this is the antenna parameter we can measure directly.

Equation (2.1) is valid if the probe and/or the DUT are perfectly matched in polarization; polarization mismatch can be quantified and properly inserted in the equation. In general, it is advisable to use well-matched antennas and align the polarization of the system when performing these measurements.

The measured coupling or S21 of the probe and DUT is directly linked to the gain of the two antennas and the distance between them as shown in (2.1). It is worth noticing, that, at a separation distance of just one wavelength, the path loss is already –22 dB. Each time the separation distance increases by a factor of 2, the path loss increases by 6 dB. The examples of free-space path loss/attenuation shown in Figure 2.2 indicate that such attenuations increase with the frequency and the distance.

A widely used gain measurement technique in both FF and NF is the *gain-substitution* or *gain-transfer technique* [3, 4]. This technique requires two measurements, the first being a measurement with a calibrated (known gain) reference antenna, as shown in Figure 2.3. Any calibrated antenna with known gain or efficiency characteristics at the desired frequency can be used as a reference.

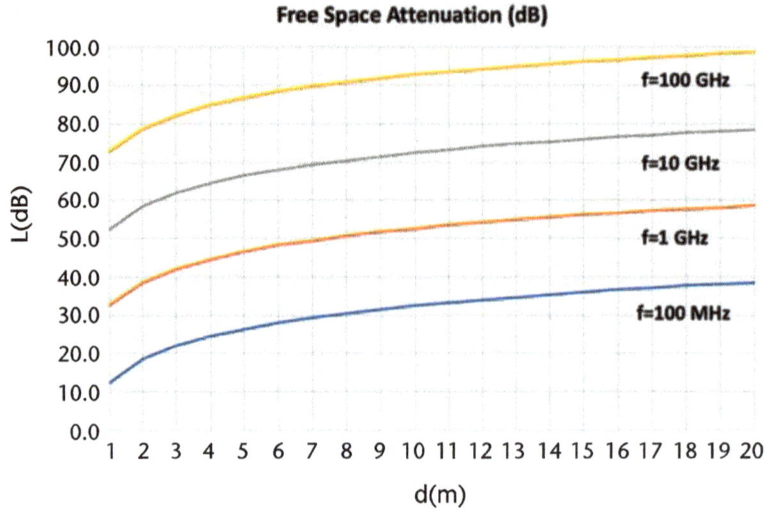

Figure 2.2 Free-space path loss attenuation depending on the distance and frequency.

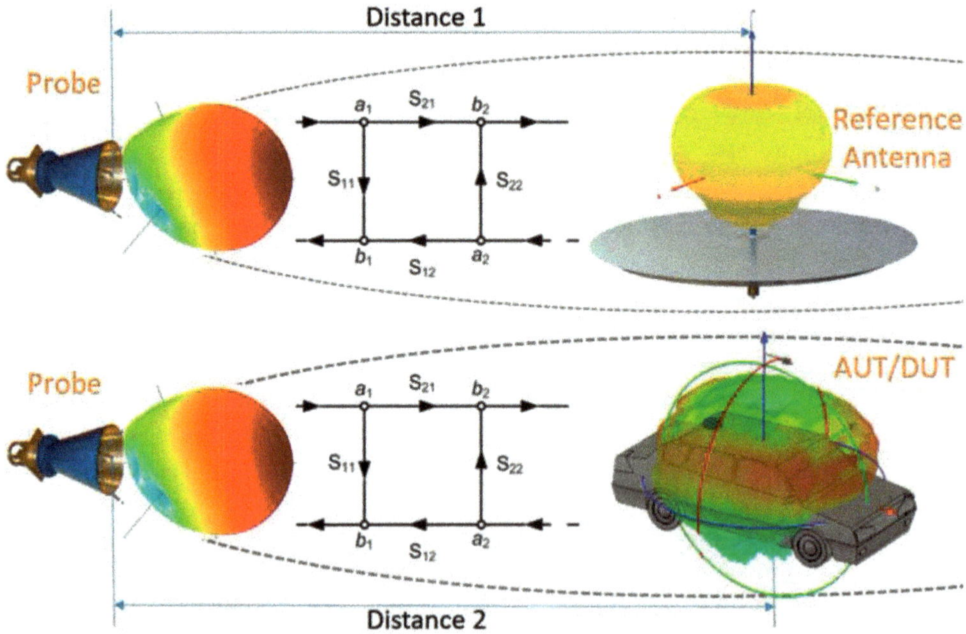

Figure 2.3 System calibration by the gain-substitution or gain-transfer technique. A reference antenna with well-known performance characteristics is used to calibrate the system.

The gain of the unknown DUT is expressed in terms of the known gain of the reference antenna G_{REF} and the measured probe antenna coupling in the two measurements as shown in (2.2) to (2.4).

$$S_{21\,REF}^{dB} = G_{Probe}^{dB} + G_{REF}^{dB} - L_{freespace,REF}^{dB} \tag{2.2}$$

$$S^{dB}_{21\,DUT} = G^{dB}_{Probe} + G^{dB}_{DUT} - L^{dB}_{freespace,DUT} \tag{2.3}$$

$$G^{dB}_{DUT} = S^{dB}_{21\,DUT} - S^{dB}_{21\,REF} + G^{dB}_{REF} - \Delta L^{dB}_{freespace,DUT-REF} \tag{2.4}$$

Often the term $\Delta L^{dB}_{freespace,DUT-REF}$ is considered negligible if the two measurements are performed at the same distance. The term C^{dB}_{Cal} is often called the *system calibration* term, as it gives the relationship between power in/out and power out/in of the probe with an isotropic radiator after system calibration. An isotropic radiator is an ideal, lossless antenna that radiates power equally in all directions and therefore has a 0-dBi gain.

$$C^{dB}_{CAL} = S^{dB}_{21REF} - G^{dB}_{REF} = Power^{dB}_{Probe} - Power^{dB}_{DUT\ Isotropic} \tag{2.5}$$

After the system is calibrated, the gain pattern of the unknown DUT can be determined in all spatial directions using the system calibration term C^{dB}_{Cal} with the measured coupling values:

$$G^{dB}_{DUT} = S^{dB}_{21DUT} - C^{dB}_{CAL} \tag{2.6}$$

The system calibration in (2.5) can also be performed using the known efficiency ε_{REF} of a reference antenna given that the 3-D integral of the antenna gain is the antenna efficiency. The following can be used to determine C^{dB}_{Cal}:

$$\varepsilon_{REF} = \frac{1}{4\pi}\iint_{\theta,\phi} G_{REF}(\theta,\phi)\sin(\theta)\,d\theta d\phi \tag{2.7}$$

$$C^{dB}_{CAL} = \left(\frac{1}{4\pi}\iint_{\theta,\phi} S_{21REF}(\theta,\phi)\sin(\theta)\,d\theta d\phi\right)^{dB} - \varepsilon^{dB}_{REF} \tag{2.8}$$

The calibration using efficiency in (2.8) requires a full 3-D or partial spherical measurement of the calibration antenna/device. For automotive applications, typically only half of the sphere is measured [11, 13]. The advantage of this approach is that efficiency is less impacted by spurious radiation in the measurement setup, given that these contributions are averaged over the full measurement sphere. Often it is easier, or more convenient, to find reference antennas with accurate efficiency information, whereas boresight or peak gain information may be less reliable. It is worth highlighting that in most automotive antenna measurement systems, only the radiation within the upper hemisphere is measured. In such cases, the so-called upper hemisphere efficiency (UHE) should be used instead of the full 3-D efficiency ε to achieve a more accurate calibration of the system. The UHE can be obtained using (2.7), truncating the integration domain at the horizon. This is discussed in further detail in Chapter 5.

It is good practice to use reference antennas that are as similar as possible to the antenna to be measured. Most automotive antennas are low directive devices, and as such, they can interact significantly with the measurement system. As described in Section 2.1.2, monocones and monopoles mounted on ground planes are typical calibration antennas for automotive systems.

The *system calibration* term C_{Cal}^{dB} in (2.5) is the relationship between power in/out of the probe and power out/in of a theoretical transmitting or receive isotropic DUT. It can be rewritten as:

$$C_{CAL}^{dB} = Power_{Probe}^{dB} - Power_{DUT\ Isotropic}^{dB} \tag{2.9}$$

As will be discussed later, for a radiating DUT, the isotropic radiated power, $Power_{DUT\ Isotropic}^{dB}$ is also the effective isotropic radiated power also known as EIRP. This relationship can be used to determine EIRP as:

$$EIRP^{dB} = Power_{DUT\ Isotropic}^{dB} = Power_{Probe}^{dB} - C_{CAL}^{dB} \tag{2.10}$$

Antenna and device performance terms are all derived from gain, efficiency, and transmitted/received power. The *system calibration* is also sometimes called *OTA system calibration,* and measurement setups such as the one shown in Figure 2.3 are often referred to as a *calibrated OTA setup*. The purpose of *OTA system calibration* is to determine the isotropic power "into" or "out of" the unknown DUT antenna, for a given measurement distance. With a calibrated OTA setup the measured power parameters are all relative to isotropic radiation. In OTA testing, spatial power quantities related to radiated power and device sensitivity are typically measured to characterize the transmitting and receiving properties of a device. To accomplish this, an external device capable of establishing a communication and control link with the AUT/DUT is required. This external device can be a wireless/mobile terminal or base-station antenna. The need of a control link makes OTA testing slightly more complicated than traditional antenna testing.

2.1.2 Calibration Antenna

It is considered good practice to calibrate the system with a reference antenna with pattern properties similar to the unknown DUT or antenna. This, of course, is difficult if the DUT is truly unknown. However, as the purpose of most automotive antennas is to provide full or partial coverage in the upper hemisphere, monocone antennas on a limited ground plane (GP), as shown in the top part of Figure 2.4, are considered a good choice as a calibration standard. The calibration antenna must have well-known performance in terms of gain, efficiency, and/or pattern. For low-frequency standards, such as the two antennas working in the 70–220-MHz and 220–700-MHz frequency ranges shown in Figure 2.4, the possibilities to calibrate in a reference facility are very limited. An alternative for determining the accurate performance of calibration antennas at FM frequencies is to perform measurements on accurately scaled models. Recommendations for how to define such scaled models and their measurements are reported in Chapter 5. While the dimensions and dielectric material properties are readily controlled during manufacturing, the conductivity and, thus, the losses cannot easily be scaled. As ohmic losses are proportional to the conductivity, we can expect the losses in higher-frequency antennas to be higher than the scaled model at lower frequencies. The measured losses of a

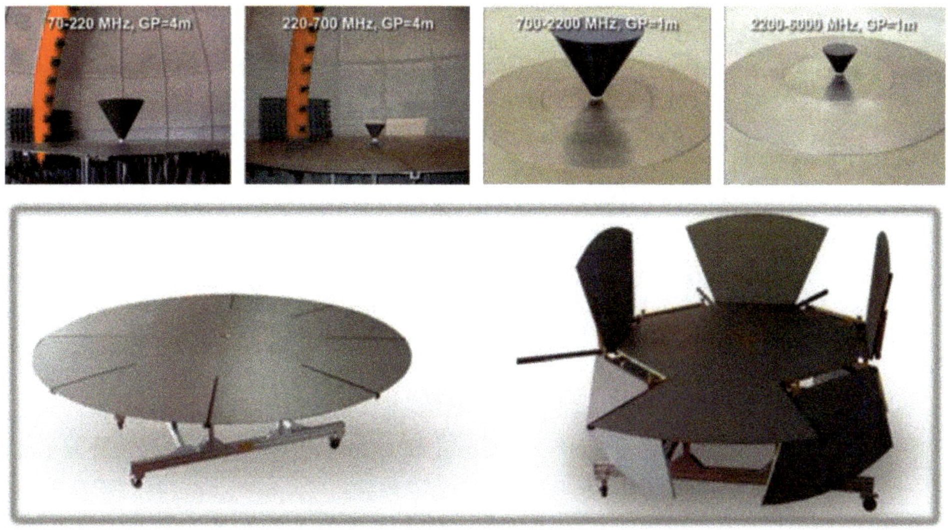

Figure 2.4 Example of monocone antennas on a limited ground plane for automotive system calibration from 70 MHz to 6 GHz, with the top figure depicting ground planes of different sizes used at different bands and the bottom depicting a deployable mechanism used for large ground planes.

higher-frequency antenna thus constitute the upper bound on antenna losses of the scaled models at lower frequencies.

An example of a scaled antenna model to determine the reference data of monocone antennas on a limited ground plane is shown in Figure 2.5. The antenna at the bottom covers the band from 70 to 220 MHz and has a 4m-diameter ground plane that makes it difficult to measure in a standard reference facility. The antenna at the top is an exact 10:1 scaled model (hence with a 40-cm-diameter ground

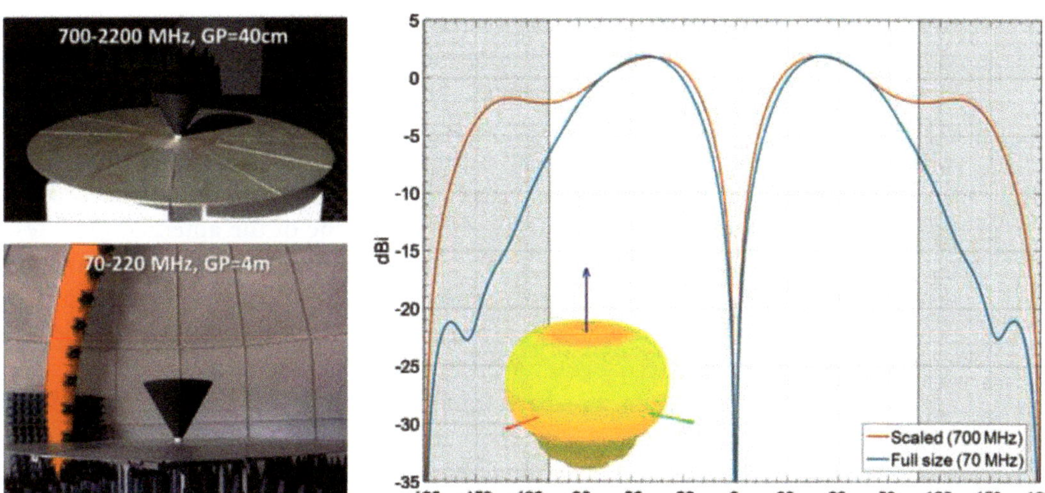

Figure 2.5 A 10:1 scaled model of a calibration antenna, with a 700–2,200-MHz monocone on a 40-cm-diameter GP versus a 70–220-MHz monocone on a 4-m-diameter. The two antennas radiate the same nominal 3-D radiation pattern. The gain radiation pattern cuts are obtained by measuring the scaled antenna in a full spherical reference facility and the full-size antenna in a typical truncated automotive range, respectively.

plane) covering the 700–2,200-MHz frequency range. This antenna can be easily calibrated in a standard reference facility. The losses in well-designed, nonresonant antennas are very low, in particular if dielectric materials are avoided (i.e., the losses in the 700–2,200 MHz monocone antennas have been measured to be less than 0.4 dB). Thus, the associated uncertainty of the gain and efficiency values of the 70–220-MHz antenna attained by not scaling the material conductivity is less than this value. This is considered an acceptable uncertainty for calibration antennas at these frequencies. The nominal 3-D radiation pattern of the two monocone antennas is shown in Figure 2.5. The reported gain pattern cuts are obtained from real measurements of the scaled antenna in a full spherical reference facility at 700 MHz (orange trace) and from the full-size antenna in a typical truncated automotive range at 70 MHz (blue trace). The 70-MHz measurements also show the effect of scan area truncation, not present in the 700-MHz measurements. More details of this example are reported in Chapter 5, but the good agreement of the two measured patterns in the upper hemisphere (e.g., $|\theta| < 90^\circ$) can be easily appreciated.

2.1.3 Testing Parameters

The family of antennas and devices can be roughly divided into active and passive antennas [2]. The passive antenna does not have an active circuit within its structure. An active antenna is a single-port or multiport antenna system with one or more devices integral to the antenna that influence the radiating and/or receiving characteristics. The amplifier is an example of such a device, but other powered, nonreciprocal, analog, or digital components may be employed. The passive antenna testing aims at metrics that characterize the antenna, whereas active testing aims at characterizing the antenna system as a device. To evaluate the field performance of an active DUT, the entire RF system must be tested to determine the interaction between the antenna and the structure in which it is integrated. When assessing the performance of the DUT as a system, the testing metrics are those that determine the system performance. Some examples of these types of systems are wireless communication devices such as mobile phones and base stations. Examples of traditional DUT testing parameters for both active and passive antenna testing are listed in Table 2.1.

OTA testing of wireless devices involves determination of transmitter and receiver characteristics, such as power, sensitivity, and, in some cases, data throughput. The testing includes the effect of the vehicle on the of the antenna. OTA testing of wireless devices is only slightly more complicated than traditional antenna testing. Such testing requires external equipment that is capable of setting up a downlink and uplink communication with the DUT. For testing of wireless devices such as handsets, such external equipment is often a base-station emulator or radio communication tester (RCT). For base-station testing, it is often a base band unit connected to the base station using a fiber-optical link. The purpose of the external equipment is to regulate the communication and maintain the test mode of the DUT.

By measuring two spatial power quantities, usually in spherical coordinates, such as EIRP and EIS, both transmitter and receiver properties can be characterized. These two terms are defined as follows:

Table 2.1 Examples of Common Testing Parameters for Active and Passive Antennas

Testing Parameters		*Passive*	*Active*
Pattern	Gain and directivity (ϕ, θ)	Yes	Yes
Power	TRP and EIRP (ϕ, θ)	No	Yes
Sensitivity	TIS/TRS and EIS (ϕ, θ)	No	Yes
Additional testing parameters	Device-dependent	Efficiency, polarization, ...	Throughput and channel emulation, spurious emission, error vector magnitude, channel leakage, blocking, ...

- *EIRP* (θ, ϕ): The amount of power that a theoretical isotropic antenna would emit to produce the power density observed in a given direction. EIRP can thus be determined as the total radiated power from a transmitting antenna times the antenna directivity, or the power delivered to the antenna times the antenna gain.
- *EIS* (θ, ϕ): The EIS of a receiver, including its antenna, expresses the level of minimum received power of an equivalent isotropic antenna to achieve a certain sensitivity threshold according to a preestablished bit error rate (BER).

The definitions of EIRP and EIS are based on radiated power with respect to an ideal isotropic antenna, making the definition very similar to isotropic gain. Due to the similarities in the definition, antenna gain and power parameters can be measured directly in a gain- (or efficiency-) calibrated OTA test setup as discussed in Section 2.1.1. By virtue of the calibration, the input/output isotropic power of the wireless device (and, thus, the EIRP and EIS performance parameters) can be measured directly at the input/output port of the test antenna or probe.

The metrics of the integrated power pattern or total radiated power (TRP) and total isotropic sensitivity (TIS)—sometimes also called total radiated sensitivity (TRS)—are the principal figures of merit in the international standardization groups such as 3GPP [7–9], CTIA [10], and 5GAA [11]. These terms also represent the average directional line-of-sight (LOS) performance of the transmitter and receiver, respectively.

TRP is defined as the integral of the power radiated by the DUT in all directions, over the entire sphere for both polarizations.

$$TRP = \frac{1}{4n}\iint EIRP(\theta,\phi)\sin(\theta)d\theta d\phi \tag{2.11}$$

Per the IEEE definition [2], TRP differs from the power available from the generator or power source. The difference is the antenna losses or efficiency and mismatch between the antenna and RF system and the transceivers response to the antenna. TRP and EIRP are related by the DUT directivity $D_{DUT}(\theta,\phi)$:

$$EIRP(\theta,\phi) = TRP \cdot D_{DUT}(\theta,\phi) \tag{2.12}$$

It follows that the antenna gain cannot be determined from the measured EIRP. The antenna gain, including antenna and mismatch losses, can only be determined if the antenna is measured separately. This requires that it is possible to access the antenna directly as a separate item such as if connectorized.

If $EIRP(\theta,\phi)$ and the directivity $D_{DUT}(\theta,\phi)$ of the DUT are known, TRP can be determined from (2.12). As TRP is constant, the evaluation can be performed by a single measurement in any direction (θ, ϕ). If the EIRP is measured in discrete samples on a regular angular grid around the DUT with N samples in θ and M samples in ϕ for a total of N by M measurement points and both polarization (θ, ϕ), the TRP can be approximated by the summation:

$$TRP \approx \frac{\pi}{2NM}\sum_{i=1}^{N-1}\sum_{j=0}^{M-1}\left[EIRP_{\theta}\left(\theta_i,\phi_j\right)+EIRP_{\phi}\left(\theta_i,f_{\phi}\right)\right]sin\left(\theta_i\right) \tag{2.13}$$

TIS or TRS refers to the average receiver sensitivity across an angular domain, often the entire radiated sphere, to achieve a specific error rate (ER) threshold for both polarizations, which represents the overall receiving capability of the DUT. It follows that TIS or TRS can be determined from the measured $EIS(\theta, \phi)$ and knowledge of the DUT directivity $D_{DUT}(\theta,\phi)$:

$$EIS(\theta,\phi) = \frac{1}{D_{\mathrm{DUT}}(\theta,\phi)}\ TIS\,or\,TRS \tag{2.14}$$

If the EIS is measured in discrete samples on a regular angular grid around the DUT with N samples in θ and M samples in ϕ for a total of $N*M$ measurement points and both polarization (θ, ϕ), the TIS or TRS can be approximated by the summation:

$$TIS\,or\,TRS \approx \frac{2NM}{\pi\sum_{i=1}^{N-1}\sum_{j=0}^{N-1}\left[\frac{1}{EIS_{\theta}\left(\theta_i,\phi_j\right)}+\frac{1}{EIS_{\phi}\left(\theta_i,\phi_j\right)}\right]sin\left(\theta_i\right)} \tag{2.15}$$

The reported summation formulas for TRP, TIS, and TRS are valid for regular angular grids. In case an irregular angular grid is used, a normalization factor on each angular term must be defined before the summation.

In some cases, power and sensitivity performance parameters are only measured in a partial angular view [11]. This can be implemented by changing the summation or integration limits in the expressions for power TRP and sensitivity TIS and TRS. In such cases, partial radiated powers (PRPs) or partial isotropic sensitivities (PISs) are computed as shown in (2.16).

$$PRP = \frac{1}{4n}\int_{0}^{2\pi}\int_{\theta_1}^{\theta_2} EIRP(\theta,\phi)\ sin(\theta)d\theta d\phi \tag{2.16}$$

where θ_1 and θ_2 are the angular limits in elevation of the partial angular view. Examples of such partial measurements are: upper-hemisphere radiated power/upper-hemisphere isotropic sensitivity (UHRP/UHIS) $\theta_1 = 0°$, $\theta_2 = 90°$), near-75-degrees partial radiated power/ near-75-degrees partial isotropic sensitivity (N75PRP/N75PIS) $\theta_1 = 60°$, $\theta_2 = 90°$), and near-horizon partial radiated power/near-horizon partial isotropic sensitivity (NHPRP/NHPIS) $\theta_1 = 60°$, $\theta_2 = 120°$) as illustrated in Figure 2.6. The choice of partial angular performance parameters to be determined in a given measurement configuration depends on the application/service of the device. For cellular bands and vehicle-to-everything (V2X) communication the NHPRP or N75PRP is often used due to the nature and application range of the communication service. For global navigation satellite system (GNSS) applications, the UHRP is often used.

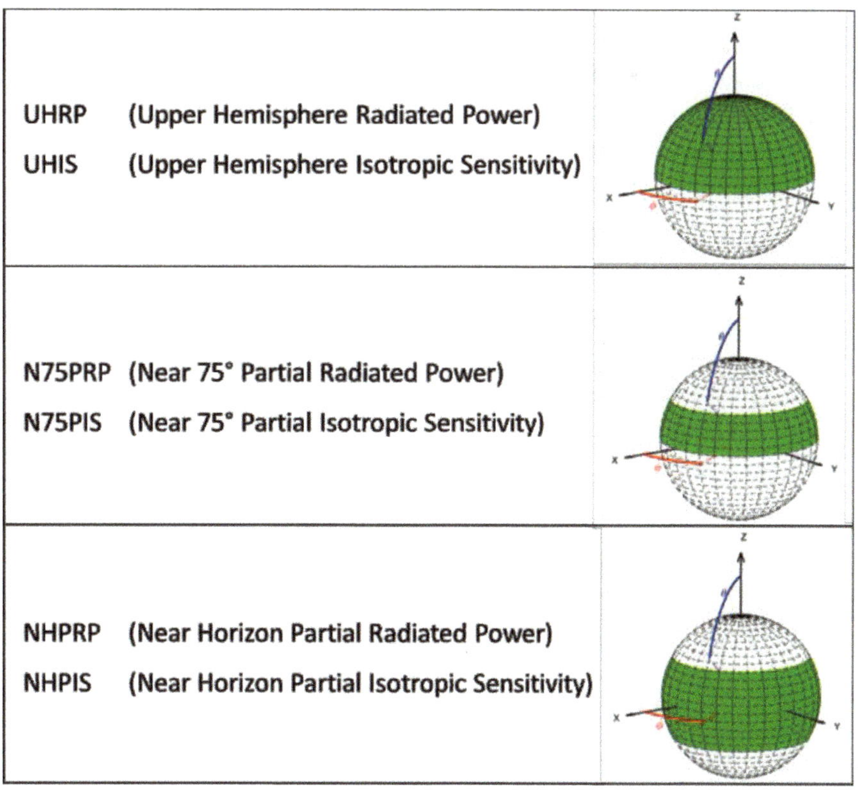

Figure 2.6 The integral OTA power and sensitivity performance metrics used in the evaluation of each measurement method.

2.2 NF versus FF Measurements

In general, antenna measurements system solutions can be divided in three categories, as shown in Figure 2.7 and briefly described as follows.

Direct FF: Where the distance between the antenna under test and the antenna probe is larger than a specified value so that a quasiplane wave condition is met. The main advantage of the FF is that the measurements are faster and do not require any complex field transformation. For automotive applications, only some frequency bands for electrically small antennas can be measured in FF.

NF: Where the acquisition is performed at a closer distance, and a transformation algorithm is applied to calculate the FF pattern or other parameters. The three classical systems are spherical, planar, and cylindrical. NF to FF transformation algorithms for these geometries are now well-established and commercially available. Currently, more complex, but sometimes convenient, scanning configurations are also possible using more sophisticated NF to FF transformation algorithms.

Indirect FF: Where a plane wave (FF condition) is obtained using specific hardware. The classical approach is the use of compact antenna test ranges based on parabolic reflectors. They use the properties of the parabola to transform a spherical phase front to a planar phase front. Recently, array antennas have also been designed to generate plane waves; they are called plane wave generators.

In direct FF antenna measurements, two conditions are employed to derive the minimum measurement distance R, depending on whether the DUT is electrically small or large. Traditionally, the DUT is defined as an antenna and vehicle enclosed within a sphere of diameter D. These criteria are maintained for automotive antenna measurements. With these two conditions we assure that the errors in the direct measurement of the electromagnetic fields are bounded within certain limits. First, the *electrical condition* stipulates that the measurements distance R must be larger than λ. Second, the geometrical condition stipulates that the measurements distance should be much larger than the largest linear dimension of the antenna.

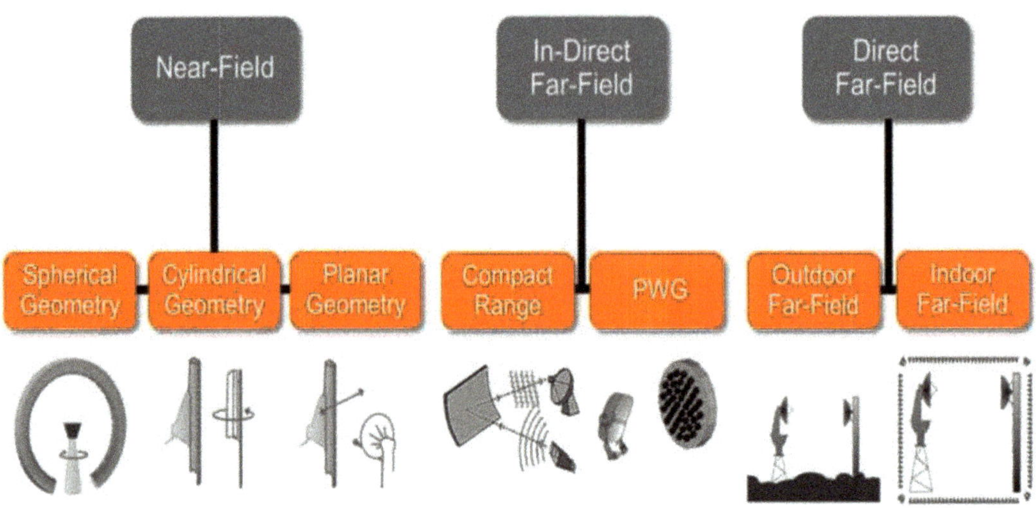

Figure 2.7 Antenna measurement systems solutions.

$R_{FF} >> \lambda$, and, $R_{FF} > 10D$ normally used for electrically small DUT and;

$R_{FF} > 2D^2 / \lambda$, where D is the maximum diameter of the DUT.

The first condition is to ensure that the reactive fields are low enough with respect to the radiated power. The second condition is to ensure a plane wave, or quasi-plane wave, illumination. More specifically, it can be shown that when the measurement distance is $2D^2/\lambda$, the maximum phase deviation between the geometric center of the spherical wave front and the edge of the wave front is 22.5°. For measurements in which the amplitude and phase accuracy is relevant, positioning the antenna of the DUT at the geometric center of the spherical wavefront will improve accuracy. For most of the frequency bands and applications, the second condition is the most relevant, but for low-frequency bands, the first one needs to be considered. Both conditions are plotted in Figure 2.8. The reader can observe that below 500 MHz, FF distance means more than 6m between antennas and that over 500 MHz, the distance depends on the size of the DUT. If the vehicle, or a portion of the vehicle (e.g., 1 meter), is considered, FF distance means more than 6m at frequencies above 1 GHz.

Considering the FF distance $R_{FF} = 2D^2 / \lambda$ when measuring antennas installed on vehicles could lead to prohibitive measurement distances if the whole vehicle dimension is considered in the equation. Examples of measurement distances and associated free-space path loss or pathloss, $L_{freespace}$ for a typical vehicle of diameter 5m are reported in Table 2.2 for different frequencies. At higher frequencies, the FF criteria could lead to very large measurement distances that are impossible to implement. Moreover, the corresponding attenuations could be significantly high, compromising the measurement dynamic range. At higher frequencies, it is thus a good practice to consider only a small portion of the vehicle around the fed an-

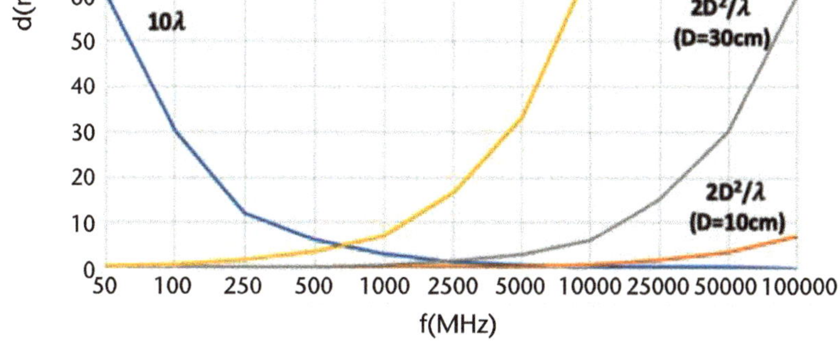

Figure 2.8 FF distances for different frequencies and antenna dimensions.

Table 2.2 FF Distances According to Conventional $2D^2/\lambda$ Criterion and Associated Free-Space Patch Losses for a Typical 5-m-Diameter Vehicle

Frequency	*FF Distance (2D2/λ)*	*Lfreespace*
70 MHz	12m	31 dB
400 MHz	67m	61 dB
3 GHz	500m	96 dB
6 GHz	1 km	108 dB
30 GHz	5 km	136 dB
60 GHz	10 km	148 dB

tenna (e.g., 10λ) as the maximum DUT dimension, leading to shorter measurement distances.

This concept can be better explained in a more formal way by evaluating the $L_{freespace}$ at the FF distance $R_{FF} = 2D^2 / \lambda$:

$$L_{freespace}^{FF} = \left(\frac{8\pi D^2}{\lambda^2}\right)^2 \tag{2.17}$$

Consider the gain of an antenna G as a function of its effective area A_e; we obtain

$$G = \frac{4\pi A_e}{\lambda^2} = \frac{2\pi\ \eta_a D^2}{\lambda^2} \tag{2.18}$$

where the last equation is obtained assuming a squared aperture and introducing the term η_a as the aperture efficiency. Expressing D^2 in terms of G and substituting it into $L_{freespace}^{FF}$ yields:

$$L_{freespace}^{FF} = \left(\frac{4G}{\eta_a}\right)^2 \tag{2.19}$$

Hence, $L_{freespace}^{FF}$ increases with the square of the gain of the antenna. Moreover, the lower the antenna aperture efficiency the higher the $L_{freespace}^{FF}$. Indeed, a highly efficient aperture antenna can reach a certain gain with a smaller physical area (and hence a smaller D) than a low-efficient antenna, leading to a shorter FF distance. Small omnidirectional antennas installed on large structures are examples of radiating systems with poor aperture efficiency. For example, depending on the operational frequency, antennas mounted on vehicles might interact with only a small portion of the car structure. The noninteracting part can make the efficiency very small. For example, a 6-GHz antenna installed on a medium-size vehicle could lead to an aperture efficiency in the range of 1%–10%. Such efficiencies would in turn lead to 40–20-dB extra attenuations of the measured signal if the entire vehicle is considered in the definition of D.

At higher frequencies, only a reduced portion of the vehicle around the antenna significantly contributes to the final radiation. It is therefore reasonable and convenient to consider only that representative portion of the vehicle as the maximum dimension of the device, relaxing the FF measurement distance and hence increasing the dynamic range.

Examples of measurement errors due to the finite measurement distance are illustrated in Figure 2.9. A typical 5-m vehicle with four different antennas installed in different positions on the body has been considered using full-wave simulation at 700 MHz, 2,170 MHz, and 5,500 MHz. Measurements at different distances have been emulated. For each simulated scenario, the gain and the N75PRP errors have been evaluated. At 700 MHz the FF distance considering the whole 5-m vehicle is about 116m. At this distance gain and N75PRP, measurements will have negligible errors. Considering a maximum acceptable error of 0.5 dB, it can be observed that the measurement distance can be reduced to approximately 21m (18% of the FF distance), corresponding to a maximum vehicle diameter of $D = 21^m$ (or 5λ). At 2,170 MHz the FF distance is 361m, but acceptable results are obtained already at 28m (7.7% of the FF distance) where the equivalent diameter of the car is 1.4m (or 10λ). Finally, at 5,500 MHz the FF distance is 916m but accuracy better than 0.5 dB is achieved at 24m (2.6% of the FF distance) corresponding to a maximum vehicle diameter of D = 0.8m (or 15λ).

With these examples, it has been shown that above relatively high frequencies (~2–3 GHz), it is usually not necessary to consider the entire vehicle when defining the FF test distance. Only a portion, roughly 10λ, around the antenna will contribute significantly to the final radiation. Consequently, reduced test distances can be considered. Nevertheless, it is also observed that at strongly reduced distances

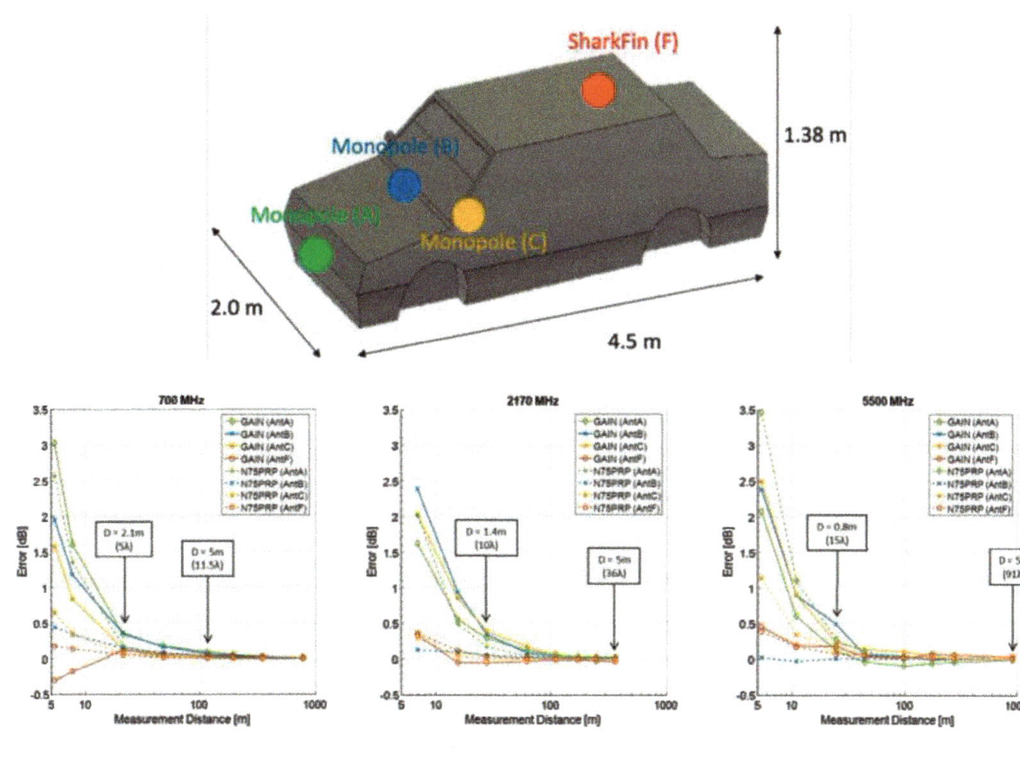

Figure 2.9 Examples of measurement errors due to the finite test distance.

like 5m, the measurement uncertainty is considerably compromised. To measure at such reduced distances, compensation techniques like the parallax correction (described in Chapter 6) or a NF measurement approach should be considered.

In NF systems, measurements are performed at distances much smaller than R_{FF}, where $L_{freespace}$ is much lower leading to higher measurement dynamic ranges. Compactness is another advantage of the NF systems, as it makes it possible to use relatively small, controlled environments or anechoic chambers thereby reducing the cost of the measurement system. Moreover, the NFFF transformation algorithms applied to the measured data allow for the enforcement of an ideal plane wave (FF) condition by propagating the field at infinite distance.

On the other hand, the main disadvantage of a NF system is the sampling requirement needed to correctly perform the NFFF transformation and avoid processing errors like aliasing. Such a requirement, especially in the case of electrically large DUT, could lead to long measurement times. The sampling criteria for spherical NF measurement depends on the electrical size of the DUT. Considering a classical spherical rollover-azimuth system, the number of azimuth scans is calculated as $D/2$:

$$N_{scans} = kD_{min} / 2 + 10 \tag{2.20}$$

where R_{min} is the radius of the minimum sphere enclosing the DUT, and k is the wave number. Figure 2.10 shows N_{scans} for different antenna sizes and frequencies. The angular sampling step can be determined as:

$$Angular\,sampling\,step^{\circ} = \frac{180^{\circ}}{N_{scans}} \tag{2.21}$$

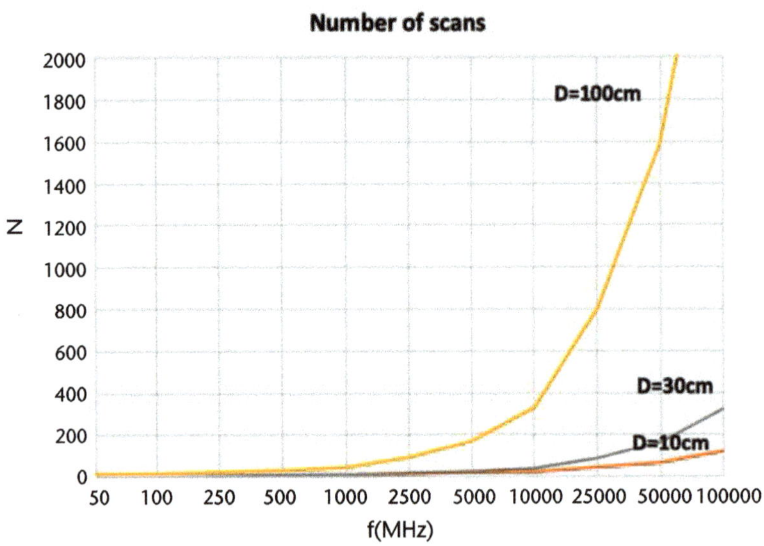

Figure 2.10 Number of scans for NF measurements.

Table 2.3 shows the required angular sampling and total number of sampling points for a 5-m vehicle ($R_{min} = 25m$) at different frequencies. As can be seen the number of sampling points can be prohibitive, especially at higher frequencies.

These examples highlight the sampling requirement as one of the main limitations of the NF techniques. However, recent advanced techniques have been proposed to reduce the number of samples and hence improve the measurement time. One of these techniques takes advantage of the fact that at higher frequencies, only a limited portion of the vehicle around the source antenna contribute significantly to the whole radiation; thus, the number of samples can be reduced drastically. Such techniques are discussed in Chapter 3.

In general, both NF and FF techniques are needed to cover all frequencies for automotive applications. Direct FF measurements are often mandatory for antennas working at very low frequencies. NF measurements are often the only viable solution to get a full characterization of the vehicles.

2.3 Introduction to Near Transformation Theory

For FF measurements (either direct or indirect FF), the electromagnetic field is a plane wave with the following characteristics:

- The electric and magnetic fields are mutually perpendicular, and both are perpendicular to the direction of propagation.
- There is a relation between the electric and magnetic field, in amplitude and phase.
- In the case of direct FF, the wave varies with the radial distance as $\frac{1}{r} \cdot e^{-jk_o r}$. In the case of an indirect FF (compact antenna test range or plane wave generator), the amplitude is constant in the quiet zone, and the variation is $e^{-jk_o r}$.

The measurement of the two tangential components is sufficient for the characterization of the electromagnetic field at any distance (keeping the FF condition). Strictly speaking, this is only true at an infinite distance, but the error is bounded if FF conditions are satisfied.

Table 2.3 Required Angular Sampling Step with Frequency and Number of Points to Measure a Full 3-D Spherical Pattern of a 5-m Vehicle

Frequency	*Angular Sampling Step (180°/N_{scans})*	*Number of Points*
70 MHz	13°	374
400 MHz	6°	1,916
3 GHz	1°	55,832
6 GHz	0.5°	210,159
30 GHz	0.1°	4,997,835
60 GHz	0.01°	1,9865,073

In a FF spherical system, the electric field is then:

$$\vec{E}(\theta,\phi,r) = \frac{1}{r} \cdot e^{-jk_o r} \cdot \left[E_\theta(\theta,\phi)\hat{\theta} + E_\varphi(\theta,\phi)\hat{\phi} \right] \tag{2.22}$$

and the magnetic field can be calculated knowing the electric field if it is necessary.

In the NF, this is not true, since both the electric and magnetic field have tangential and normal components to the surface. However, the knowledge of the two tangential components (normal to propagation vector) is still sufficient for the calculation of the electric and magnetic fields. The reasoning is based on the equivalent theorem; Figure 2.11 summarizes the process. Let's assume an antenna radiating the electromagnetic fields $\vec{E}_{rad}$ and $\vec{H}_{rad}$ out of the measurement sphere.

This electromagnetic field has the same solution, out of the sphere, if we replace the radiating antenna with any electromagnetic field inside the sphere, $\vec{E}_{int}$ and $\vec{H}_{int}$ and some surface electric and equivalent currents in the surface, whose values are defined by:

$$\vec{J}_s = \hat{n} \times \left(\vec{H}_{rad,s} - \vec{H}_{int,s} \right) \tag{2.23}$$

$$\vec{M}_s = -\hat{n} \times \left(\vec{E}_{rad,s} - \vec{E}_{int,s} \right) \tag{2.24}$$

where the subindex s is the value of the field in the sphere. In fact, (2.23) and (2.24) are valid for any value of the electric and magnetic fields $\vec{E}_{int}$ and $\vec{H}_{int}$. Love's equivalent principle fixes those values to zero (null electric and magnetic fields). Also, it is still valid if we force having a perfect electric conductor inside the sphere. In this case, we know that the tangential component of the electric current is null,

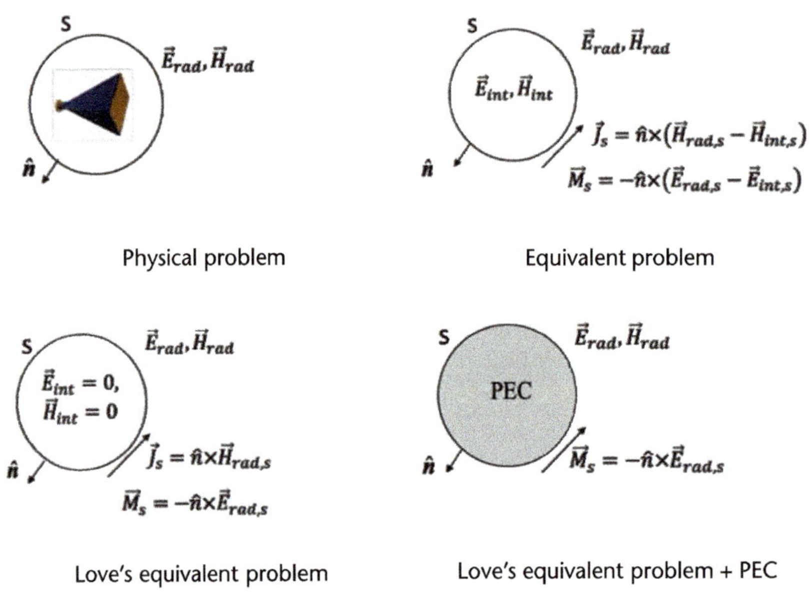

Figure 2.11 Equivalent principles to justify the NFFF transformation theory.

and therefore, we only need the equivalent magnetic current—that is, the measurement of the tangential components of the electric field, $E_\theta(\theta,\phi)$ and $E_\varphi(\theta,\phi)$. Once we know these components, the radiated field out of that sphere can be calculated.

NFFF transformation algorithms are based on the equivalent principle, which states that, from the measurement of the (amplitude and phase) electric field components tangential to a generic scanning surface, the electric and magnetic fields can be computed everywhere outside that surface. To simplify the problem, spherical modes are used for the calculation of the radiated field in case of the spherical NF acquisitions. This is explained in Chapter 3.

2.4 Radiofrequency System

The radiofrequency system is a key aspect in the measurement systems. Figure 2.12 shows a general setup for an antenna measurement. Depending on the frequency range, dynamic range, and available equipment, this setup can change. In this introductory chapter, we want to show some of the main aspects.

The radiofrequency setup must meet the following basic criteria:

- It should not saturate the receiver in the maximum of the radiation pattern.
- It should have a sufficient signal-to-noise ratio for the minimum values (e.g., nulls of the pattern and cross-polar).
- It should be able to measure in a very wide frequency band. Usually, the setup is slightly different depending on the frequency.

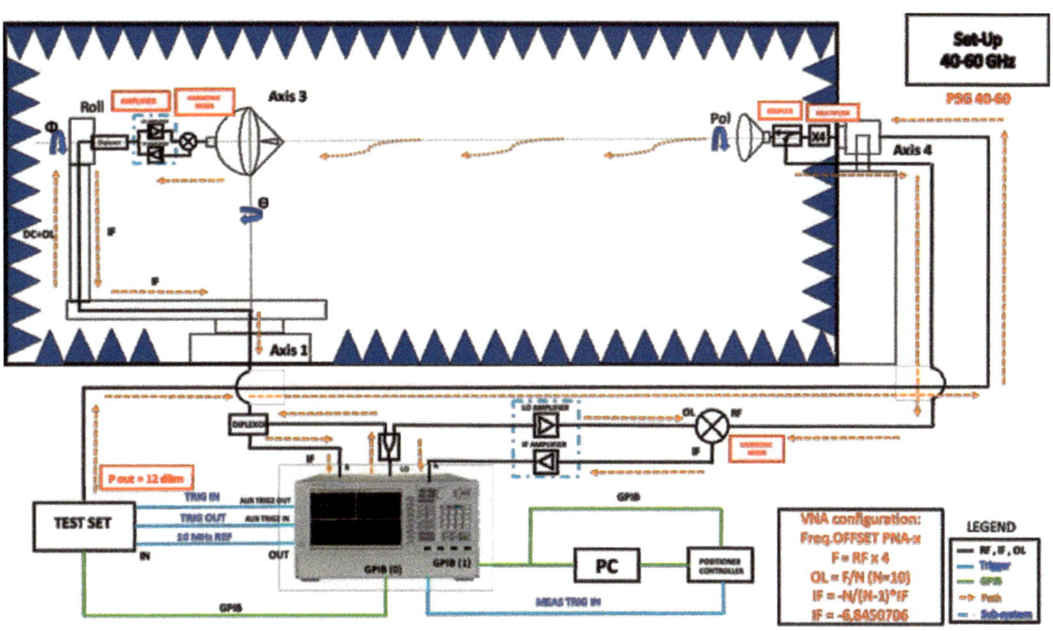

Figure 2.12 Standard antenna measurement setup.

This is a difficult compromise, but some general aspects are considered. The most important equipment is the vector network analyzer (VNA), with a RF transmitter and an amplitude and phase receiver. VNAs can cover most of the frequency spectrum in general antenna measurements; however, they are often limited at the highest frequencies. At these frequencies, multipliers can be used to enable the measurements. For instance, in this setup, the multiplier (×4) is placed closed to the measurement probe. This system is working by transmitting from the probe and receiving by the DUT. The probe is often a horn, due to low losses, stability, and good performance. Ridge horns are used for wideband applications where corrugated or choke conical horns are used to simplify the spherical NF/FF transformation as they have less azimuthal spherical modes due to their symmetry (see Chapter 3). In some cases, open-ended waveguides are also used. Chapters 3 and 5 explain the recent advancement in NF/FF transformation allowing compensation for any multimode horns.

The standard measurement configuration allows for DUT transmitting or receiving. In the case of passive antennas, the measurement can easily be performed in both ways. Due to reciprocity the measured result will be the same. However, for active antennas, or integrated antennas with the RF subsystem together the antenna subsystem, the right configuration should be selected. It is important to have the multiplier close to the transmitting antenna to reduce path losses. Minimizing path losses, in cables or due to free-space attenuation, is essential to guaranteeing a sufficient dynamic range in the measurement. The free-space attenuation is important at millimeter and submillimeter frequencies. To compensate for the losses of the cables, an amplifier can be placed after the multiplier, if used, or just after the transmission cable. The possible drift, usually due to thermal variations, is compensated for in the receiver since two signals are usually compared. The received signal captured by the DUT is compared with a reference signal. This reference signal is extracted from the directional coupler placed after the multiplier. In this way, it is possible to measure the phase with low uncertainty, since the drift originated before this element is compensated for.

When multipliers are used, in both paths, the receiver one and the reference one, harmonic mixers are used to convert to an intermediate frequency. These harmonic mixers are located close to the antennas to reduce the cable losses. When harmonic mixers are used, a local oscillator (LO) signal must be inserted in the harmonic mixer. The vector network analyzer will generate this LO signal.

For some measurement configurations (e.g., outdoor FF or OTA measurements), it is not possible to extract a reference signal. In those cases, a fixed-reference antenna can be placed to obtain a reference signal. This is in place of extracting the reference from the directional coupler as is usually done. In some cases, the internal reference of the VNA is used, but the uncertainty in the phase measurement can be higher, mainly in higher frequencies.

When the dynamic range is deemed insufficient, maybe at higher frequencies and in configurations with high free-space losses or other reasons, an amplifier can be located in the reception path. This solution is not optimal, since drift generated by this amplifier is not compensated for in the reference path. However, the measurement system design is always a compromise among different sources of measurement error and uncertainties.

Finally, the positioners are equipped with rotary joints and slip rings. Their function is to allow the RF cables (rotary joints) and control or power cables (slip rings) to rotate infinitely without damaging the cables.

Figure 2.12 is an example of an antenna measurement setup, but different configurations must be designed depending on the factors mentioned in this section.

References

[1] Bryant, J. H., and H. Hertz, *The Beginning of Microwaves,* Piscataway, NJ: IEEE Service Center, 1988.

[2] IEEE Standard Definitions of Terms for Antennas, IEEE Std 145-2013.

[3] IEEE Recommended Practice for Antenna Measurements, IEEE Std 149-2021 (Revision of IEEE Std 149-1977 Standard Test Procedures for Antennas).

[4] IEEE Recommended Practice for Near-Field Antenna Measurements, IEEE Std 1720-2012.

[5] Graham, A., and P. O. Iversen, "Rapid Spherical Near Field Antenna Test System for Vehicle Mounted Antennas," *AMTA 2004,* Stone Mountain Park, Georgia, October 17–22, 2004.

[6] Dooghe, S., et al., "Spherical Near-Field Antenna Test Range for Automotive Testing from 70 MHz to 6 GHz," *Proc. EuCAP 2006,* Nice, France, Nov. 2006.

[7] 3GPP TR 25.914, "Measurement of Radio Performances for UMTS Terminals in Speech Mode."

[8] 3GPP TS 34.114, "User Equipment (UE) / Mobile Station (MS) Over the Air (OTA) Antenna Performance—Conformance Testing."

[9] 3GPP TS 37.114, "User-Equipment (UE) and Mobile Station (MS) GSM, UTRA, and E-UTRA Over the Air Performance Requirements."

[10] CTIA OTA Test Plan version 3.9.1, May 2020, www.ctia.org.

[11] 5GAA VATM (Vehicular Antenna Test Method) Technical report version 1.0, April 2021.

[12] Friis, H. T., "A Note on a Simple Transmission Formula," *Proceedings of the I.R.E. and Waves and Electrons,* May 1946, pp 254–256.

[13] Saccardi, F., et al., "Accurate Calibration of Truncated Spherical Near Field Systems with Different Ground Floors Using the Substitution Technique," *AMTA 2019,* San Diego, CA, October 6–11, 2019.

CHAPTER 3

NFFF Transformations

Francesco Saccardi, Fernando Rodríguez Varela, Manuel Sierra Castañer, and Lars J. Foged

NF measurements are a valuable alternative to FF methods. FF methods can require large measurement distances. NF techniques allow the antenna to be tested at a shorter distance, in an anechoic chamber with a controlled environment. A NF system requires the NF-to-FF (NF/FF) transformation to be performed to obtain the antenna radiation pattern and related electrical properties.

The NF/FF transformation uses the uniqueness theorem, which states that in a volume, there is a set of electric and magnetic currents that produce a specific electromagnetic field on a surface enclosing these currents. Therefore, the characterization of electromagnetic fields over a closed surface allows all the antenna radiation characteristics to be determined, including the FF radiation pattern.

Figure 3.1 depicts the NF/FF transformation process typically followed in antenna measurements. In the first step, the electric and magnetic currents of the DUT are obtained from the NF, which has been determined by solving an inverse problem. The FF is then computed with what is called a direct problem. Both inverse and direct problems can be solved using the relationships between electromagnetic currents and fields governed by Maxwell equations.

For an efficient implementation of field transformations, the electromagnetic fields are expanded into a set of linearly independent wave objects that are orthogonal with respect to each other. These wave objects are called modal expansions, and they are derived solving Maxwell equations in a canonical coordinate system: cartesian, cylindrical, and spherical. Each of these expansions is suitable for application on planar, cylindrical, and spherical NF measurement systems, respectively. They enable an efficient mathematical evaluation of the inverse and direct problems of the NF/FF.

For cylindrical and planar setups, the measurement surface suffers from an inherent truncation. This limits its use to high-directive antennas to ensure that most of the radiated power can be measured. For low-directive antennas, the spherical setup is better suited, because the radiation can be measured in all directions, minimizing the measurement uncertainty. Most communication antennas installed

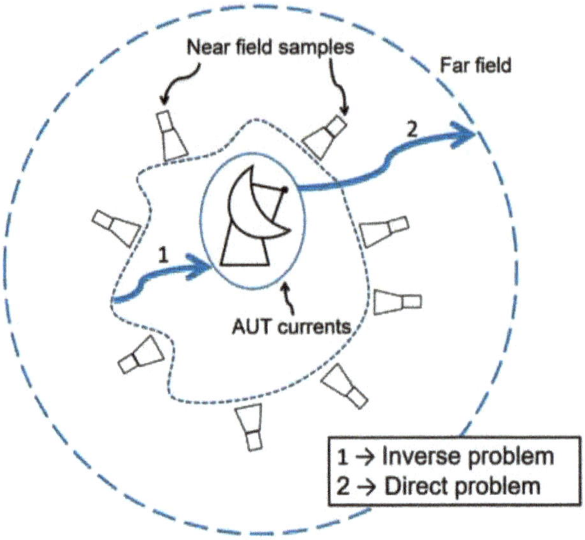

Figure 3.1 Schematic depiction of the NFFF transformation principle. ©2018 UPM. (Reprinted with permission.) [69].

on vehicles have a low-directive pattern to ensure full coverage. In this case, the spherical NF (SNF) setup is normally the preferred choice for accurate automotive antenna testing.

This chapter introduces an extensive review of the main NF/FF transformation methods for automotive measurements in SNF setups. The application of spherical modal expansion to automotive environments is discussed, covering topics such as sampling, truncation errors, measurements over conducting ground floors, and probe correction. In Section 3.2 the capabilities of NF processing using equivalent current formulation are explored. Last, an introduction to new and emerging post-processing techniques is given to offer a perspective of future trends in automotive antenna testing. This chapter refers to relative power or directivity normalization, disregarding the absolute value of the gain pattern, which is addressed in Chapter 5.

3.1 NF/FF Transformation based on Spherical Wave Expansion

This section introduces the basic concepts of spherical NF/FF transformation relevant to automotive measurements using modal expansions. The spherical wave expansion (SWE) is a powerful tool for spherical field transformation and the evaluation of the AUT performance. Traditional post-processing techniques based on the SWE are introduced, and innovative and emerging applications such as full-probe compensation and down sampling in translated measurements are reviewed.

3.1.1 SWE

SNF measurements are postprocessed, solving Maxwell's equation in a spherical coordinate system. These solutions, known as spherical wave functions, include a series of properties that facilitate efficient and stable field transformation algorithms. An extensive bibliography [1–4] has been developed analyzing the mathematical formulation of the spherical wave functions in full detail, so this analysis will not be repeated here. In this section we will present the basic concepts of SNF formulation to provide an understanding of conventional and advanced post-processing techniques for automotive environments.

Figure 3.2 depicts the geometry of a spherical coordinate system typically used in SNF measurements, with the AUT located at the origin. The measured NFs are referred to this coordinate system, and to postprocess them, the mathematical equations governing the electromagnetic fields must be derived. To facilitate the mathematical formulation, only the source-free region of space will be considered. This region corresponds to $r > R_{min}$, which is the radius of the smallest sphere that encloses the AUT/DUT, known as the minimum sphere (see the green circle in Figure 3.2). The spherical wave functions are derived by solving the well-known scalar wave equation for the region outside the minimum sphere:

$$\left(\nabla^2 + k^2\right) f(r,\theta,\phi) = 0 \tag{3.1}$$

with k being the propagation constant and $f(r, \theta, \phi)$ the function to be solved in spherical coordinates. The time convention $e^{j\omega t}$ is assumed and suppressed.

It may be shown [2] that there exist two families of vectorial generating functions that satisfy (3.1) given by the expression:

$$\vec{F}^{(c)}_{1mn}(r,\theta,\phi) = \frac{1}{\sqrt{2\pi}} \frac{1}{\sqrt{n(n+1)}} \left(-\frac{m}{|m|}\right)^m \left\{ z_n^{(c)}(kr) \frac{jm\bar{P}_n^{|m|}(\cos\theta)}{\sin\theta} e^{jm\phi}\hat{\theta} - z_n^{(c)}(kr) \frac{d\bar{P}_n^{|m|}(\cos\theta)}{d\theta} e^{jm\varphi}\hat{\phi} \right\} \tag{3.2}$$

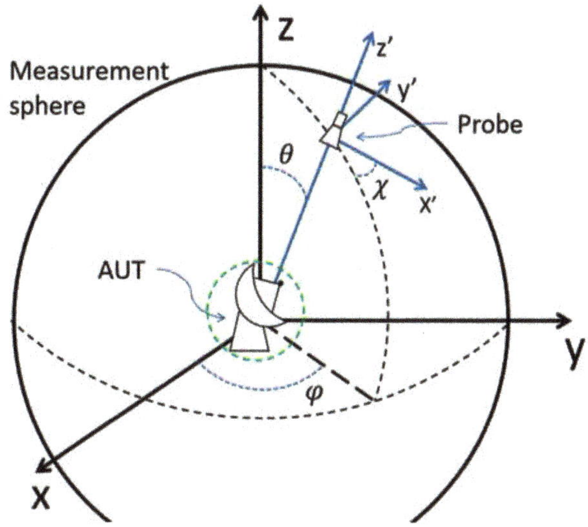

Figure 3.2 Geometry and coordinate system of a SNF measurement system. ©2020 IEEE. (Reprinted with permission.) [70].

$$\vec{F}^{(c)}_{2mn}(r,\theta,\phi) = \frac{1}{\sqrt{2\pi}}\frac{1}{\sqrt{n(n+1)}}\left(-\frac{m}{|m|}\right)^m$$

$$\left\{\begin{array}{l}\dfrac{n(n+1)}{kr}z_n^{(c)}(kr)\overline{P}_n^{|m|}(\cos\theta)e^{m\phi}\hat{r}+\dfrac{1}{kr}\dfrac{d\left\{z_n^{(c)}(kr)\right\}}{d\{kr\}}\dfrac{d\overline{P}_n^{|m|}(\cos\theta)}{d\theta}e^{jm\phi}\hat{\theta}+\\ \dfrac{1}{kr}\dfrac{d\left\{z_n^{(c)}(kr)\right\}}{d\{kr\}}\dfrac{jm\overline{P}_n^{|m|}(\cos\theta)}{\sin\theta}e^{jm\phi}\hat{\theta}\end{array}\right\} \tag{3.3}$$

where $\overline{P}_n^{|m|}$ is the associated Legendre function [5], and the operator $\frac{d(\;)}{d(\;)}$ denotes the derivative. $z_n^{(c)}(kr)$ is a radial function that expresses the traveling nature of the propagating wave, depending on the index c. For $c = 1$ and $c = 2$, $z_n^{(c)}(kr)$, becomes the spherical Bessel and Neumann functions [5], respectively, which indicates standing waves. In the case of $c = 3$ and $c = 4$ the radial function becomes the spherical Hankel function of the first and second kind, representing inward and outward waves, respectively, for the assumed time convention. In SNF measurements, it is assumed that all sources are contained on the minimum sphere and that there are no mutual interactions with the probe. Therefore, only outward waves propagating from the origin exist, and $c = 4$ is considered from now on.

The radial behavior of the spherical Hankel function of the first kind is depicted in Figure 3.3 for several orders. In all cases two regions with different propagating conditions can be distinguished. The first region corresponds to $kr < n$, and it shows a very steep response where the spherical waves experience a very high attenuation for small radial distances. This is the equivalent of the cut-off region for a waveguide. In the second region ($kr > n$) the radial decay is of 20 dB per

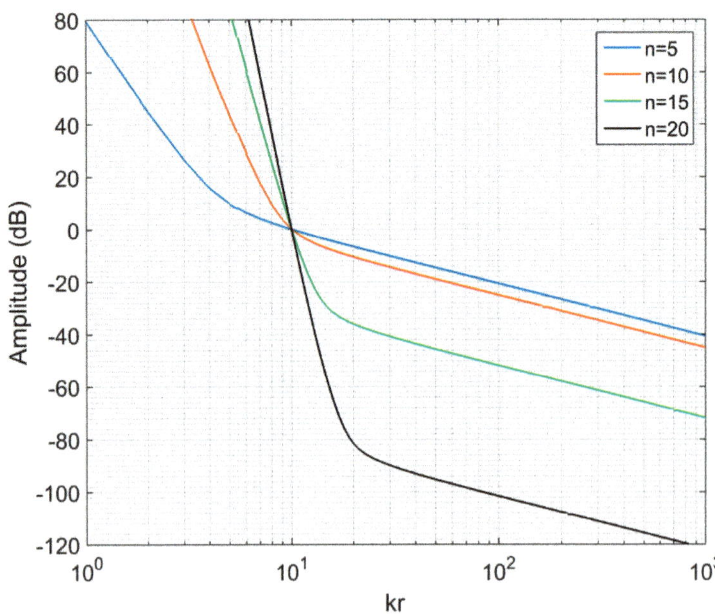

Figure 3.3 Radial behavior of the spherical wave functions for various values of n. For comparison the amplitude is normalized to 0 dB at the surface of an assumed minimum sphere with $kR_{min} = 10$.

decade corresponding to the well-known $1/(kr)^2$ factor of propagating spherical waves. If we consider that the field propagation starts at $r = R_{min}$, the spherical waves with indexes $n > kr_{min}$ will start to propagate in the cut-off region. These are denoted as evanescent waves because, due to the strong attenuation, their contribution to the radiated field is negligible.

The angular dependence of the spherical wave functions can be seen in Figure 3.4, where their amplitude for several indexes has been depicted. It can be seen that the indexes m and n control the number of oscillations along the ϕ and θ coordinates, respectively. Although it is not shown in Figure 3.4, the index s controls the polarization nature of the propagating wave, which can be either transverse electric (TE) in the case of for $s = 1$ or transverse magnetic (TM) for $s = 2$.

The electric field at any point of the space outside the minimum sphere may be written as a superposition of all propagating spherical wave functions weighted by complex coefficients:

$$\vec{E}(r,\theta,\phi) = \frac{k}{\sqrt{\eta}} \sum_{s=1}^{2} \sum_{n=1}^{N} \sum_{m=-n}^{n} Q_{smn} \vec{F}_{smn}^{(4)}(r,\theta,\phi) \tag{3.4}$$

Equation (3.4) is known as SWE [1]. In the SWE expression, Q_{min} are the spherical wave coefficients (SWCs), and N is an integer truncation constant so that only the propagating waves are considered. The value of this constant is given by:

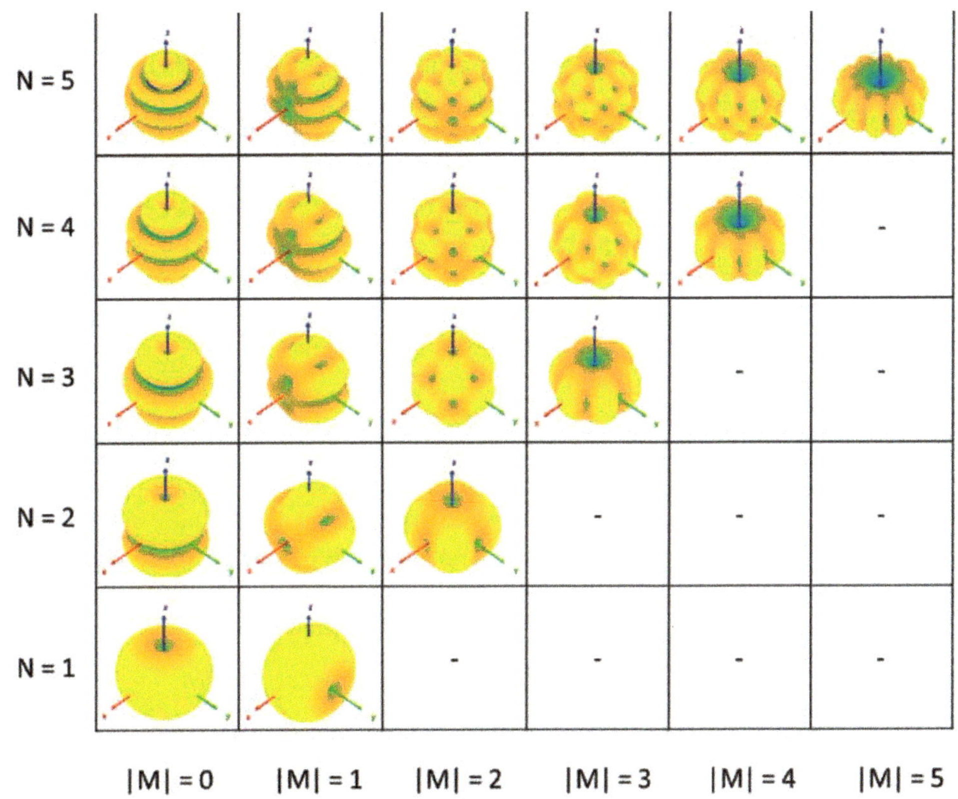

Figure 3.4 Angular behavior of the spherical wave functions.

$$N = \lfloor kR_{min} \rfloor + n_{safety} \tag{3.5}$$

where the square brackets indicate the floor function, and n_{safety} is a safety constant to ensure that we do not neglect spherical waves, which contribute significantly to the radiated field; this typically takes a value of 10. A more sophisticated expression of n_{safety}, defined according to the required accuracy of the SWE, is given in [6].

The SWCs represent the amplitude and phase contribution of each spherical wave function to the total AUT radiated field. Considering the introduced truncation constant, there is a total of $2N(N + 2)$ SWC, and each antenna has its own set different to those of any other antenna, so they become an effective equivalent representation of the AUT/DUT.

The SWE is a key tool to perform the NF/FF transformation process in spherical geometries. The spherical NF/FF transformation is a two-step procedure that starts with the measured NF at a given distance $r = R_M$. First, the field is expanded in spherical waves to compute the SWC. In the second step, the FF is evaluated from the SWC using the spherical wave summation (SWS) of (3.4). The first step is not a straightforward task, because it requires special considerations and additional formulation. These topics will be covered in the next sections.

3.1.1.1 Computation of the SWC

The easiest and most intuitive way of calculating the SWC is to rewrite (3.4) as a linear system of the form **b** = **Ax**, where **b** is a column vector with **i** measurement NF points, $\boldsymbol{x}$ is another column vector with $\boldsymbol{j}$ unknown SWC, and $\boldsymbol{A}$ is a $\boldsymbol{i} \times \boldsymbol{j}$ matrix containing the SWE basis functions. The solution of the system by means of $\boldsymbol{x}$ can be simply performed by standard inversion methods. Although relatively simple, this approach is computationally inefficient, and for large problems it may require lot of memory resources and be quite time-consuming.

To compute the SWC in a computationally efficient way, it is convenient to express the measured signal over the spherical surface in terms of the *transmission formula* [1] shown in (3.6). This formula expresses the measured signal $w(r, \chi, \theta, \phi)$ as the coupling between the AUT/DUT and the measurement probe.

Figure 3.2 shows a generic measurement scenario, with the probe receiving the field radiated by the AUT, at a point in space (r, θ, ϕ), with a relative orientation angle χ, which defines the measurement polarization. If the probe is a few wavelengths away from the AUT, the interactions between both can be neglected. The coupling between AUT and probe can be expressed as a scalar product of the SWC of both antennas if they are referred to the same coordinate system. To derive a useful formula, the SWC of the probe is translated to the AUT coordinate system using translation and rotation of the spherical waves. This leads to the well-known transmission formula of spherical NF measurements:

$$w(r,\chi,\theta,\phi) = \sum_{s=1}^{2}\sum_{n=1}^{N}\sum_{m=-n}^{n}\sum_{\mu=-V}^{V} Q_{smn} e^{jm\phi} d^{n}_{\mu m}(\theta) e^{j\mu\chi} P_{s\mu n}(kr) \tag{3.6}$$

$$P_{s\mu n}(kr) = \sum_{\sigma=1}^{2} \sum_{\substack{v=|\mu| \\ v \neq 0}}^{V} C_{\sigma\mu v}^{sn}(kr) R_{\sigma\mu v} \tag{3.7}$$

with $d_{\mu m}^{n}(\theta)$ and $C_{\sigma\mu\varpi}^{sn}(kr)$ being rotation [7] and translation [8] coefficients, respectively, and $R_{s\mu v}$ the probe SWC. The terms $P_{s\mu n}(kr)$ are known as the *probe response constants*, and it is convenient to isolate them because they only depend on the measuring probe and measurement distance.

The transmission formula is a very powerful tool because, if certain requirements are met, the SWC can be obtained in a computationally efficient way using a fast Fourier transform (FFT) technique and can compensate for the probe effect. The FFT-based (or Wacker) inversion method [1, 9] of the transmission formula to compute the SWC is briefly described as follows.

Let us assume that an ideal χ-oriented *Hertzian* dipole antenna is used as probe. As described in [1], this antenna has only two nonzero SWCs (i.e., $Q_{\sigma=2,\mu=1,v=1}$ and $Q_{\sigma=2,\mu=-1,v=1}$), which are used to precompute the probe response constants shown in (3.7). The SWCs of the AUT/DUT are then found by inverting the four summations in (3.6), which is done by exploiting the orthogonality properties of the rotation terms [1]:

$$\int_0^{\pi} d_{\mu m}^{n}(\theta) d_{\mu m}^{n'}(\theta) \sin\theta d\theta = \frac{2}{2n+1}\delta_{nn'} \tag{3.8}$$

$$\int_0^{2\pi} e^{jm\phi} e^{-jm'\phi} d\theta = 2\pi\delta_{mm'} \tag{3.9}$$

By applying scalar products of the measured field $w(r, \chi, \theta, \phi)$ with, $e^{-j\mu\chi}$, $e^{-jm\phi}$, and $d_{\mu m}^{n}(\theta)$, the SWC can be extracted from the summations in (3.6). In particular, note the following:

- The summation μ has only two terms ($\mu = \pm 1$), because an ideal Hertzian dipole is assumed, and it can be solved exploiting the orthogonality of $e^{j\mu\chi}$. This translates into the application of a FFT of the two acquired field components measured with the probe χ orientations (typically $\chi = 0°$ and $\chi = 90°$).
- The summation in m is inverted in a similar way using the orthogonality of $e^{j\mu\chi}$ (FFT along ϕ variable). Because this summation involves a maximum number of $2M$ terms, this is the number of samples required in ϕ.
- The summation in n involves the rotation term $d_{\mu m}^{n}(\theta)$, which is expressed as a Fourier series [1]. To properly apply this data transformation, N samples in the θ variable are required.
- The summation in s involves only two terms, so it can be inverted by solving a linear system of equations provided that there are two polarizations measurements.

In summary, the requirements for a successful inversion of the transmission formula using the FFT-based method are the following:

- The NF must be measured in a closed spherical surface $\theta \in [0,\pi] \phi \in [0,2\pi]$ of constant radius *r*.
- The sampling along θ and ϕ must be evenly spaced.
- Two orthogonal field orientations must be measured with a first-order ($\mu = \pm 1$) probe.

In the above description the Hertzian dipole has been assumed to be a probe. Since it is the simplest radiating element, with this assumption the effect due to the probe pattern is not compensated for. Nevertheless, the Wacker inversion method does support the probe pattern correction if the probe modes with indexes $\mu \neq \pm 1$ are assumed negligible (while no assumption on the v and s mode orders is required). Probes fulfilling such requirements are called *first-order probes*, and they are characterized by a rotationally symmetric azimuthal radiation pattern (ϕ-dependency is simply in the form of $sin(\phi)$ and $cos(\theta)$). When an actual probe is considered, the Wacker inversion method is typically called the first-order probe compensation (FOPC) technique.

Examples of first-order probes are shown on the left side of Figure 3.5 [10, 11]. On the right side the typical behavior of the azimuthal spherical wave spectrum [see (3.5) and Section 3.1.1.3 for more details] of a first-order probe over frequency is depicted. The dominant $|\mu| = 1$ modes are shown in blue. The thin colored traces represent the individual higher-order azimuthal modes (e.g., individual spherical modes with $|\mu| \neq 1$), which as seen, are well below –45 dB. Usually, –35/–40 dB is the threshold to consider a probe a first-order one [12]. The thicker orange trace is obtained combining all the higher order modes together, and is below –40 dB.

State-of-the-art first-order probes can achieve a maximum bandwidth of one octave. The design and manufacturing of first-order probes on a larger bandwidth is a very challenging task, and probes may exhibit higher-order modes.

It should be noted that the application of the FOPC to SNF measurements performed with probes having nonnegligible higher-order μ-modes could lead to residual errors unless the probe is sufficiently electrically small and/or the angle with which the probe illuminates the AUT/DUT is also sufficiently small. Otherwise, higher-order probe compensation techniques should be adopted as detailed in Section 3.1.3.2.

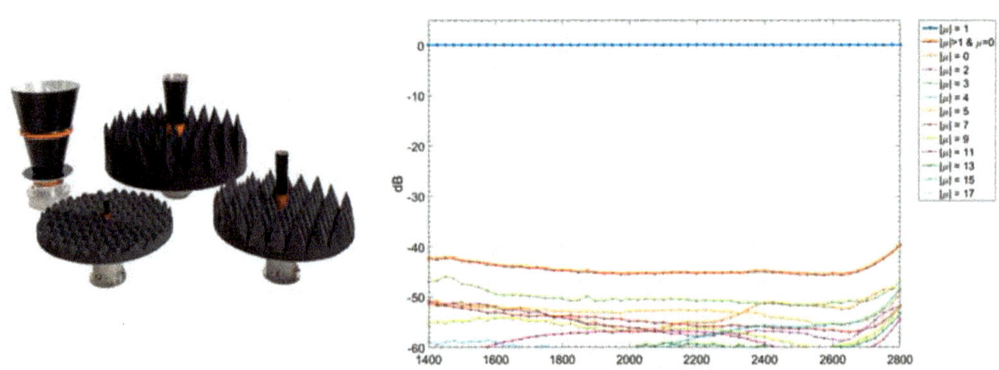

Figure 3.5 Example of first-order probes and a typical spherical wave spectrum.

3.1.1.2 Sampling Criteria

As seen in Section 3.1.1.1, and in general when dealing with data transformations, a sufficient number of sampling points should be acquired in the initial domain to avoid the well-known aliasing effect [13]. In particular, the sampling rate, according to the Nyquist-Shannon sampling theorem for signal processing, is proportional to the inverse of twice the bandwidth of the signal to be transformed. The same criterion also applies to the spatial sampling of electromagnetic waves whose rate must be at least half-wavelength ($\lambda/2$). This directly applies to planar NF (PNF) measurements [13] where the field is measured on a plane in front of the antenna, and the above-mentioned half-wavelength sampling is used along both scanning axes. For SNF measurements the NF/FF transformation is performed with the SWE, and, as seen in (3.5), the minimum number of SWCs to be computed (N) depends on the radius of the AUT/DUT minimum sphere. Intuitively, the minimum number of samples to be measured should be at least equal to the number of SWCs to be computed. Hence, considering a uniform sampling along the θ-axis (with θ = [0°, 180°]) and the ϕ-axis (with ϕ = [0°, 360°]) the minimum angular sampling rate is given by the angular steps (in radians) reported in (3.10).

$$\Delta\theta = \Delta\phi = \frac{\pi}{N} = \frac{\pi}{kR_{min} + n_{safety}} \tag{3.10}$$

The sampling requirements in SNF measurements can be relaxed when the AUT/DUT dimensions are smaller in the xy plane than in the xz or yz plane, as in the base station antenna shown in Figure 3.6. In these cases, the radiated field will exhibit less variation along the ϕ-axis than in the θ-axis. Consequently, the sampling along the ϕ-axis can be relaxed. In particular, in these scenarios, the m-summation in (3.4) can be truncated at $m = \pm M$, where M is defined considering the radius of the minimum cylinder (C_{min}) enclosing the AUT/DUT as shown in (3.11).

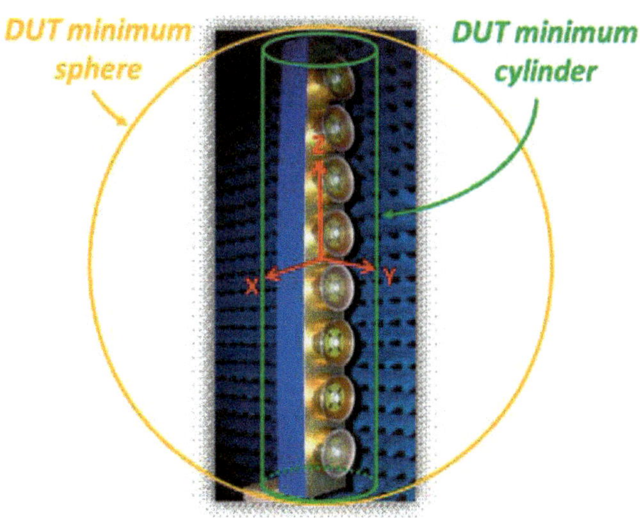

Figure 3.6 Example of AUT minimum sphere and AUT minimum cylinder in the case of a base station antenna.

$$M = \lfloor kC_{min} \rfloor + m_{safety} \tag{3.11}$$

Equation (3.11) is similar to (3.5) and always leads to $M \leq N$. Hence, when C_{min} is smaller than R_{min} the ϕ sampling step can be defined as

$$\Delta\phi = \frac{\pi}{M} = \frac{\pi}{\lfloor kC_{min} \rfloor + m_{safety}} > \Delta\theta \tag{3.12}$$

It is interesting to note that from (3.10) and (3.12), neglecting the safety factors, we can obtain the equations

$$\Delta\theta\, R_{min} = \frac{\pi}{k} = \frac{\lambda}{2} \tag{3.13}$$

$$\Delta\phi\, C_{min} = \frac{\pi}{k} = \frac{\lambda}{2} \tag{3.14}$$

which show that for SNF measurements the minimum sampling rate is given by the conventional half-wavelength spacing. The difference with respect to PNF measurements is that the minimum sampling should be enforced considering the AUT minimum sphere/cylinder instead of the spherical scanning surface.

Following the same argument, it is possible to define an *equivalent sampling area* based on the applied angular sampling step. In particular, the equivalent sampling sphere (ESS) and the equivalent sampling cylinder (ESC) are the virtual sphere/cylinder where the half-wavelength sampling is met, and they are given by (3.15) and (3.16).

$$R_{ESS} = \frac{\lambda}{2\Delta\theta} = \frac{\pi}{k\Delta\theta} \tag{3.15}$$

$$R_{ESC} = \frac{\lambda}{2\Delta\phi} = \frac{\pi}{k\Delta\phi} \tag{3.16}$$

where R_{ESS} is the radius of the ESS, and R_{ESC} is the radius of the ESC. It should be noted that if a denser sampling with respect to the minimum one is considered (oversampling), the ESS and the ESC will be larger than the actual AUT/DUT minimum sphere and cylinder. In such a case, higher-order spherical modes can be computed, and the so-called *modal filtering* can be applied to improve the signal-to-noise ratio [14] and to mitigate the effect of possible unwanted contributions present in the measurements (*echo suppression* [15]).

Finally, note that the ESS and the ESC cannot be larger than the radius of the measurement spherical surface (R_M). This is a direct consequence of the cut-off properties of the spherical wave functions, which do not allow modes with n and m indexes higher than kR_M to propagate (see Figure 3.3).

3.1.1.3 Representation and Examples of the SWC

For a better understanding of the SWE, some examples of SWC associated with different AUT/DUT are reported in this section.

The first antenna considered is the MVG SH600 dual-ridge horn shown in Figure 3.7. This antenna operates over a wide frequency range, 0.6–9 GHz and is typically used for gain calibration purposes. The directivity radiation pattern at 3 GHz obtained with full-wave simulation is also shown in Figure 3.7. The dimension of the antenna aperture is 199 × 356 mm (E-plane x H-plane) while its height is 295 mm. Considering the origin coordinate system in the geometrical center of the antenna, the radius of the minimum sphere enclosing the antenna is $R_{min} = 252$ mm (approximately 25λ at 3 GHz). The radius of the minimum cylinder enclosing the antenna is $C_{min} = 204$ mm (approximately at 3 GHz). According to (3.5) and (3.11) the minimum number of modes needed to represent such an antenna is $N = 25$ and $M = 22$ (obtained by setting $n_{safety} = m_{safety} = 10$). According to (3.10) and (3.12), the minimum sampling required to measure such an antenna is approximately $\Delta\theta = 7.2°$ and $\Delta\phi = 8.2°$.

The computed SWCs of this simulated horn are shown in Figure 3.8. The $Q^{(4)}_{1min}$ modes (or simply Q_1) are shown on the top-left while the $Q^{(4)}_{2min}$ modes (or simply Q_2) are shown on the top-right. Both mode maps are normalized with respect to the total radiated power given by

$$P_{tot} = \frac{1}{2}\sum_s\sum_m\sum_n \left|Q^{(4)}_{smn}\right|^2 \tag{3.17}$$

and expressed in decibels $\left(10\log_{10}\left(\left|Q^{(4)}_{smn}\right|2 / P_{tot}\right)\right)$. As can be seen, modes with indexes $N > 25$ and $|M| > 22$ are, as expected, well below the considered dynamic.

A more compact way of representing the spherical modal spectrum of an antenna is obtained using the $P_{\hat{n}}$ and $P_{|\hat{m}|}$ (or simply $P_{\hat{m}}$) power spectrum curves defined by (3.18) and (3.19), respectively:

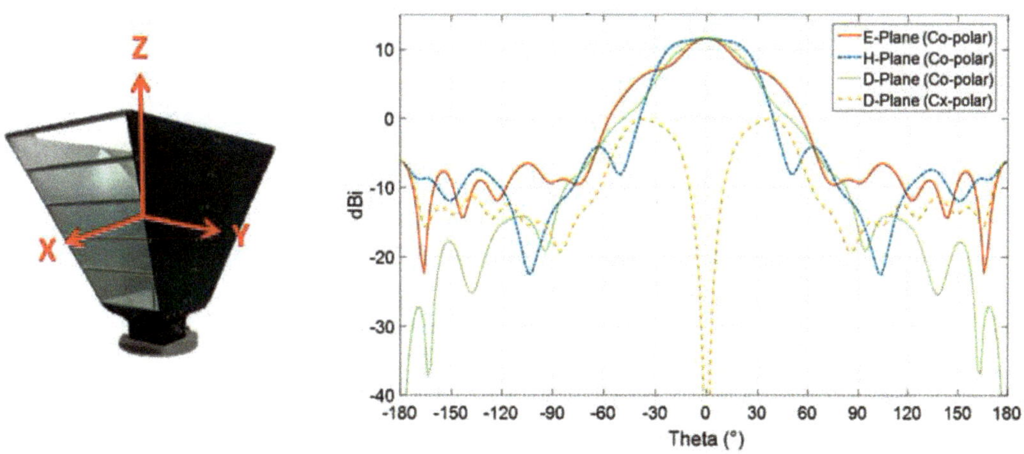

Figure 3.7 Directivity radiation pattern cuts of the SH600 dual-ridge horn at 3 GHz.

$$P_n = \frac{1}{P_{tot}} \sum_{s=1}^{2} \sum_{m=-n}^{n} \left| Q_{smn}^{(4)} \right|^2 \tag{3.18}$$

$$P_{|m|} = \frac{1}{P_{tot}} \sum_{s=1}^{2} \sum_{n=|m|}^{N} \left(\left| Q_{s,m,n}^{(4)} \right|^2 + \left| Q_{s,-m,n}^{(4)} \right|^2 \right) \tag{3.19}$$

As P_n and P_m are obtained by the squared summation of the *s*-indexes and of all the *m*- or *n*-indexes, respectively, one-dimensional plots like those shown at the bottom of Figure 3.8 are obtained. The P_n power spectrum gives a compact description of the radial extension of the antenna. In this case it can be observed that modes with index $n > 23$ are, as expected, more than 60 dB below peak, meaning that a proper convergence of the *n*-series of the SWE is reached. The P_m power spectrum is useful for a compact description of the azimuthal (ϕ) behavior of the antenna. As can be seen, a convergence of the series is reached before the predicted $M = 20$ minimum index, meaning that the edges of the antenna do not contribute significantly to the azimuthal behavior of the radiate field. In other words, the *effective* minimum cylinder is slightly smaller than the geometrical one.

We now consider the same horn antenna but placed in different position, offset in the reference system. When an antenna is translated away from the origin of the coordinate system, its minimum sphere/cylinder is increased and, consequently, the number of modes needed to describe the device and the associated number of sampling points is higher.

For example, let us consider the horn antenna with its geometrical center at $z_t = +5\lambda$, $z_t = +10\lambda$, and $z_t = +15\lambda$. The radius of the antenna minimum sphere is increased from 25λ to 7.5λ, 12.5λ, and 17.5λ, respectively. While a translation along the *z*-axis, the radius of minimum cylinder is unchanged. The P_n and P_m power

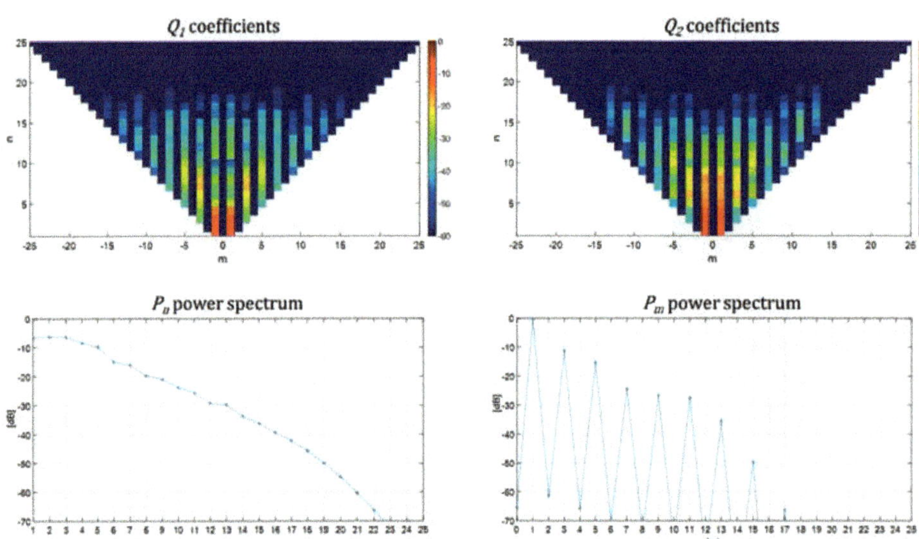

Figure 3.8 Simulated SH600 dual-ridge antenna at 3 GHz: 2-D representation of the SWC (top) and 1-D representation with the P_n and P_m power spectra (bottom).

spectra of the horn antenna at the different locations along the z-axis are shown at the top of Figure 3.9. The increase of the antenna minimum sphere is reflected directly on the spectrum whose modes are spread along the n-indexes. Moreover, it can be observed that a cut-off of the spectrum is present at approximately $N_{cutoff} = [kd]$ with d being the translation distance from the origin. With $d = z_t$, we in fact have $N_{cutoff} = 32$, $N_{cutoff} = 63$, and $N_{cutoff} = 94$, respectively, for the three considered z-translations. The translation along the z-axis does not modify the azimuthal symmetry of the antenna (same minimum cylinder); hence, the P_m power spectra does not vary.

Similarly, we now consider the horn antenna with its geometrical center translated along the x-axis, namely $x_t = +5\lambda$, $x_t = +10\lambda$, and $x_t = +15\lambda$. As in the previous case, the radius of the antenna minimum sphere is increased from 2.5λ to 7.5λ, 12.5λ, and 17.5λ, respectively. In this case, the radius of minimum cylinder is increased and is approximately as large as the one of the minimum sphere. The translation along the x-axis modifies both the radial extent and the azimuthal symmetry of the antenna. Hence, the spherical modes are spread along both the n- and m-indexes after the translation. These effects are clearly visible in the P_n and P_m power spectra shown at the bottom of Figure 3.9. A similar behavior of the spectrum would be observed with any translation in the xy-plane.

The second example is a simulation of an automotive environment shown in Figure 3.10, where a patch antenna operating at 2.6 GHz has been placed on the hood of a sedan car with dimensions of 4.7 × 1.8 × 1.4m. Two coordinate systems have been defined. The first one is located approximately at the center of the car. The second coordinate system is placed at the center of the patch given that it corresponds to the theoretical phase center of the antenna. In practice, most of the radiation comes from the patch, but due to presence of electric currents coupled to the car body, nonnegligible contributions can come from the car structure.

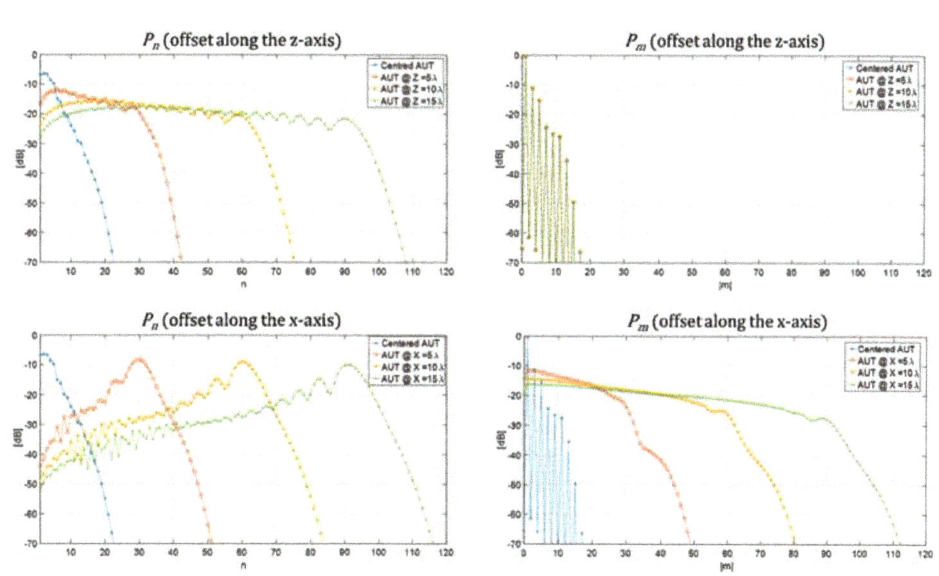

Figure 3.9 P_n and P_m power spectrum comparison of the simulated SH600 at 3 GHz with different offset.

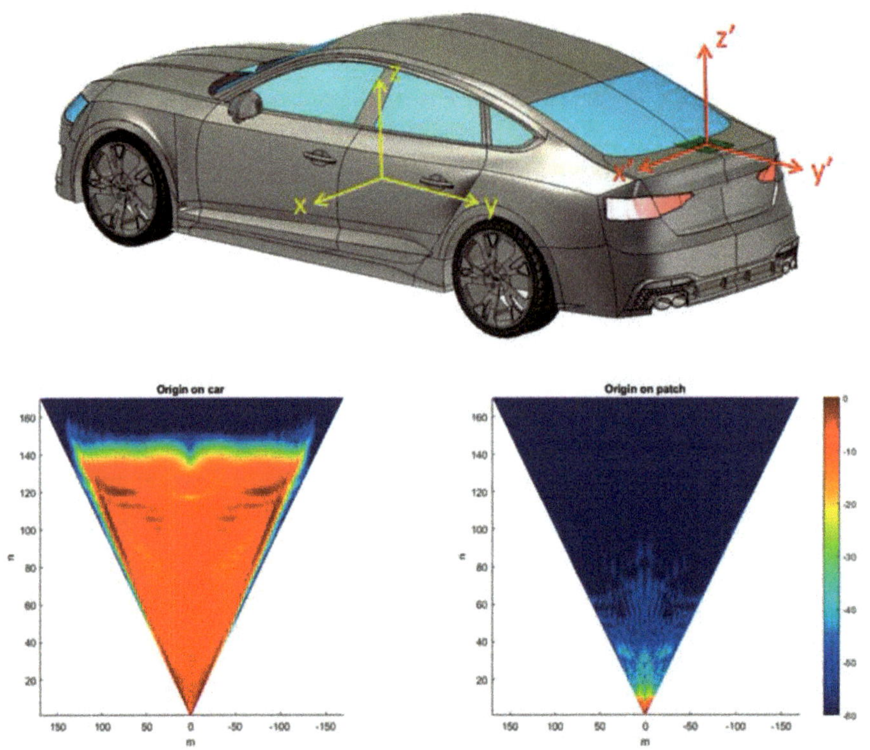

Figure 3.10 Simulated patch antenna at 2.6 GHz mounted to the rear part of a sedan car, and associated SWCs referred to the reference system center on the car (left) and on the patch (right).

The SWCs of the field radiated by the patch have been computed in the two different coordinate systems and are also depicted in Figure 3.10. When the SWCs are referred to the car center (plot on the left), the patch is placed far from the origin, increasing the modal content up to approximately n = 150 for power levels higher than –60 dB. This truncation number corresponds to a minimum sphere of 1.9m, which is roughly the distance of the patch from the origin. Instead, when the coordinate system is on the patch (plot on the right), the SWCs are much more concentrated on the low-order modes. In fact, for values of higher than 20, the power of the SWCs is below –35 dB, which corresponds to a minimum sphere of similar dimensions of the patch. However, from this point, the power decay is very slow, reaching values lower than –60 dB only after n = 100. This residual power comes from the radiating currents coupled to the car, which increases the effective minimum sphere.

To better analyze the influence of the coupling currents, the car has been removed from the simulation, and the SWCs have been recomputed. Figure 3.11 compares the P_n power spectra of the different scenarios. The blue and orange traces are the power spectra of the patch antenna installed on the vehicle, respectively, when the coordinate system is centered on the car and on the antenna. Similarly, the black and green traces refer to the same coordinates, but without the vehicle. It is interesting to see the different decays between the car and patch coordinate system. In the former, the power spectrum exhibits a cut-off behavior around

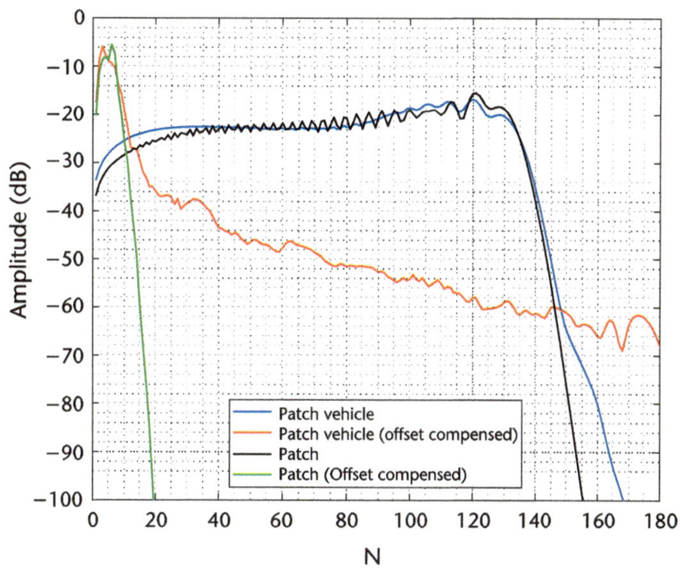

Figure 3.11 P_n spectrum of the patch antenna in free space and installed on the car for the reference system centered on the car (blue and back traces) and the one centered on the patch (orange and green traces).

$n = 150$, which corresponds to a minimum sphere that encloses the complete radiating structure. For the patch coordinate system, the power is concentrated in the first harmonics, but the coupled currents are generating a spectrum with residual power even after $n = 180$ (orange trace), because the minimum sphere required to enclose the complete structure is now bigger. If we are interested in accurate NF measurements, the car coordinate system may be the optimal choice, because it allows all the significant radiation to be captured, with approximately 150 radial modes. However, we can sacrifice some accuracy by using the patch coordinate system and reducing the maximum n-mode index to 50. In this case we would be neglecting the harmonics with a power below –45 dB, but the sampling rate can be reduced by a factor of 3 on each angular dimension. This concept, called the *local measurement approach,* is detailed in Section 3.1.3.3.

3.1.2 Truncated SNF Measurements

The SWE theory introduced in Section 3.1.1 assumes a SNF data acquisition performed over the whole, untruncated, sphere. In practice, SNF measurements are often performed on a portion of the scanning sphere for various reasons. For example, measurements of electrically large antennas such as arrays or reflectors could lead to a prohibitive scanning time, so it is convenient to measure only a portion of the sphere, where the radiated power is concentrated. In this case the NF/FF transformation can be performed by setting the missing samples to zero [zero-padding (ZP)]. For high-gain antennas, the ZP does not introduce significant errors since it is a good approximation of the actual radiated field, being very low on the angular regions far from the main beam [16]. In other situations, the full spherical acquisition may be unfeasible due to the physical size of the test object, which could be too big and/or too heavy. Automotive antenna measurements fall into this category.

As discussed in Chapters 4 and 5, in SNF automotive systems the spherical scan is typically truncated at or close to the horizon ($\theta = 90°$). The scanning surface can be terminated on a floor covered by absorbing materials [17] or on a conductive/metallic floor [18, 19]. The first solution is intended to emulate free-space conditions while the other approximates a perfect electric conductor (PEC). Such solutions have different advantages and disadvantages involving different aspects [20]. For example, the PEC-based solution reduces the downtime between measurement setup because the vehicle is simple parked in the middle of the system, while more time is usually needed to set up an absorber-based measurement because of the arrangement of the absorber around the car. On the other hand, the emulation of the performances of the vehicle over an arbitrary ground [21, 22] is strongly simplified in the case of absorber-based systems than in the case of PEC-based ones. These two aspects are detailed in Chapters 5 and 8; here we focus on the differences between these two solutions in terms of uncertainty introduced in the NF/FF transformation.

When truncated absorber-based systems are used, the ZP is normally used to fill the missing field samples. Since low/medium directivity antennas are usually installed on the vehicles, this operation could introduce abrupt field discontinuity, which, in turn, could generate the so-called truncation errors [23–28]. As in Section 3.1.2.1, such errors are more pronounced at lower frequencies, and proper advanced processing based on the SWE can be used to mitigate them.

PEC-based measurements are normally NF/FF processed using image theory [29, 30], which is used to enforce the PEC boundary condition. This allows a "virtual" untruncated spherical measurement surface to be emulated where the field on the backward hemisphere is a mirrored replica of the one on the forward hemisphere. Some residual truncation might be present close to the floor interface, because, for example, mechanical constraints might make it impossible to properly measure at these points. However, such systems are usually more immune to truncation errors than absorber-base systems.

3.1.2.1 Free-Space Truncation

As discussed above, the truncated SNF acquired in absorber-based (free-space) systems are normally extrapolated with a simple ZP before the NF/FF transformation. To illustrate the effect of ZP on an omnidirectional device, a z-oriented sleeve dipole at 1.9 GHz is considered. Figure 3.12 shows the SWC of the dipole obtained from full spherical acquisition (top-left) and from hemispherical acquisition (bottom-left). The discontinuity introduced by the ZP creates high-order modes and modifies the low-order mode distribution. The effect on the reconstructed pattern is the generation of unwanted ripples all over the pattern as illustrated by the directivity pattern comparison also shown in Figure 3.12.

It should be noted that in this example the omni-like antenna is truncated at an angle that corresponds to its maximum radiation ($\theta = 90°$), generating a strong field discontinuity when the ZP is applied. Hence, this example represents a worst-case scenario. Other measurement scenarios with higher directive antennas and/or smaller truncated areas at lower field levels are less sensitive to these types of errors.

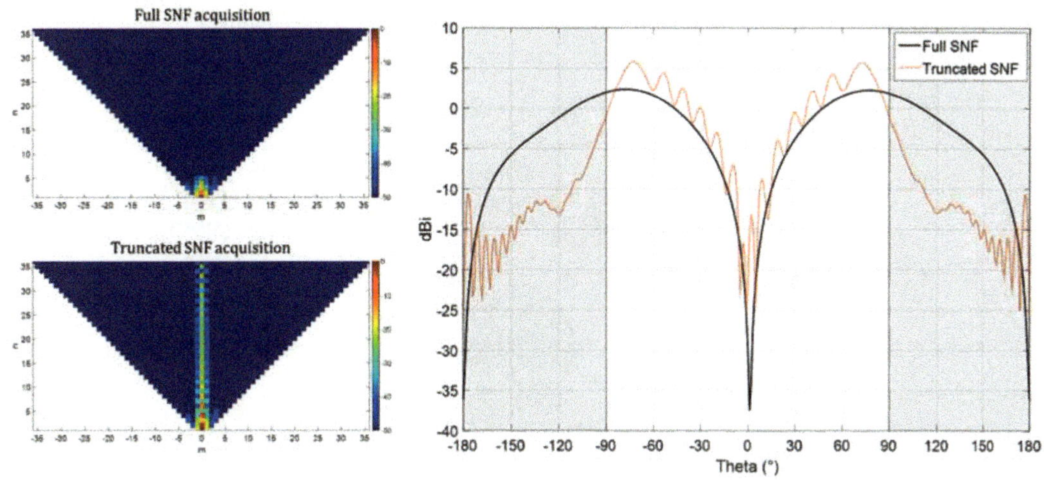

Figure 3.12 Example of truncation error for a sleeve dipole at 1.9 GHz: SWCs from SNF measurement over a full sphere (top-left), from truncated acquisition at $\theta = 90°$ (bottom-left) and associated directivity patterns (right).

It is also important to point out that truncation errors usually have a more significant impact at lower frequencies. In fact, as illustrated above, to represent the field discontinuity introduced by the truncation in the spherical wave domain, many SWCs are required (ideally an infinite number). Due to the cut-off properties of the SWE [1] the highest SWC order is given by

$$N_{max} = k R_M = \frac{2\pi f}{c} R_M \tag{3.20}$$

where f is the frequency and R_M is the measurement radius. For a given and finite R_M, the total number of computable spherical modes reduces when the frequency is decreased. Truncation errors are thus more important at lower frequencies, because the discontinuity introduced by the ZP cannot be represented by a smaller number of modes.

Different truncation error mitigation techniques have been proposed in literature [16, 23–28]. An effective and computationally efficient technique based on the SWE, called iterative modal filtering (IMF), is introduced as follows [24–27].

The block diagram of the IMF technique is shown in Figure 3.13. The ZP is first applied to the truncated area of the measured SNF and then the SWE is computed. Low-pass filtering (modal filtering) is then applied to the computed SWC. As observed in Figure 3.12, higher-order SWCs are generated when a truncated dataset is considered. Nevertheless, it is known that the AUT/DUT itself cannot radiate modes with indices higher than $N = \lfloor kR_{min} \rfloor + n_{safety}$ as already explained in Section 3.1.1. Hence, modes higher than N can be set to zero. It should be noted that this operation is simply based on the knowledge of the AUT/DUT minimum sphere (R_{min}). From the filtered SWC, the spherical NF at the same measurement radius is recomputed on the full sphere using (3.4). The computed NF samples in the truncated region are an improved estimate of the missing samples. In the last step of the procedure, these extrapolated

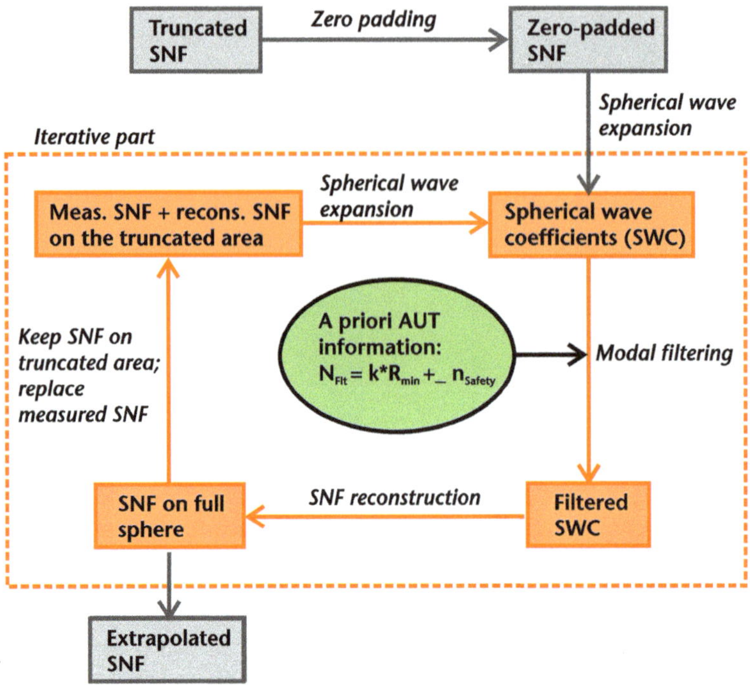

Figure 3.13 Block diagram of the IMF technique. (Reprinted with permission.) [27].

samples are kept while the recomputed NF samples in the untruncated region are replaced with the real measured data, not affected by errors due to the truncation. This process is repeated until a convergence is reached. The convergence of the algorithm can be controlled by evaluating at each iteration the deviation between the measured and the reconstructed field in the untruncated region. When a certain threshold level defined by the user is reached the iterative process is terminated.

It should be noted that to apply modal filtering, more samples are required than those associated with the AUT/DUT minimum sphere defined in Section 3.1.1. These are needed to compute higher-order SWCs and filter them out. The acquisition of additional samples is usually called oversampling (OS).

As an example, we consider the SMC700 monocone antenna mounted on a 40-cm circular ground plane (GP40) shown in Figure 3.14. This antenna is a 10-time scaled version of the SMC70 monocone antenna with a 4-m ground plane normally used as gain reference to calibrate automotive measurement systems at low frequencies (70–220 MHz). The known SWCs of the SMC700 with GP40 at 1,200 MHz have been used to emulate a simple, but realistic, measurement scenario with truncation. The reference directivity pattern obtained directly from the SWCs is shown in Figure 3.14 (solid black trace). From the same SWCs, a truncated SNF acquisition with measurement radius $R_M = 6\text{m}$ and samples in θ ranging from 0° to 100° with a constant angular increment of 2.5° (OS = 4.8) is emulated. This scenario is equivalent to the measurement of a 10-time bigger antenna (SMC70 with GP400) at 120 MHz measured in a system with $R_M = 6\text{m}$, with the same 2.5° angular spacing of the samples and with the same truncation (see scaled model technique described in [31]). The directivity pattern obtained from the NF/FF transformation of the emulated truncated SNF with ZP is also shown in

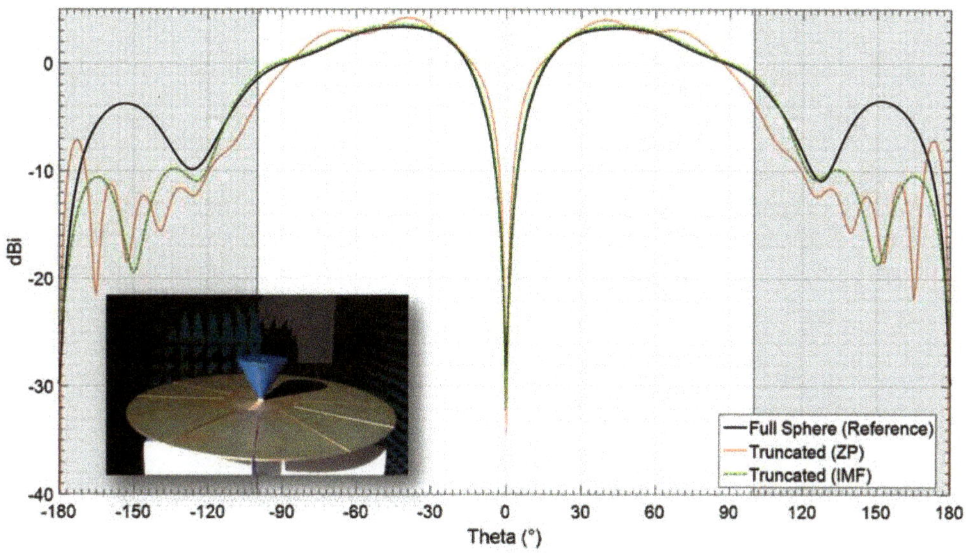

Figure 3.14 SMC700 monocone antennas with 40-cm circular ground plane: Directivity radiation pattern comparison at 1,200 MHz from SNF measurements without truncation and with truncation at (J = [−100°, 100°]) processed with the ZP and with the IMF.

Figure 3.14. As in the example of the z-oriented sleeve dipole shown in Figure 3.12, ripple is generated all over the pattern. Nevertheless, this ripple is less pronounced in this case because the truncation is applied at an angle where the field radiated by the monocone is more attenuated, and hence the discontinuity introduced by the ZP is less sharp. The dashed green trace in Figure 3.14 is the pattern obtained with the aforementioned IMF technique where the modal filtering is applied at the AUT minimum sphere (R_{min} = 20 cm). The ripple at angles corresponding to the measured samples (ϑ = [−100°, 100°]) is strongly attenuated. It should be noted that in this case a good pattern extrapolation is obtained even up to $\vartheta = \pm 120°$, hence slightly outside the untruncated area. This result has been obtained with 45 iterations of the IMF technique.

3.1.2.2 PEC Truncation

When acquisition is performed over a good conductor the floor can be assumed to be an infinite PEC. These types of measurement systems are called PEC-based systems. According to image theory [29–30], the radiation in the upper hemisphere of any device radiating above a PEC is equivalent to the same device in free space with its image superimposed. Truncated SNF measurements over a PEC can be effectively extrapolated by mirroring the measured field with respect to the conductive floor interface as shown in Figure 3.15.

If the conductive floor coincides with the horizontal plane intersecting the center of the measurement sphere (i.e., plane at $z = 0$, as shown in Figure 3.15), the field can be simply extrapolated with the PEC boundary condition shown in (3.21):

$$E_\theta^{Full}(\theta,\phi) = E_\theta(\theta,\phi) + E_\theta(180° - \theta,\phi)$$

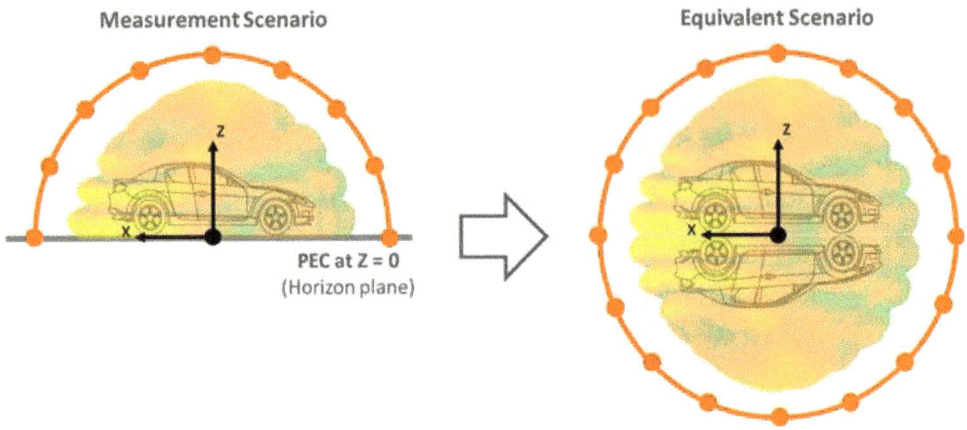

Figure 3.15 Schematic illustration of the PEC-based measurements: measurement (left) and equivalent scenario (right).

$$E_\phi^{Full}(\theta,\phi) = E_\phi(\theta,\phi) - E_\phi(180^\circ - \theta,\phi) \tag{3.21}$$

where $\left(E_\theta^{Full}, E_\phi^{Full}\right)$ are the full (measured plus extrapolated) field components and (E_θ, E_ϕ) are the measured field components with $\theta = [0{,}90°]$ and $\phi = [0{,}360°]$.

Alternatively, as described in [18, 19], the PEC boundary condition can be enforced directly on the spherical wave spectrum by computing only the SWCs for which the sum of all the mode indexes, $s + |m| + n$, is odd. Due to the properties of the spherical waves, any SWC not satisfying this condition is strictly zero [30].

It should be noted that sampling the field in close proximity to the PEC interface (i.e., close to $\theta = 90°$) is usually not feasible due to mechanical limitations, with the probe colliding on the floor, or is affected by higher uncertainty. For this reason, a full hemispherical acquisition is not possible and depending on the intensity of the field at the horizon, truncation errors could also occur in PEC-based measurements. To mitigate these errors, the missing samples in proximity to the PEC are normally extrapolated in a different way depending on the field component. For the component perpendicular to the PEC (E_θ) a linear extrapolation or a repetition of the last available sample is applied. Instead, the missing samples of the parallel component (E_ϕ) are set to zero because of the properties of the PEC boundary condition [30].

As an example, let us consider again the z-oriented dipole at 1.9 GHz used in Section 3.1.2.1 and an infinite PEC placed at $z = 0$. The dipole is displaced so that its center is at $(x, y, z) = (1.0, 0, 0.5)$m. The radiation pattern is shown in Figure 3.16 (black trace). The effect of the PEC can be clearly identified by the lobes that modify the conventional radiation of a dipole. The radius of the minimum sphere including the dipole (and its image) is $R_{min} = 1.12$m, resulting in a maximum mode index of $N = 55$. The SWC (Q_2 modes) of the dipole over the PEC are shown on the top-left side of Figure 3.16. The mode cut-off is around the expected n-index; moreover it can be seen that, due to the presence of the PEC, half of the modes are zeros. From that SWC, a simple PEC-based measurement scenario is generated, emulating the measured hemispherical SNF at $R_M = 5$m. A sampling

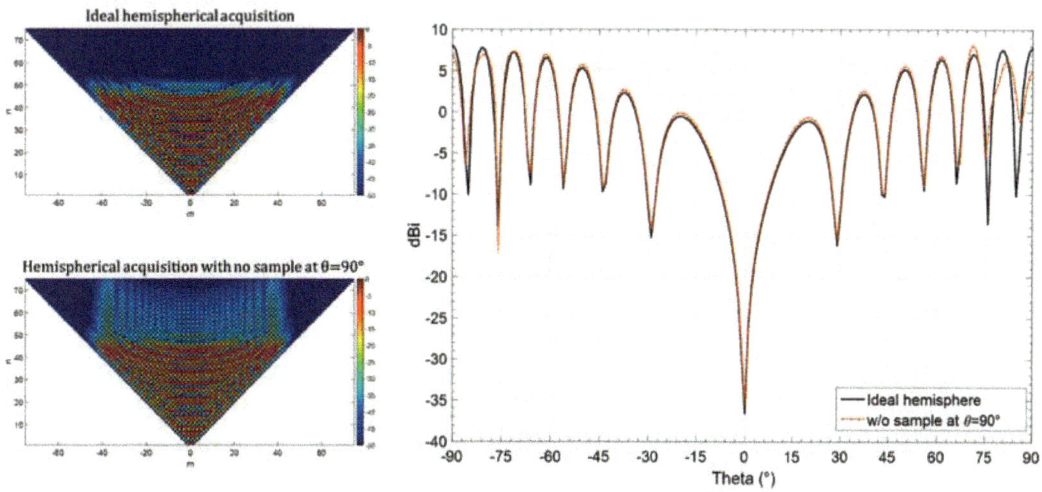

Figure 3.16 *z*-oriented dipole at 1.9 GHz placed at (x, y, z) = (1.0, 0, 0.5) over a PEC at $z = 0$: SWCs from full hemispherical acquisition (top-left), from hemispherical acquisition with missing sample at $\vartheta = 90°$ (bottom-right) and associated radiation pattern.

step of 2.4° along both the scanning coordinates has been considered. Finally, to emulate a realistic scenario, the samples at $\vartheta = 90°$ have been removed from the emulated acquisition. From that data, the NF/FF transformation has then been computed extrapolating the data at $\vartheta = 90°$ and on the backward hemisphere with the procedure described above. The obtained SWCs are shown on the bottom-left side of Figure 3.16. The effect of the missing sample close to the PEC is clearly identified by the generation of the higher-order modes. This truncation effect is also reflected on the radiation pattern as shown by the dashed red trace.

Due to the difficulties of measuring in close proximity to the PEC interface, the PEC-based systems are also not entirely immune to truncation errors. Nevertheless, since there are fewer missing samples in comparison to a free-space system, the effect due to truncation is normally small or negligible. Moreover, error mitigation techniques like the IMF technique introduced in Section 3.1.2.1 could be exploited in PEC-based systems to improve the measurement accuracy.

The techniques presented for data processing of the measurements over conductive floor are valid under the assumption that the floor interface coincides with the horizontal plane intersecting the center of the measurement sphere (i.e., $z = 0$). In some cases, for the sake of convenience or mechanical constraints, PEC-based automotive measurement systems have the floor at an arbitrary height. In such cases advanced and generalized processing tools should be used as described in Section 3.1.3.5.

3.1.3 Advanced SWE Techniques

The techniques discussed so far offer efficient and reliable ways to acquire the AUT/DUT SWCs from the measured field and perform the NF/FF transformation in different scenarios. To use these techniques, some requirements must be fulfilled, but these requirements could be too strict or even impossible to meet in some measurement scenarios. For example, as described in Section 3.1.1.1, first-order probes are needed to accurately compute the SWCs using the FFT-based technique but could limit the bandwidth of the measurement. In addition, the sampling requirements to measure a full vehicle at frequencies like 5–6 GHz, could lead to angular sampling steps in the order of 0.5°, which could be difficult to implement and/or lead to a long acquisition time. Section 3.1.2.2 notes that to properly handle PEC-based measurements, the PEC interface should coincide with the center of the measurement sphere, which is not always the case.

It should be noted that in many practical cases, if some particular measurement conditions are met, the above-mentioned requirements could be relaxed without significant loss of accuracy. For example, for a small vehicle-mounted antenna centered in the reference system, the effect due to the probe can usually be neglected and the sampling rate can be reduced considering only a significant portion of the vehicle around the antenna (i.e., local measurement approach).

On the other hand, to cope with more generic measurement situations where the antenna could be arbitrarily placed on the vehicle, advanced and generalized field expansion techniques may need to be considered. This section introduces a generalized method to compute the SWC, consisting of solving a linear system of equations. This method is extremely powerful because the forementioned limitations of the conventional SWE based method are overcome, namely, the probe compensation using generic probes, the sampling requirements for offset mounted antennas, and the handling of the PEC floor at arbitrary heights.

3.1.3.1 Generalized SWE Method

To formulate the following set of equations, the matrix notation of the transmission formula is used. This is done by arranging the measured samples in a column vector $\vec{w}$:

$$\vec{w} = \begin{pmatrix} w(r_1,\chi_1,\theta_1,\phi_1) \\ w(r_2,\chi_2,\theta_2,\phi_2) \\ \vdots \\ w(r_K,\chi_K,\theta_K,\phi_K) \end{pmatrix} = \begin{pmatrix} \sum Q_{smn} e^{j\mu\phi_1} d^n_{\mu m}(\theta_1) e^{j\mu\chi_1} P_{s\mu n}(kr_1) \\ \sum Q_{smn} e^{j\mu\phi_2} d^n_{\mu m}(\theta_2) e^{j\mu\chi_1} P_{s\mu n}(kr_2) \\ \vdots \\ \sum Q_{smn} e^{j\mu\phi_K} d^n_{\mu m}(\theta_K) e^{j\mu\chi_1} P_{s\mu n}(kr_K) \end{pmatrix} \tag{3.22}$$

and then expanding the right-hand side of (3.22) in a matrix vector product:

$$\begin{pmatrix} w(r_1,\chi_1,\theta_1,\phi_1) \\ w(r_2,\chi_2,\theta_2,\phi_2) \\ \vdots \\ w(r_K,\chi_K,\theta_K,\phi_K) \end{pmatrix} = \begin{pmatrix} V_{1,-1,1}(\vec{r}_1)\, V_{1,0,1}(\vec{r}_1) \cdots V_{2,N,N}(\vec{r}_1) \\ V_{1,-1,1}(\vec{r}_2)\, V_{1,0,1}(\vec{r}_2) \cdots V_{2,N,N}(\vec{r}_2) \\ \vdots \\ V_{1,-1,1}(\vec{r}_K)\, V_{1,0,1}(\vec{r}_K) \cdots V_{2,N,N}(\vec{r}_K) \end{pmatrix} \begin{pmatrix} Q_{1,-1,1} \\ Q_{1,0,1} \\ \vdots \\ Q_{2,N,N} \end{pmatrix} \tag{3.23}$$

where $V_{smn}(\vec{r}_k)$ is the compact form of the partial summation $\sum_\mu e^{j\mu\phi} d^n_{\mu m}(\theta) e^{j\mu\chi} P_{s\mu n}(kr)$.

Equation (3.23) becomes a linear system of equations where the unknowns are denoted as:

$$\vec{w} = V\vec{q} \tag{3.24}$$

with $\vec{q}$ being the vector of unknowns containing the SWC, and the coupling matrix. The number of unknowns is $2N(N+2)$, and the number of equations is equal to the number of measurement samples. These two numbers will be generally different making V not square, so the least-squares version of (3.24) is used:

$$V^H\vec{w} = V^H V\vec{q} \tag{3.25}$$

where $\vec{q}$ is found using an iterative matrix inversion algorithm such as conjugate gradient (CG) [32], which is more efficient and stable than a direct inversion.

Using this approach, we have high flexibility on the parameters of the input field, because the matrix formulation supports arbitrary locations of the measured points, probe type, and polarizations. The downside is that a linear system of equations must be solved, which is less computationally efficient, and some care must be taken to ensure a proper conditioning of the system matrix ($V^H V$). Experimental results suggest that when using standard methods like CG to solve (3.25), the required sampling rates are similar to those needed for the Wacker algorithm described in Section 3.1.1.1. More advanced solvers like *l1-minimization* can be used to obtain significant reductions on the required number of measured samples, but this topic is still under investigation [33].

3.1.3.2 Full Probe Compensation

The probe's effect on SNF measurements depends on many factors. First, an electrically large probe could distort the measured pattern because the NF is not sampled at a single point in space; it is the result integrated over the probe aperture. This distortion is accentuated for an electrically large AUT/DUT because of the rapidly varying field. On the other hand, when electrically small antennas and/or probes are considered, the probe effect tends to be negligible. Another important factor is the AUT-probe view angle (α), illustrated in Figure 3.17, and defined as $\alpha = \tan^{-1}\left(\frac{R_{min}}{R_M}\right)$. For relatively small values of α (e.g., $\alpha < \pm 10°$) the probe compensation in SNF measurements can often be neglected without significant loss of accuracy.

Most vehicle-installed antennas are electrically small, but they are arbitrarily mounted on the car structure giving rise to possible large AUT-probe view angles, which could make the probe effect nonnegligible.

The FOPC, or Wacker, approach described in Section 3.1.1.1 can be used to compensate for the probe effect, but as already pointed out, it requires the probe to radiate only $\mu = \pm 1$ azimuthal modes, reducing its applicability, especially for large measurement bandwidths (e.g., more than an octave).

The generalized SWE method introduced in Section 3.1.3.1 can be exploited to relax the constraint related to the probe modal content and hence apply a *full* or *higher-order probe compensation* [32–37]. The clear advantage is that wideband

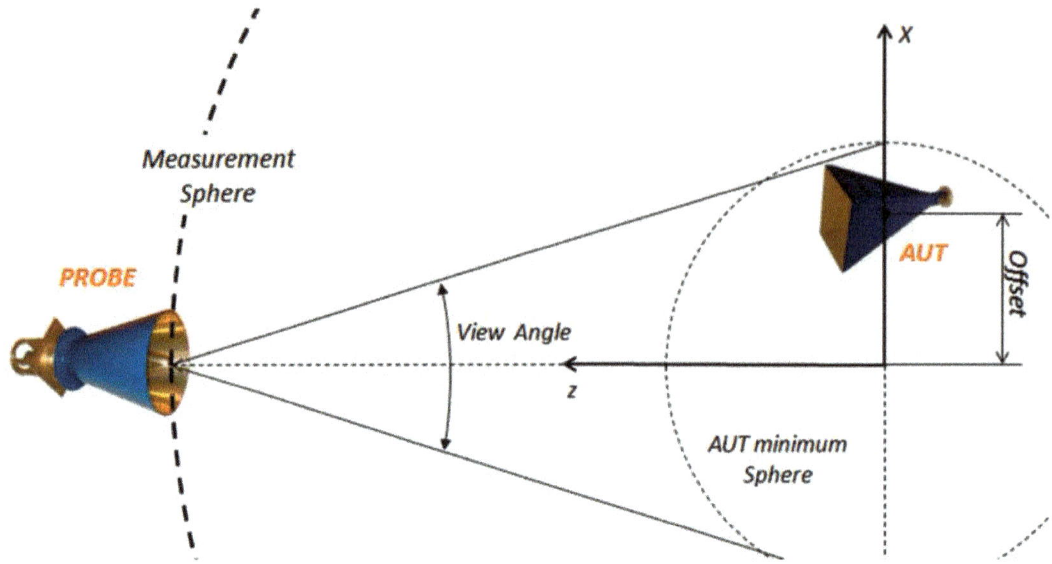

Figure 3.17 Illustration of the AUT-probe view angle.

antennas with more than a decade of bandwidth can conveniently be selected as the probe improving the efficiency of the measurements (i.e., less probes to be used).

In addition to the mathematical formulation reported in Section 3.1.3.1, the full PC approach can be easily understood by considering the SWE in (3.4). This expression does not allow the probe effect to be accounted for unless the spherical wave functions are modified to properly include it. Ad-hoc spherical wave functions can be defined with the transmission formula (3.6), as described in [35]. The projection of the measured field over the ad-hoc expansion base allows the full probe compensation to be achieved.

The drawback of this approach is that the new basis expansion is not orthogonal anymore; hence a matrix inversion approach must be used instead of an FFT. More specifically, in this case the orthogonality of the spherical wave functions is lost only along the θ-coordinates, while is kept along the ϕ-coordinates. The full-probe correction approach can hence be applied considering a hybrid FFT/matrix inversion approach as described in [32].

An example of application of the forementioned full PC approach is shown in Figure 3.18, where the SNF measurements of a x-band standard gain horn is shown [35]. The antenna has been measured in an offset configuration with a higher-order probe (the quad-ridge horn shown in Figure 3.18). The spherical modes of this probe at 12 GHz are shown on the left side. As can be seen, the probe spectrum has nonnegligible higher-order azimuthal modes up to approximately $m = 30$ (see trace in red). The AUT radiation patterns obtained by processing the measured SNF without the PC (black trace), with the FOPC (green trace), and with the full PC (orange trace) are shown on the right. The reference pattern (blue trace) is obtained by measuring the same antenna with a conventional first order and electrically small probe. There are significant deviations from reference when no PCs or FOPCs are considered due to the higher-order modes of the probe not being accounted for in the NF/FF processing. On the other hand, for both the co-polar and the cx-polar patterns, a very good agreement is achieved when the full PC is applied.

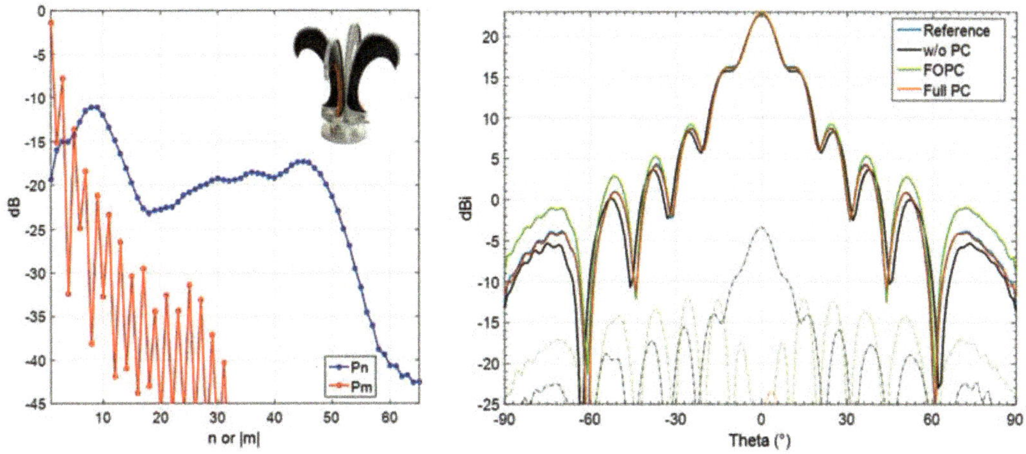

Figure 3.18 Example of application of the full probe compensation technique. Standard gain horn at 12 GHz measured by a quad-ridge horn, higher-order probe. Solid and dashed traces are respectively the co-polar and cx-polar radiation patterns.

Specific examples relative to the application of the full PC approach to automotive measurements can be found in Sections 5.3.1 and 5.3.2.

It is finally highlighted that the full PC approach also supports the use of dual-polarized probes. Dual-polarized probes are often considered to measure two orthogonal field components at the same time to reduce the acquisition time by approximately half. With FOPC the pattern radiated by the second port of the probe is assumed to be a 90° rotated replica of the first port. This assumption is not always appropriate, especially when wideband and complex probes are considered. With this approach, two sets of probe coefficients are considered (one for each port of the probe) and the corresponding probe response constants are computed as shown in (3.26) $P^{p}_{s\mu n}(kr)$, where p is used to refer to the two ports of the probe). It should be noted that the $e^{j\mu\chi}$ term, needed to rotate the probe around its axis, is now not needed because the probe rotation is already included in the second set of probe coefficients. This probe correction approach is usually called the *dual-polarized PC* technique [37].

$$w^{p}(r,\theta,\phi) = \sum_{\substack{smn \\ \mu}} Q^{(4)}_{smn} e^{jm\phi} d^{n}_{\mu m}(\theta) P^{p}_{s\mu n}(kr) \tag{3.26}$$

3.1.3.3 Down-Sampling in Translated SNF Measurements

As described in Section 3.1.1.2, the sampling rate required for SNF measurements depends on the electrical size of the minimum sphere enclosing the DUT. Because the minimum sphere is defined with its center at the origin of the coordinate system, a common practice is to place the geometrical center of the DUT at that origin so that the minimum sphere is kept to a minimum, thus reducing the number of required NF samples and the measurement time.

SNF measurements of a common 5-m vehicle (i.e., R_{min} = 2.5m) at high frequencies such as 6 GHz requires an angular sampling rate of approximately 0.5°.

As described in Chapter 5, some measurement systems do not allow this dense sampling. In other instances, this level of sampling can be realized, but it could lead to an excessive acquisition time.

To overcome these limitations the so-called *local measurement approach* can be used together with a proper definition of the coordinate system to apply the SWE.

The local measurement approach assumes that, in general, only a relatively small portion of the structure around the antenna significantly contributes to the final radiation. This concept is illustrated in Figure 3.19, where the full wave simulation of vehicle fed by a 5.9-GHz monopole antenna is considered. The simulated radiation pattern, accounting for the coupling of the antenna with the whole car structure, is depicted in black in the plots at the bottom of the figure. By placing the coordinate system on the antenna, the SWE has been applied and truncated to n-indexes associated to spheres of radius 5λ, 10λ, 15λ, and 20λ. The corresponding radiation pattern cuts, along the car's longest dimension, are also shown in Figure 3.19. The solid traces are the patterns associated with the considered portion of the spheres, while the dotted ones are associated with the filtered portions. Considering the 5λ-sphere, the ripple caused by the interaction with the car body is filtered out, but the overall pattern envelope is maintained. As expected, when larger spheres around the antenna are considered, more accurate reconstructions of the radiation pattern are achieved. Nevertheless, it is observed that sufficient accuracies can be achieved by considering only a small portion of the vehicle around the antenna. The measurement accuracy of the local approach is quantified in Table 3.1, where the gain and TRP errors, computed for each sphere, are reported. It can be concluded that considering an equivalent sphere of 10–15λ radius around the antenna would allow a measurement accuracy suitable for most applications to be achieved.

The local measurement approach can sometimes be implemented by placing the vehicle in the measurement system so that the position of the fed antenna corresponds to the center of the measurement sphere. Unfortunately, this is often

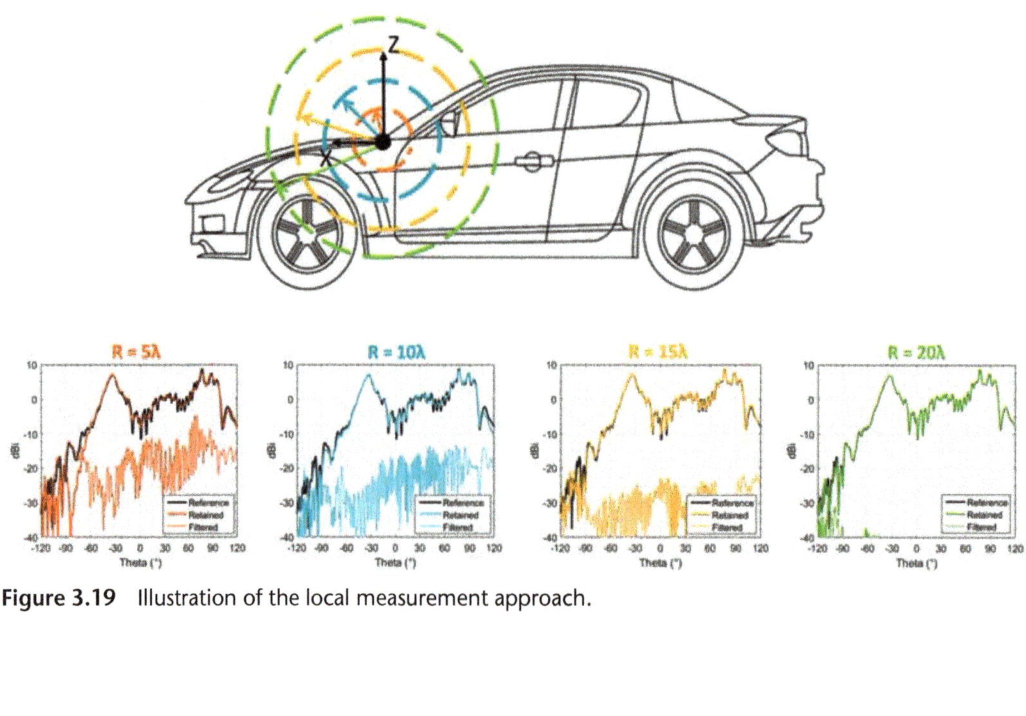

Figure 3.19 Illustration of the local measurement approach.

Table 3.1 Gain and TRP Errors Obtained with the Local Measurement Approach

Radius (λ)	*Radius (m)*	*Gain Error (dB)*	*TRP Error (dB)*
5	0.25	1.0	0.2
10	0.50	0.6	0.1
15	0.76	0.2	<0.0
20	1.02	<0.0	<0.0

unfeasible because of space constraints. Moreover, it is usually desirable to perform the measurements without repositioning the vehicle for each antenna location. Keeping the vehicle in the same position reduces the downtime between measurements and also improves the measurement accuracy (e.g., especially for diversity evaluations).

To overcome this limitation, it is convenient to define a local coordinate system (x',y',z') in the center of the antenna during the application of the SWE, as shown in Figure 3.20. The SWE with an arbitrary defined coordinate system was first introduced in [38] and is typically called translated-SWE (or simply TSWE). The choice of an optimal reference system allows the size of the minimum sphere to be kept to a minimum, reducing the sampling requirements because a smaller truncation index $N_t < N$ can be used (i.e., less SWCs are needed to represent the radiated field as observed in Figures 3.10 and 3.11).

From Figure 3.20 it is evident that the local reference system defines an offset measurement sphere that complicates the relative distribution of the NF samples, which are at different distances from the new origin and are not evenly spaced anymore. Consequently, the FFT-based expansion cannot be applied, and the

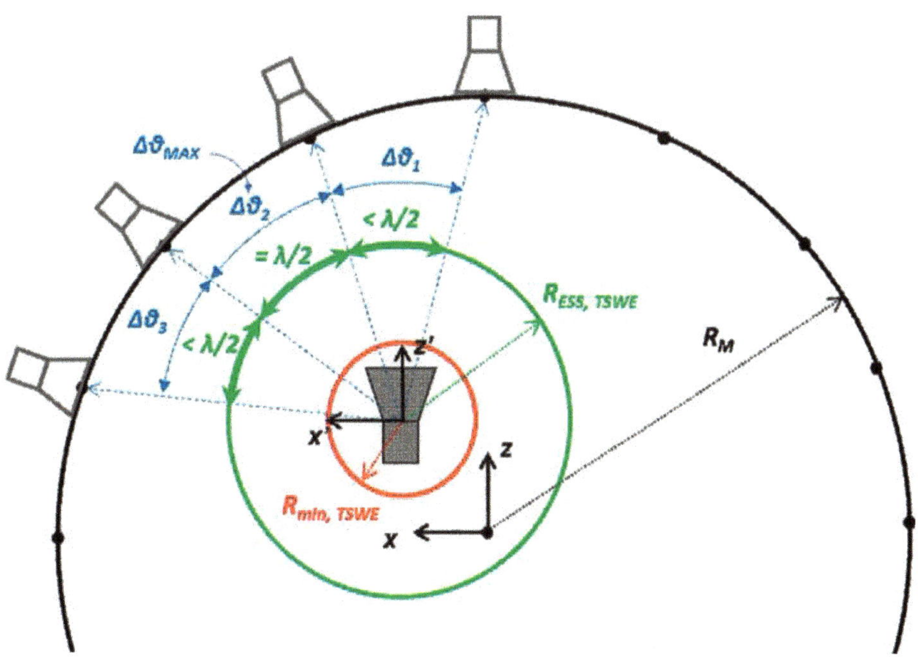

Figure 3.20 Illustration of the TSWE concept: Definition of the local coordinate system on the antenna and sampling implications.

generalized SWE approach described in Section 3.1.3.1 must be applied instead. Like the full PC approach, the generalized SWE is now used by redefining the spherical wave functions with respect to the translated coordinate system.

The downside of the TSWE is the increase in the complexity of the expansion, which must be performed with a matrix inversion. The complexity can be significantly reduced when the local reference system is translated only along the *z*-direction. In fact, in this case the constant spacing of the azimuthal NF samples is kept, and, like the full PC technique, a hybrid FFT/matrix inversion approach can be adopted [38]. Instead, when a generic *xyz*-translation is considered, the regular spacing of the sampling is lost both along the θ- and ϕ-axis; hence, a full-matrix inversion must be applied as done in [39]. It should be noted that a generic *xyz*-translation is typically found in vehicle-mounted antennas. An approximate solution of the TSWE that is able to efficiently deal with this type of automotive measurement scenario has been proposed in [40].

It is also important to specify that the nonuniformity of the measured samples in the local coordinate system has some implications on the equivalent sampling sphere (ESS) and on the sampling requirements. From the constant angular step $\Delta\theta$, relative to the original coordinate system, nonequally spaced samples $(\Delta\theta_1 \neq \Delta\theta_2 \neq \Delta\theta_3 \neq \cdots)$ are generated when the local coordinates are defined. (See Figure 3.20.) As already defined in (3.15), the radius of the ESS relative to the original coordinates is $R_{ESS,SWE} = \lambda / (2\Delta\theta)$. In the local coordinates, the maximum angular step $(\Delta\theta_{MAX})$ must be considered to ensure a maximum spacing between the samples of half-wavelength:

$$R_{ESS,\,TSWE} = \frac{\lambda}{2\,\Delta\theta_{MAX}} = \alpha R_{ESS,SWE} \tag{3.27}$$

$$\alpha = \frac{\Delta\theta}{\Delta\theta_{MAX}} \approx \left(1 - \frac{|offset|}{R_M}\right) \leq 1 \tag{3.28}$$

When the TSWE is applied, the ESS is reduced by the α-factor defined in (3.28), which depends on the offset (Euclidian distance between original and local coordinates) and the measurement radius. The α-factor is purely geometric, and it corresponds to the oversampling to be applied to the SNF offset measurements processed with the TSWE. The sampling step to measure a certain minimum sphere arbitrary displaced in space $\left(R_{min,TSWE}\right)$ is given by

$$\Delta\theta_{TSWE} \leq \alpha\left(\frac{\pi}{k\,R_{min,\,TSWE} + n_{safety}}\right) \tag{3.29}$$

To demonstrate the capabilities of the TSWE technique we consider again the simulated automotive scenario at 2.6 GHz presented in Section 3.1.1.3. (See Figure 3.21.) First, we considered the conventional, full SNF measurement with the origin at the center of the car. As previously discussed, this means that the patch antenna, mounted on the rear part of the car, is considerably offset, and requires SWC up to $n = 150$. Under such conditions, a standard SNF measurement requires a sampling (θ, ϕ) grid of 151 × 300 points ($\Delta\theta = \Delta\phi = 1.2°$).

With TSWE, a local coordinate system can be defined on the antenna center, and the truncation number can be significantly reduced. The equivalent minimum sphere must account for the currents coupled to the car body. Using the results of the local measurement approach of Figure 3.19, a minimum sphere radius of 11λ is selected as a compromise between accuracy and reduction of measurement data. This results in a truncation number $N = 80$, so using TSWE the NF can be sampled with a reduced grid of 101×200 points. This accounts for the α-factor previously introduced, considering the offset and measurement radii are 2m and 9m, respectively. Therefore, a sampling reduction of more than 50% is achieved, which will result in a significant reduction in measurement time.

Figure 3.21 depicts the radiation pattern computed from the fully sampled grid using standard NF/FF along with the pattern obtained with the down-sampling and the TSWE technique. The agreement between the two methods is excellent, showing the potential of the TSWE method. The differences, on the order of –40 dB, are due to the local measurement assumption and the intrinsic filtering of part of the vehicle, as described above. (See the example in Figure 3.19.)

3.1.3.4 Down-Sampling and Probe Compensation in Translated SNF Measurements

The transmission formula introduced in Section 3.1.1.1 assumes that the probe is always pointing to the origin of the coordinate system, which is usually maintained in conventional SNF measurements. However, when the TSWE is applied and the local coordinate system is defined, this condition is broken. The application of the probe compensation in an undersampled measurement based on the TSWE is an additional challenge because the orientation of the probe changes at the different sampling points. To address this issue, three different techniques have been proposed.

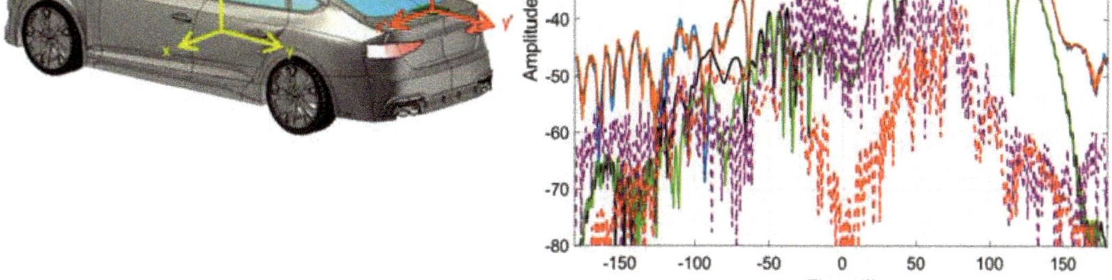

Figure 3.21 Comparison of the transformed FF using the standard transformation algorithm with 160×321 samples (reference) and the proposed technique with 101×200. ©2021 IEEE. (Reprinted with permission.) [71].

The first approach is based on a further generalization of the matrix V in (3.10) to incorporate the necessary angular rotations to model the probe orientation on each measurement point [41]. This is equivalent to measuring the NF with a different probe on each point whose probe response constants are given by:

$$P_{s\mu n}(kr,\alpha,\beta,\gamma)=\sum_{\sigma=1}^{2}\sum_{\substack{\nu=|\mu| \\ \nu\neq 0}}^{V}C_{\sigma\mu\nu}^{sn}(kr)\sum_{\mu'=-\mu}^{\mu}e^{j\mu\alpha}d_{\mu'\mu}^{\nu}(\beta)e^{j\mu'\gamma}R_{\sigma\mu'\nu} \tag{3.30}$$

where α, β, γ are the rotation angles of the probe with respect to the new coordinate system. Once the new probe constants are incorporated in the system matrix, the linear system of equations can be solved without any further considerations. Therefore, with this approach a more generalized probe correction is achieved but at the expense of an increased computational time to evaluate (3.30) for each measurement point.

The second approach is a two-stage procedure where the effect of the probe is first neglected, by assuming an ideal Hertzian dipole as the probe and the application of the TSWE [42]. To compensate for the probe influence, the SWCs are then translated to the original coordinate system, and then propagated to the measurement sphere to retrieve a "virtual" acquisition with full sampling (i.e., interpolation). The second step consists of retrieving the probe-corrected SWCs from this virtual acquisition. Because we now have a full sampling grid, the conventional FOPC or the full PC introduced in Section 3.1.3.2 can be applied as described above. This approach shows an improved computational performance because it avoids the use of complicated probe rotations required in the first method.

The third approach is like the previous one, because it alternates between the translated and original coordinate systems, but multiple times [43]. On each step of the matrix iterative inversion algorithm, the matrix-vector products are replaced by fast operators that make the calculations "on the fly," thus avoiding the need to compute and store V. These fast operators are derived by noting that V can be defined as the product of two matrices:

$$V=(P)(C) \tag{3.31}$$

where C performs a translation of the SWC from the translated coordinate system to the original one, and P is a matrix that calculates the field on the measurement sphere. The use of matrices is just for notation purposes, because in practice the translation of the coordinate system can be done very efficiently by propagating the SWC to the FF, adding a phase shift, and back-propagating to the SWC domain. After this translation, we are back to the original coordinate system, where the samples are uniformly spaced, and the probe is always pointing to the center. Therefore, matrix P can be replaced by a fast evaluation of the transmission formula using FFT in both θ and ϕ. This approach avoids the use of matrices obtaining a very high computational efficiency, but the translation-propagation approach must be performed at each step of the iterative inversion algorithm, which is complex to implement.

3.1.3.5 SWE with Conductive Floor at Arbitrary Height

As described in Section 3.1.2.2, automotive measurements are often performed over conductive floors because of the easy placement of the vehicle and the significant mitigation of the truncation errors during the NF/FF transformation. The latter is achieved with proper enforcement of the PEC boundary condition, which can easily be applied with a mirroring of the field if the floor coincides with the center of the measurement sphere (i.e., PEC at $z = 0$, as in the top-left image of Figure 3.22).

In some measurement scenarios, due to mechanical/geometrical constraints the conductive floors do not coincide with the center of sphere (i.e., PEC at $z \neq 0$, as in the bottom-left image in Figure 3.22). As discussed in Chapter 5, situations like this can be found when absorber-based measurement systems are used as PEC-based by removing the absorbing material from the floor. The TSWE described in Section 3.1.3.3 can be used to manage these types of measurement scenarios to properly enforce the PEC boundary condition. The TSWE can be used to define the local coordinate system on the PEC interface as shown in the bottom-left illustration of Figure 3.22. The SWCs are then directly computed with the PEC boundary condition applied. Due to the properties of the spherical waves, the PEC condition is applied by computing only the SWCs for which the sum of all the mode indexes, $s + |m| + n$, is odd. Any SWC not satisfying this condition is strictly zero [18, 30].

It should be noted that with this type of application of TSWE, the translation of the reference system is always along the z-axis, meaning that the expansion can be efficiently computed with the FFT/matrix inversion approach (i.e., the FFT applied along the ϕ-axis).

An example of application of the TSWE to automotive measurements with the conductive floor at an arbitrary height is shown on the right side of Figure 3.22 [44]. The elevation radiation pattern of a patch antenna at 500 MHz mounted on the roof of a vehicle is shown. The blue trace, obtained by measuring the car in a

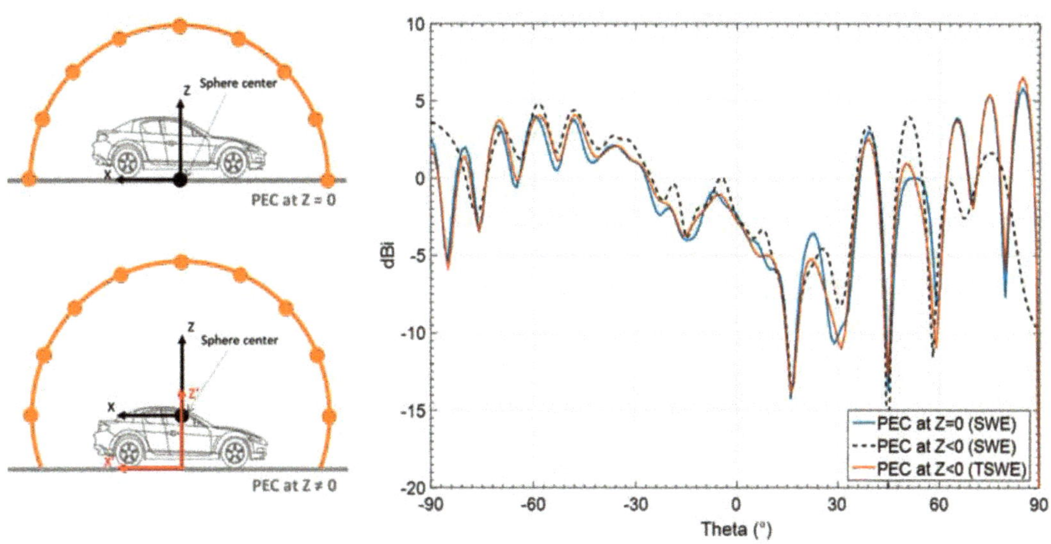

Figure 3.22 Application of the TSWE technique to PEC-based measurements with the floor at an arbitrary height.

conventional PEC-based scenario with the floor at $z = 0$, is considered as a reference. The same vehicle was then measured in a similar system, but with the conductive floor at $z = -132$m. The application of the conventional SWE to this second measurement shows nonnegligible errors, especially at scanning angles close to the PEC interface (see black dashed trace). Using the TSWE permits the proper application of the PEC boundary condition, and the resulting radiation pattern (orange trace) is in good agreement with the reference.

3.2 Equivalent Current Technique

This section introduces the basic concepts of the equivalent current (EQC) technique, which essentially consists of the computation of equivalent sources with known locations that radiate a given field (called the goal field). This technique has attracted considerable interest [45] given its appealing advantages over the well-known wave expansion techniques, most notably a simple and unified treatment without binds to any coordinate system.

In view of the equivalence principle, several formulations of the source reconstruction problem are possible, and some may prove advantageous in particular applications [46]; for example, it has been shown that for diagnostic applications, Love's form of equivalence is preferred since the resulting equivalent sources bear a direct relationship with the actual source of the target field, providing useful information on it [46]. Interestingly, Love's form provides additional advantages such as numerical stability and higher accuracy for NF/NF transformation [47].

3.2.1 Formulation

The equivalence principle [3] allows substituting all radiators/scatterers within a volume Ω by a distribution of equivalent currents on Ω's boundary (a closed surface) denoted by Σ_R. (See Figure 3.23.) Field equivalence outside Σ_R results from the enforcement that the equivalent currents radiate the same tangential field as measured at a surface Σ_M (the measurement range):

$$\hat{n} \times E(r) = \hat{n} \times \left[-\eta_0 \mathcal{L}\left(J_{eq}; r\right) + \mathcal{K}\left(M_{eq}; r\right) \right] \qquad r \in \Sigma_M \tag{3.32}$$

where:

$$\mathcal{L}\left(J_{eq}; r\right) = jk_0 \int_{\Sigma_R} \left[J_{eq}(r') + \frac{1}{k_0^2} \nabla \nabla'_s \cdot J_{eq}(r) \right] g(r, r') ds' \tag{3.33}$$

$$\mathcal{K}\left(M_{eq}; r\right) = \int_{\Sigma_R} M_{eq}(r') \times \nabla g(r, r') ds' \tag{3.34}$$

$$g(r, r') = \frac{e^{-jk_0 |r - r'|}}{4\pi |r - r'|} \tag{3.35}$$

The solution of (3.32) is not unique [3]; therefore, uniqueness needs to be enforced.

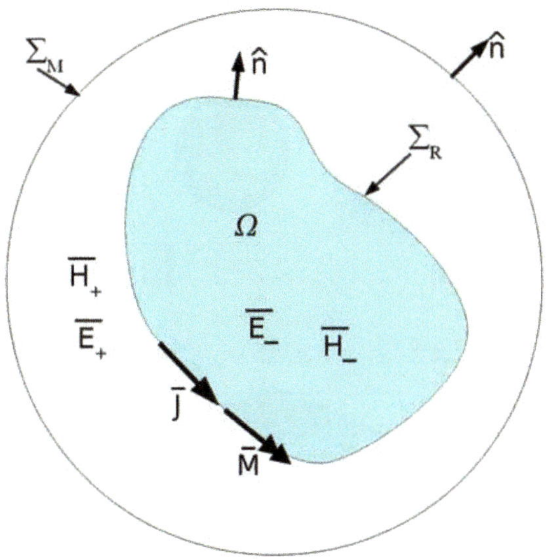

Figure 3.23 Depiction of the equivalence theorem. ©2011 IEEE. (Reprinted with permission.) [72].

Love's formulation results in currents proportional to actual fields and thus is the most appropriate for diagnostics, where fields give direct feedback on the presence of sources and/or certain materials beneath the surface (e.g., metal conductor). Love's form results from enforcement of the following boundary integral identities [46].

$$\hat{n}\times\left[-\eta_0\mathcal{L}\left(J^{Love};r\right)+\mathcal{K}\left(\mathrm{M}^{Love};r\right)\right]=-\frac{1}{2}M^{Love}(r) \quad r\epsilon\Sigma_R \tag{3.36}$$

$$\hat{n}\times\left[-\frac{1}{\eta_0}\mathcal{L}\left(\mathrm{M}^{Love};r\right)+\mathcal{K}\left(J^{Love};r\right)\right]=\frac{1}{2}J^{Love}(r) \quad r\epsilon\Sigma_R \tag{3.37}$$

Equations (3.31) to (3.37) are solved via the standard method of moments (MoM), employing discretization of the surface ΣR in terms of the Rao-Wilton-Glisson (RWG) basis. For testing, Dirac delta functions are used for (3.32) while the Galerkin method is used for (3.36) and (3.37) with proper treatment of the singularities.

The EQC technique has recently implemented a multifrequency formulation based on frequency interpolation of data [48] and a fast technique based on a nested skeletonization scheme (NSS) [49] to better manage electrically large cases in terms of computational time and occupied memory, as for large applications in automotive.

3.2.2 Features

This section discusses the many capabilities of the versatile EQC technique.

From the measured NF, the EQC technique allows for the computation of the equivalent electric and magnetic currents on an arbitrarily shaped reconstruction surface fully enclosing the antenna/DUT. An example of the workflow is shown is shown in Figure 3.24, where the DUT is the antenna mounted on the vehicle.

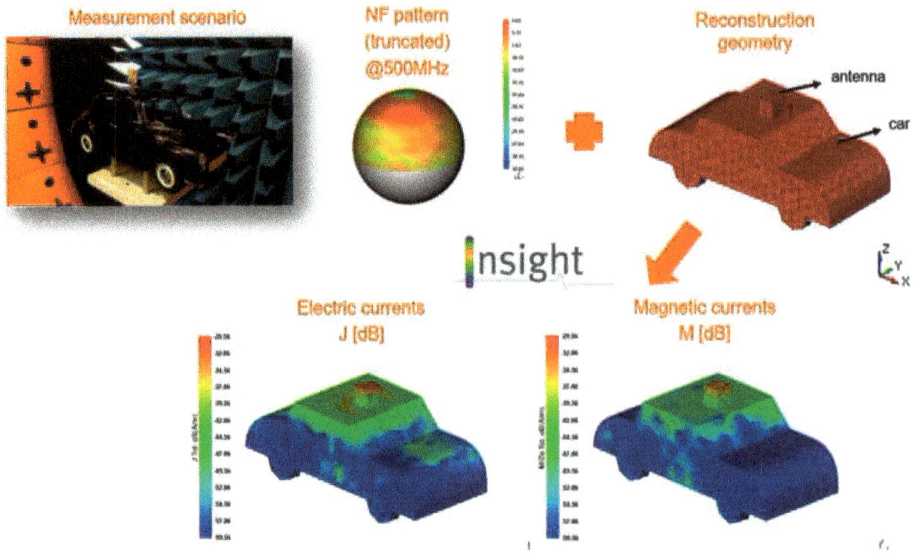

Figure 3.24 Workflow calculations of the equivalent currents.

According to the equivalence principle, once the EQCs are known, the electric and magnetic field can be computed everywhere outside the reconstruction geometry. The EQC technique can thus be used as an NF/FF transformation tool applicable to any kind of scanning surfaces, including the canonical ones (planar, cylindrical, and spherical), emerging scanning techniques like plane-polar and spiral scanning, and irregular ones (for example, generated from drone acquisitions). Besides its versatility, when the EQC technique is used as the NF/FF transformation of partial NF acquisitions, it is also more robust against truncation errors than other expansion techniques mentioned here. (See [26, 27] for more details.) On the other hand, the main drawback of the EQC technique is its computational cost. Unlike the SWE, cylindrical wave expansion (CWE), and plane wave expansion (PWE), the EQC expansion is not an FFT-based method. Its implementation is based on the inverse MoM, which is more computationally demanding.

In addition to the usage of the EQC method as an NF/FF transformation tool, this technique allows the following:

- Advanced diagnostics of the AUT/DUT to determine root causes of faults, such as find antenna array element defects, excitation errors, loose connections, and sources of asymmetry/unwanted coupling/depolarization [50–52];
- Determination of interactions between the antenna and its surroundings (e.g., positioner, mounting structure, and fixture) or detection of antenna feeding leakage in measurement systems [53];
- Artificially remove unwanted currents from the measurement (i.e., currents from cables and fixtures) and monitoring for improvement in the antenna radiation pattern [54];
- Evaluation of NF and/or power densities very close to the radiating antenna surface with NF-to-NF transformation [55].

- Troubleshooting electromagnetic compatibility (EMC) shielding, electomagnetic sensitivity (EMS) investigation, and pinpointing sources (bright spots) on the shielding structure allowing for the identification of unwanted radiation sources [56];
- Extrapolation of truncated areas to reduce errors [26, 27];
- The use of measurement in numerical computations of the antenna in a larger EM problem [57].

The last feature is widely used in complex environments or antenna placement analyses requiring knowledge of the exact model of antenna to be used with the final structure (e.g., vehicles, aircrafts, satellites, and ships) in computational electromagnetic (CEM) solvers. Very often, the antenna comes from third parties that might not be allowed to share details of their design. Furthermore, the numerical model of the antenna does not take into account manufacturing tolerances which might create discrepancies between the modeled and the realized device. The EQC computed from measurements of the antenna allows a link to be established between measured data and the simulations, enabling accurate antenna placement analysis. By doing this, the antenna is seen as a "black-box," and no additional information is required.

Examples of automotive measurements postprocessed with the EQC method are reported in Chapter 7, which focuses in more detail on the diagnostics, filtering/removal of unwanted currents from the measurement, NF/FF transformations, extrapolation of truncated areas, and links between measurements and simulations.

3.3 New Methods for NF-to-FF Transformations

Standard SNF measurements are implemented using angular positioners to acquire the samples in a regular (θ, ϕ) grid. The most common sampling scheme is the scan-and-step mode, where the positioner is rotated continuously in one axis (θ or ϕ), while the radiofrequency equipment is triggered to acquire the samples on the desired angular coordinates. Once the scan is complete, the positioner steps to the next point on the other axis (ϕ or θ, respectively) to perform a second scan. This process is repeated until the complete solid angle is covered. The resulting grid from this process is denoted as equiangular sampling since it is equally spaced in both angular coordinates. The simplicity to implement it, combined with the possibility of using very efficient FFT techniques for its post-processing, have made the equiangular grid the most popular sampling grid in SNF measurements.

In recent years there has been an increasing trend to use alternative sampling schemes for SNF measurements. This stems from the following two factors:

- The development of nonredundant sampling distributions so the minimum number of points is measured to minimize the scanning times;
- The use of scanning paths (like spirals) that minimize the measurement time so that the positioner covers all points as fast as possible.

This section continues to review a series of methods that employ alternative sampling schemes.

3.3.1 Sparse Recovery of SWC

A signal is known to be sparse when it contains a small number of nonzero elements compared to its dimension. The SWCs of conventional antennas are usually sparse, due to the symmetries exhibited by their radiation patterns, which cancel most of the possible wave components from the complete set.

The sparsity of the SWCs can be exploited using *l1-minimization* techniques in the framework of compressed sensing [58], which is able to retrieve the solution vector with a small number of equations compared to the number of unknowns. A common sparse recovery solver is the basis pursuit denoising [59]:

$$\min_{x} x_1 s.t. Px - w_2 \leq \sigma \tag{3.38}$$

where the parameter controls the tolerance of the fitting.

Multiple studies exist [60–63] on the applicability of sparse SWC reconstruction over different measurement scenarios and theoretical bounds for the minimum number of samples required for a successful transformation. One of the most important of the parameters that affect the performance of these methods is the sampling grid. To this end, alternate sampling grids to the equiangular have been proposed, including spiral, igloo or random sampling grids, which have shown improved performance in sparse retrieval solvers.

Theoretically, the use of sparse spherical measurements allows a considerable reduction in the number of measurement samples. Measurement sample savings of levels up to 90% have been reported [62], but this is not directly translated on a measurement time reduction because the obtained sampling grids cannot be acquired with the same speed as the equiangular one. The length of the scanning path and the number of times that the positioner needs to step should be considered when designing the sampling grid. This is a complicated task, and, in practice, the savings in measurement time are more modest due to these mechanical considerations [63].

3.3.2 Optimal Sampling Interpolation

The field radiated by a given AUT can be represented by a finite set of spherical waves because it is a band-limited signal. Band-limited signals can be reconstructed accurately in arbitrary points of their domain if they have been previously characterized in a finite set of samples. This property is exploited in SNF measurement by studying the radiation properties of the electromagnetic radiation to design a grid that captures all degrees of freedom with the minimum number of samples [64, 65]. Then, an optimal sampling interpolation (OSI) is applied to recover the NF in the required additional points so that conventional FF transformation techniques can be applied.

The nonredundant sampling representation is derived by defining first the so-called reduced voltage measured on the probe over a curve defined by an opportune parametrization $\vec{r} = \vec{r}(\eta)$:

$$\tilde{w}(\eta) = w(\eta)e^{j\psi(\eta)} \tag{3.39}$$

with $w(\eta)$ being the probe voltage evaluated on the parametrized curve and $\psi(\eta)$ a phase function that reduces the highly oscillatory behavior intrinsic to the AUT phase response. This phase function depends on the AUT geometry and the curve parametrization.

The reduced voltage can be retrieved via the OSI expansion:

$$\tilde{w}(\eta) = \sum_{n} \tilde{w}(\eta_n)\Omega_N(\eta - \eta_n)D_{N'}(\eta - \eta_n) \tag{3.40}$$

where $\vec{w}(\eta_n)$ are intermediate samples of the reduced voltage and Ω_N and $D_{N'}$ interpolating functions. The summation in n can be truncated to just consider the neighbor samples of the point of interpolation depending on the bandwidth of the reduced voltage.

The introduced formulation makes it possible to design an optimal sampling scheme and a subsequent interpolator for a given curve parametrization. A popular choice is a spiral arrangement, because all samples can be acquired by a continuous movement of both angular positioners leading to faster measurements [66–68]. With this approach, the obtained errors are lower than the sparse retrieval approaches because OSI is an analytically exact and robust approach. However, the measurement time reductions are less significant.

References

[1] Hansen, J. E., "Spherical Near Field Antenna Measurements," IEE Electromagnetic Series, Peter Peregrinus Ltd., 1988.

[2] Stratton, J. A., *Electromagnetic Theory*, New York, NY: McGraw-Hill, 1941.

[3] Harrington, R. F., *Time Harmonic Electromagnetic Fields*, New York, NY: McGraw-Hill, 1961.

[4] Rudge, A. W., et al., *The Handbook of Antenna Design* (Vols. 1 & 2), Peter Peregrinus, 1986. (Chapter 8: Antenna Measurement by J. Appel-Hansen, E. S. Gillespie, T. G. Hickman & J. D. Dyson), London.

[5] Abramowitz, M., and I. A. Stegun (eds.), *Handbook of Mathematical Functions with Formulas, Graphs, and Mathematical Tables*, New York: Dover Publications, 1964, p. 439, eq. 10.2.21.

[6] Jensen, F., and A. Frandsen., "On the Number of Modes in Spherical Wave Expansions," *AMTA 26th Annual Symposium (AMTA2004)*, Stone Mountain Park, GA, October 17–22, 2004.

[7] Edmonds, A. R., *Angular Momentum in Quantum Mechanics*, Princeton, NJ: Princeton University Press, 1974.

[8] Bruning, J., and Y. Lo, "Multiple Scattering of EM Waves by Spheres, Part I—Multipole Expansion and Ray-Optical Solutions," *IEEE Trans. Antennas Propag.*, Vol. 19, No. 3, May 197, pp. 378–90.

[9] Wacker, P. F., "A New Data Reduction Scheme Makes Practical the Determination of Patterns of Electrically Large Antennas from Near-Field Spherical Scans," in *Conference on Precision Electromagnetic Measurements*, 1974

[10] Foged, L. J., A. Giacomini, and R. Morbidini, "Probe Performnce Limitation Due to Excitation Errors in External Beam Forming Network," *AMTA 2011*, Colorado.

[11] Foged, J. A., A. Giacomini, and S. Pivnenko, "Wide Band Dual Polarized Probes for Near and Farfield Measurement Systems," *AMTA 2007*, November 4–9, 2007, St. Louis, MO.

[12] Jerauld, J., "Insights Into Spherical Near Field Probe Correction Gained from Examining The Probe Response Constants," *AMTA 2016*, Austin, Texas.

[13] Newell, A. C., "Error Analysis Techniques for Planar Near-Field Measurements," *IEEE Trans. Antennas Prop.*, Vol. 36, June 1988, pp. 754–768.

[14] Foged, L. J., M. Faliero, and P. O. Iversen, "Random Noise in Spherical Near Field Systems," *AMTA*, Salt Lake City, Utah, November 2009.

[15] Foged, L. J., et al., "Echo Suppresion by Spatial Filtering Techniques in Advanced Planar and Spherical NF Antenna Measurements," AMTA Symposium, October 2012, Seattle, Washington.

[16] Cano-Fácila, F. J., S. Pivnenko, and M. Sierra-Castañer, "Reduction of Truncation Errors in Planar, Cylindrical and Partial Spherical Near-Field Antenna Measurements," *International Journal of Antennas and Propagation*, Volume 2012 (2012).

[17] Noren, P., L. J. Foged, and P. Garreau, "State of the Art Spherical Near-Field Antenna Test Systems for Full Vehicle Testing," *2012 6th European Conference on Antennas and Propagation (EUCAP)*, 2012, pp. 2244–2248.

[18] Mauermayer, R. A. M., and T. F. Eibert, "Spherical Field Transformation Above Perfectly Electrically Conducting Ground Planes," in *IEEE Transactions on Antennas and Propagation*, Vol. 66, No. 3, March 2018, pp. 1465–1478.

[19] Cornelius, R., D. Heberling, and D. Pototzki, "Spherical Near-Field Far-Field Transformation with Infinite Ground Plane At Arbitrary Height z," *EuCAP 2017*, 19–24 March, Paris, France.

[20] Saccardi, F., et al., "Experimental Comparison of Vehicular Antenna Measurements Performed over Different Floors," *EuCAP 2020*, Copenhagen, Denmark, 15–20 March, 2020.

[21] Saccardi, F., et al., "Estimation of Real Ground Effect in Absorber-Based Automotive Measurements: Experimental Validation," *APS 2020*, 5–10 July 2020, Montréal, Canada.

[22] Saccardi, F., et al., "Virtual Drive Testing Based on Automotive Antenna Measurements for Evaluation of Vehicle-to-X Communication Performances," *AMTA 2019*, October 6–11, San Diego, CA.

[23] Foged, L. J., et al., "Truncation Impact on Measured Radiation Pattern in Spherical Near Field Antenna Test Ranges," *IEEE APS Symposium and USNC-URSI National Radio Science Meeting*, Boston, MA, July 8–12, 2001.

[24] Martini, E., S. Maci, and L. J. Foged, "Spherical Near Field Measurements with Truncated Scan Area," *5th European Conference on Antennas and Propagation*, EuCAP 2011, Rome, Italy, 11–15 April 2011.

[25] Martini, E., S. Maci, and L. J. Foged, "Reduction of Truncation Errors in Spherical Near Field Measurements", *AMTA 2010*, 10–15 October, Atlanta, Georgia.

[26] Foged, L. J., et al., "Comparative Investigation of Methods to Reduce Truncation Errors in Partial Spherical Near-Field Antenna Measurements," *EuCAP 2012*, 26–30 March, Prague, Czech Republic.

[27] Saccardi, F., et al., "Truncation Error Mitigation in Free-Space Automotive Partial Spherical Near Field Measurements," *AMTA 2017*, 15–20 October, Atlanta, GA.

[28] Wittmann, R. C., C. F. Stubenrauch, and M. H. Francis, "Using Truncated Data Sets in Spherical Scanning Antenna Measurements," *International Journal of Antennas and Propagation*, Volume 2012 (2012).

[29] Balanis, C.A., *Advanced Engineering Electromagnetics*, Hoboken, NJ: John Wiley & Sons, 2012.

[30] Foged, L. J., et al., "Infinite Ground Plane Antenna Characterization from Limited Groundplane Measurements," *APS 2010*, 11–17 July, Toronto, ON, Canada.

[31] ANSI/IEEE Std 149-1979 "IEEE Standard Test Procedures for Antennas."

[32] Saad, Y., *Iterative Methods for Sparse Linear Systems* (second edition), Society for Industrial and Applied Mathematics, 2003.

[33] Hofmann, B., O. Neitz, and T. F. Eibert, "On the Minimum Number of Samples for Sparse Recovery in Spherical Antenna Near-Field Measurements," in *IEEE Transactions on Antennas and Propagation,* Vol. 67, No. 12, Dec. 2019, pp. 7597–7610.

[34] Laitinen, T., et al., "Theory and Practice of the FFT/Matrix Inversion Technique for Probe-Corrected Spherical Near-Field Antenna Measurements with High-Order Probes," *IEEE TAP,* Vol. 58, No. 8, August 2010.

[35] Saccardi, F., A. Giacomini, and L. J. Foged, "Probe Correction Technique of Arbitrary Order for High Accuracy Spherical Near Field Antenna Measurements," *AMTA 2016,* October 30–November 4, Austin, TX.

[36] Saccardi, F., et al., "Using Standard Wideband Antennas as Probe in Spherical Near Field Measurements with Full Probe Correction: Experimental Validation," *AMTA 2018,* 4–9 November, Williamsburg, VA.

[37] Saccardi, F., et al., "Experimental Validation of Full Probe Correction Technique using Wideband and Dual-Polarized Probes in Spherical NF Antenna Measurements," *AMTA 2021,* October 24–29, Daytona Beach, FL.

[38] Foged, L. J., et al., "Spherical Near Field Offset Measurements Using Downsampled Acquisition and Advanced NF/FF Transformation Algorithm," *EuCAP,* Davos, Switzerland, 10–15 April 2016.

[39] Saccardi, F., "Application of the Translated-SWE Algorithm for the Characterization of Antennas Installed on Cars Using a Minimum Number of Samples," *AMTA 2017,* 15–20 October, Atlanta, GA.

[40] Saccardi, F., et al., "Theory and Validation of Reduced Complexity Translated-SWE Algorithm with Applications to Automotive Measurements with Minimum Sampling," *EuCAP 2018,* 9–13 April, London, England.

[41] Cornelius, R., and D. Heberling "Spherical Wave Expansion With Arbitrary Origin for Near-Field Antenna Measurements," *IEEE Transactions on Antennas and Propagation,* Vol. 65, Issue: 8, Aug. 2017.

[42] Saccardi, F., et al., "Fully Probe Corrected Spherical Near Field Offset Measurements with Minimum Sampling Using the Translated-SWE Algorithm," *AMTA 2018,* 4–9 November, Williamsburg, VA.

[43] Rodríguez Varela, F.,B. Galocha Iragüen, and M. Sierra Castañer, "Fast Spherical Near-Field to Far-Field Transformation for Offset-Mounted Antenna Measurements," in *IEEE Antennas and Wireless Propagation Letters,* Vol. 19, No. 12, pp. 2255–2259, Dec. 2020.

[44] Saccardi, F., et al., "Experimental Validation of the Translated-SWE Technique Applied to Automotive Measurements over PEC-Floor at Arbitrary Height," *EuCAP 2020,* 15–20 March, Copenhagen, Denmark.

[45] Quijano, J. L. A., and G. Vecchi, "Improved-Accuracy Source Re¬construction on Arbitrary 3-D Surfaces," *IEEE Antennas and Wireless Propagation Letters,* Vol. 8, 2009, pp. 1046–1049.

[46] Quijano, J. L. A., and G. Vecchi, "Field and Source Equivalence in Source Reconstruction on 3D Surfaces," *Electromagnetic Waves (Pier),* 2010, Vol. 103, pp. 67–100.

[47] Quijano, J. L. A., and G. Vecchi, "Near- and Very Near-Field Accuracy in 3-D Source Reconstruction," *Antennas and Wireless Propagation Letters,* IEEE, Vol. 9, pp. 634–637, 2010.

[48] Scialacqua, L., "Equivalent Source Technique Processing of Broadband Antenna Measurements," *14th European Conference on Antennas and Propagation (EuCAP 2020—virtual format),* 2020.

[49] Scialacqua, L., et al., "Diagnostics on Electrically Large Structures by a Nested Skeletonization Scheme Enhancement of the Equivalent Current Technique," *14th European Conference on Antennas and Propagation (EuCAP 2020—virtual format),* 2020.

[50] Scialacqua, L. et al., "Practical Application of the Equivalent Source Method as an Antenna Diagnostics Tool," *AMTA Symposium,* October 2011, Englewood, Colorado.

[51] Quijano, J. L. A., et al., "3D Spatial Filtering Applications in Spherical Near Field Antenna Measurements," *AMTA 2010 Symposium,* October, Atlanta, Georgia.

[52] Foged, L. J., et al., "Application of the Dual-Equation Equivalent Current Reconstruction to Electrically Large Structures by Fast Multipole Method Enhancement," *35th Annual Meeting and Symposium—Antenna Measurement Techniques Association,* AMTA 2013, Columbus, Ohio, October 6–11, 2013.

[53] Foged, L. J., et al., "Detection and Suppression of Scattered Fields from Coplanar Micro-Probe and Positioner in Millimeter Wave On-Chip Antenna Measurements," *ISAP2016, International Symposium on Antennas and Propagation,* Ginowan, Okinawa, Japan, October 24–28, 2016.

[54] Araque Quijano, J. L., et al., "Antenna Measurement Processing for Diagnostics and Filtering Based on Integral Equations," *Antennas and Propagation Society International Symposium,* 2011, APSURSI '09. IEEE, Spokane, July 2011.

[55] L. Scialacqua, L., et al., "Power Density Measurement at 5G Millimeter-Wave Using Inverse Source Method," submitted to the *40th Annual Symposium of the Antenna Measurement Techniques Association,* AMTA, Williamsburg, Virginia, November 4–9, 2018.

[56] Foged, L. J., et al., "Using Measured Fields as Field Sources In Computational EMC," *37th Annual Symposium of the Antenna Measurement Techniques Association,* AMTA, Long Beach, CA, October 11–16, 2015.

[57] Foged, L. J., et al., "Bringing Numerical Simulation and Antenna Measurements Together," *IEEE Antennas and Propagation Society International Symposium,* July 6–11, 2014.

[58] Donoho, D. L., "Compressed Sensing," *IEEE Trans. Inf. Theory,* Vol. 52, No. 4, April 2006, pp. 1289–1306.

[59] Gill, P. R., A. Wang, and A. Molnar, "The In-Crowd Algorithm for Fast Basis Pursuit Denoising," in *IEEE Transactions on Signal Processing,* Vol. 59, No. 10, October 2011, pp. 4595–4605.

[60] Hofmann, B., O. Neitz ,and T. F. Eibert, "On the Minimum Number of Samples for Sparse Recovery in Spherical Antenna Near-Field Measurements," in *IEEE Transactions on Antennas and Propagation,* Vol. 67, No. 12, December 2019, pp. 7597–7610.

[61] Fuchs, B. , "Fast Antenna Far Field Characterization Via Sparse Spherical Harmonic Expansion," in *IEEE Tranactions on. Antennas and Propagation,* Vol. 65, No. 10, October 2017, pp. 5503–5510.

[62] Culotta-López, C., et al., "A Compressed Sampling for Spherical Near-Field Measurements," in *AMTA Proceedings,* Williamsburg, Virginia, October, 2018.

[63] Rasmus, C., *Fast Spherical Near-Field Antenna Measurement Methods,* Apprimus Verlag, Aachen, 2019.

[64] Bucci, O. M., and G. Franceschetti, "On the Spatial Bandwidth Of Scattered Fields," *IEEE Trans. Antennas Propagat.,* Vol. AP-35, December 1987, pp. 1445–1455.

[65] Bucci, O. M., C. Gennarelli, and C. Savarese, "Representation of Electromagnetic Fields over Arbitrary Surfaces by a Finite and Nonredundant Number of Samples," in *IEEE Transactions on Antennas and Propagation,* Vol. 46, No. 3, March 1998, pp. 351–359.

[66] D'Agostino, F., et al., "Experimental Assessment of an Effective Near-Field–Far-Field Transformation with Spherical Spiral Scanning for Quasi-Planar Antennas," in *IEEE Antennas and Wireless Propagation Letters,* Vol. 12, 2013, pp. 670–673.

[67] D'Agostino, F., et al., "Pattern Evaluation from NF Spherical Spiral Data in the Noncentered Quasi-Planar Antennas Case," in *IEEE Antennas and Wireless Propagation Letters,* Vol. 19, No. 12, December 2020, pp. 2275–2279.

[68] D'Agostino, F., et al., "A Nonredundant Sampling Representation Managing an Offset Mounting of an Elongated Antenna in a Spherical Near-Field Facility," in *IEEE Antennas and Wireless Propagation Letters,* Vol. 18, No. 12, December 2019, pp. 2671–2675.

[69] Rodriguez Varela, F., M. Sierra Castañer, and B. Galocha Iragüen, "Multi-Level Spherical Wave Expansion for Fast Near-Field to Far-Field Transformation," *2018 AMTA Proceedings,* pp. 1–6.

[70] Rodríguez Varela, F., B. G. Iragüen, and M. Sierra-Castañer, "Under-Sampled Spherical Near-Field Antenna Measurements with Error Estimation," in *IEEE Transactions on Antennas and Propagation,* Vol. 68, No. 8.

[71] Varela, R., B. G. Iragüen, and M. S. Castañer, "Spherical Near-Field Measurements for Offset-Mounted Antennas," *2021 15th European Conference on Antennas and Propagation (EuCAP).*

[72] Foged, L. J., et al., "Practical Application of the Equivalent Source Method as an Antenna Diagnostics Tool," *33rd Annual Symposium of the Antenna Measurement Techniques Association, AMTA,* Englewood, Colorado, October 2011.

CHAPTER 4

Chamber Design and Analysis

Andrea Giacomini, Maria Alberica Saporetti, Francesco Saccardi, and Manuel Sierra Castañer

This chapter provides readers with a basic knowledge of chamber design and analysis with particular attention to automotive applications.

The performance of a chamber is evaluated in terms of the uniformity of illumination of a certain area inside the chamber. In such an area, called the quiet zone (QZ), the radiation from the range antenna approximates a plane wave condition [1]. The AUT or DUT installed on a car is placed in the QZ, where an appropriate level of field illumination uniformity is ensured in order to guarantee the right performance of the measurement system. A well-designed chamber is, therefore, extremely important. The criteria for defining NF and FF rectangular and tapered ranges are described in this chapter, including details on absorbers and ferrite lining. Chambers typically used for automotive applications are shown and compared in the context of emulated real scenarios. In addition, available analysis methods to evaluate the chamber are outlined, and the most common procedures used to evaluate the chamber performances with measurements are presented.

4.1 Measurement Ranges

Antenna testing can be performed in indoor or outdoor ranges. There is no limitation on the kinds of measurements that can be performed in indoor facilities; NF, FF, compact ranges, electromagnetic compatibility, and radar cross-section measurements are all good candidates. Indoor measurements have several advantages over outdoor systems, including their immunity to weather conditions, the possibility to perform antenna measurements on a 24/7 basis, and, above all, the assurance of a quasi-interference-free environment with no ground effects and no electromagnetic pollution. These conditions can be successfully achieved with a well-designed anechoic chamber where such measurements must be performed.

Shielded anechoic chambers are indoor antenna ranges that, in reducing undesired echoes, emulate a free-space scenario so that the measurement of the AUT or of the DUT is not influenced by external factors, such as unwanted reflections from the walls. To this purpose internal surfaces of the room or chamber are lined with RF absorbers to help minimize reflections into the test zone [1]. A schematic of an anechoic chamber is provided in Figure 4.1; the RF absorbers are represented by blue triangles, and a probe and an AUT or DUT are present. The probe can be used in transmitting mode, and the DUT can be used in receiving mode, or vice versa. The DUT is in the QZ area. If, for example, we consider a configuration where the probe is transmitting, and the DUT is receiving, the probe will illuminate the AUT (direct field) but also the walls of the chamber. Part of the energy arriving on the walls will be reflected toward the QZ: a good anechoic chamber will ensure that the level of the reflected field in the QZ is much lower than the level of the direct field. In such conditions, the pattern of the DUT can be measured with an acceptable error contribution.

The anechoic capability of the chamber has an impact on the QZ illumination in a way that depends on the working frequency; it is determined by a combination of two main factors: the geometry of the chamber and the absorber layout. Typical geometries include rectangular and tapered chambers. The absorber layout includes the shape, the disposition, and the reflectivity property of the absorbing material. The most common absorbers are pyramidal foam absorbers. Due to their shape, multiple reflections occur at the air/material boundary. For each interaction/reflection, a portion of the wave is absorbed inside the material, thereby allowing only a very small portion of the incident wave to be reflected back into the environment. In addition to the classical foam absorbers, thin-layer absorbers (tiles) based on ferrite-loaded composites can be added to the lining in case of low-frequency measurements.

In some cases, including automotive measurements, the floor can be conductive instead of being covered by absorbers, as shown in Figure 4.2. The purpose is to emulate a specific ground environment and include its contribution in the measurement of the vehicle-installed antenna.

The chamber construction, especially for ranges that have to host vehicles, can be really expensive, considering the costs for the real estate, the absorbers, and the ferrite materials. Consequently, it is fundamental for chamber suppliers to

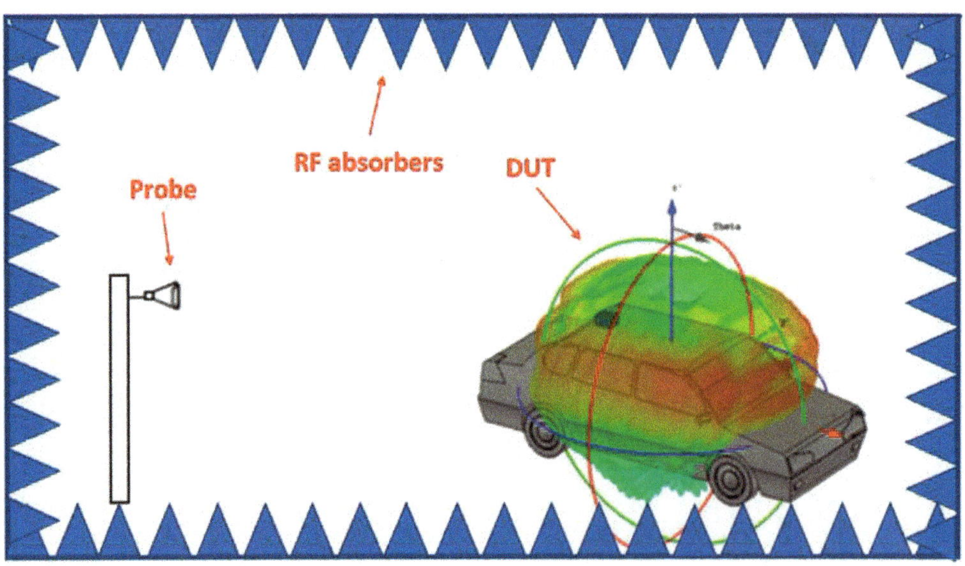

Figure 4.1 Conceptual illustration of an anechoic chamber.

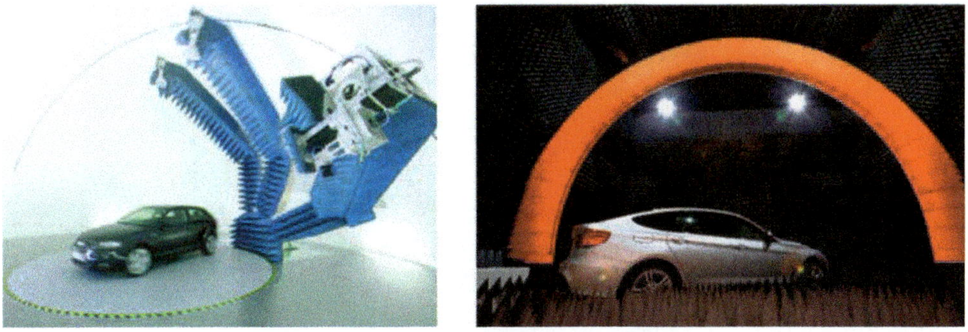

Figure 4.2 Anechoic chamber for automotive measurements: conductive floor (left) and absorbing floor (right). ©2019 IEEE. (Reprinted with permission.) [28].

adopt proper design criteria (as described in Section 4.2) to correctly estimate with analysis methods (described in Section 4.3) the performance of a certain design before building the chamber. This ensures a good trade-off between the objectives of reducing the costs and achieving good performance levels. Once the chamber has been built, the actual reflectivity in the QZ has to be measured in order to cross-check the predicted performances.

4.2 Design Criteria

This section details the criteria usually followed when designing an anechoic chamber. To this purpose, Section 4.2.1 provides a description of the absorbers and the ferrites, and Sections 4.2.2–4.2.4 illustrate the typical geometry and configuration for anechoic chambers for NF and FF measurements.

4.2.1 Absorbers and Ferrites

As already underlined, the aim of the absorber lining on the walls of the chamber is to minimize the reflections in the QZ.

RF absorbers are made of lossy material, usually foam with a certain percentage of carbon loading, that, together with a particular absorber shape, permit the penetration of the incident EM field with absent or very low reflections. The electrical thickness of the absorber (i.e., its physical thickness normalized by the wavelength) is proportional to the EM field absorption. It has been verified that, to achieve a significant attenuation, the absorber should be, at least, larger than one free-space wavelength, but for high performances, a length of several wavelengths is suggested [2].

Typical RF absorbers have a pyramidal geometry, as shown in Figure 4.3. This shape makes it possible to provide an impedance match from free space to the lossy material. However, with an increased angle of incidence, the performance of the pyramidal absorber tends to deteriorate.

Manufacturers' datasheets usually include the reflectivity of RF absorbers at normal incidence as a function of frequency, as shown in Figure 4.4, where it is evident that at lower frequencies, physically larger absorbers are needed. The number

Figure 4.3 Pyramidal RF absorbers.

Specifications

		AEP-4	AEP-6	AEP-8	AEP-12	AEP-18	AEP-24	AEP-26	AEP-36	AEP-48	AEP-60	AEP-72	AEP-96
Geometry	Right/Twisted	R	R	R	R	R	R	R	R/T	R/T	R	T	T
Height	in	4	6	8	12	18	24	26	36	48	60	72	96
	cm	10.2	15.2	20.3	30.5	45.7	61	66	91.4	122	152.4	182.9	243.8
Pyramids/Blocks	per block	144	144	64	36	16	9	9	4	4	1	1	1
Absorption @ Normal Incidence @ 30 MHz	dB								5	7	8	9	15
@ 80 MHz	dB							6	12	15	18	20	25
@ 125 MHz	dB							10	15	28	30	33	35
@ 250 MHz	dB						30	30	32	35	37	37	40
@ 500 MHz	dB					30	35	35	37	40	40	43	45
@ 1.0 GHz	dB			30	35	40	40	40	42	45	47	48	50
@ 3.0 GHz	dB	30	33	37	40	45	45	50	50	50	50	50	50
@ 6.0 GHz	dB	35	37	45	45	50	50	50	50	50	50	50	50
@ 10.0 GHz	dB	40	40	50	50	50	50	50	50	50	50	50	50
@ 15.0 GHz	dB	45	50	50	50	50	50	50	50	50	50	50	50
@ 18.0 GHz	dB	50	50	50	50	50	50	50	50	50	50	50	50
@ 40.0 GHz	dB	50	50	50	50	50	50	50	50	50	50	50	50
Max Power Handling	Watt/in²	1	1	1	1	1	1	1	1	1	1	1	1
	Watt/m²	1550	1550	1550	1550	1550	1550	1550	1550	1550	1550	1550	1550
Weight	lbs/pc.	2.5	3	4.5	6	9	13.5	15	18.5	25	30	42	60
	kg/pc.	1.1	1.4	2	2.7	4.1	6.1	6.8	8.4	11.3	13.6	19	27.2

Figure 4.4 Examples of absorption of pyramidal RF absorbers at normal incidence [3].

in the name of the absorber represents the dimension in inches. Moreover, it should be noted that performance improves when the frequency increases.

Looking at the schematic of a chamber, it is evident that there are some locations inside the chamber where the incident field will not be perpendicular to the absorbers. The reflectivity at oblique incidence can be computed based on an experimentally observed behavior [4]. The reflection of typical absorbers is approximately constant up to 45 degrees (coincident with the value at normal incidence) and linearly increases to 0 dB at grazing angle (90 degrees). Such an approximated behavior is shown in Figure 4.5, where the factor $\alpha(\theta_i)$ used to scale the reflectivity at normal incidence (*Reflectivity* $(\theta i = 0)|_{\text{db}}$, with $-A$ being the absorption at normal incidence) is plotted as a function of the incidence angle (θ_i).

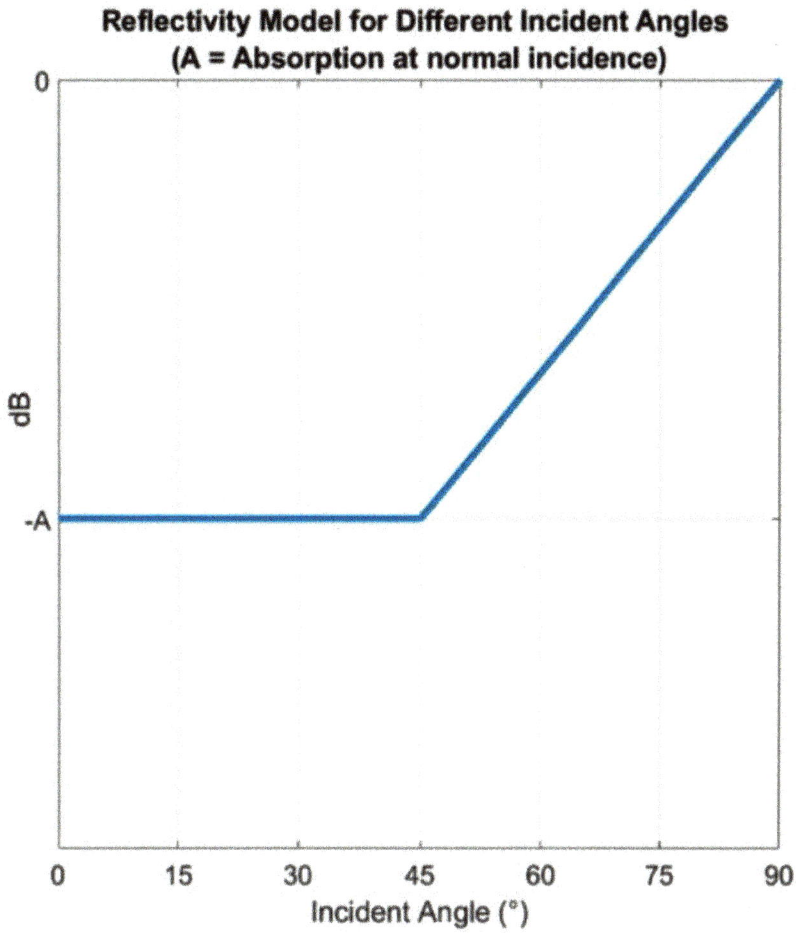

Figure 4.5 Reflectivity versus angle of incidence.

$$Reflectivity(\theta_i)|_{dB} = \alpha(\theta_i) Reflectivity(\theta_i = 0)|_{dB} \tag{4.1}$$

From the typical reflectivity versus frequency data shown in Figure 4.4 it is evident that, to achieve relatively low reflectivity levels (e.g., < -15 dB) at low frequencies like 100 MHz and below, very large and hence heavy foam absorbers are needed. To provide a sufficient reflectivity at such low frequencies, thin-layer absorbers (tiles) based on ferrite-loaded composites are often considered for the lining of the chamber walls, as shown in Figure 4.6. At low frequencies, ferrite tiles are characterized by an impedance close to the one of free space, which allows direct transmission without any significant reflection at the surface and magnetic losses, meaning that the wave is almost completely absorbed inside the material. The typical performance of ferrite tiles is depicted in Figure 4.6.

Beside the good reflectivity provided at low frequency, another advantage of the ferrite tiles is their usability inside smaller measurement environments (no extra space to host large absorbers is required) and hence the possibility to keep the cost of building relatively low. Nevertheless, since ferrite tiles are usually more expensive than conventional foam absorbers, their use must be properly evaluated. It is

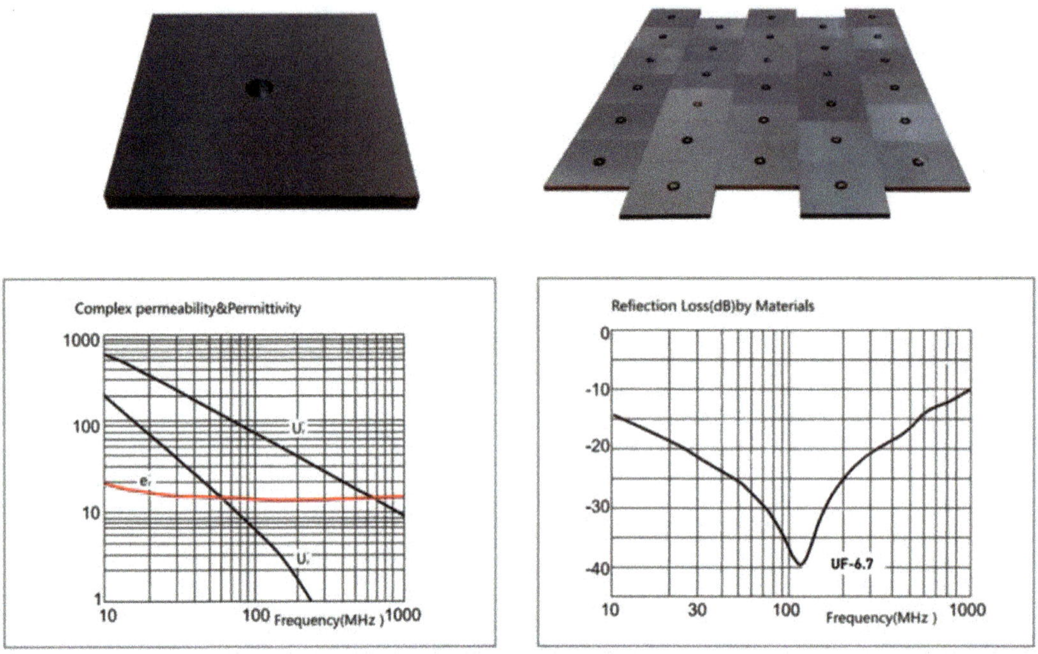

Figure 4.6 Example of ferrite tiles and their typical performance [5].

also important to consider that while ferrite tiles make it possible to achieve better performance at low frequency, they could also significantly degrade performance at higher frequencies. Accordingly, special care must be paid when ferrite tiles are used at different frequency bands in the same range.

The lining of a chamber, called the absorber layout, can include different kinds of absorbers. In these cases, the most critical areas on the chamber walls are covered by more performing absorbers or ferrite. For example, if a relevant illumination of the wall behind the AUT is estimated, this wall should be covered by absorbers larger than the rest of the chamber. This is visible in Figure 4.7, where the 24-inch absorbers cover all the walls except the end wall, which is lined by larger 36-inch absorbers. In addition, a diamond of Chebyshev absorbers (a design that achieves high performance with restrictive size constraints [6]) has been placed on the lateral walls to further reduce the reflection coming from the side walls.

4.2.2 Rectangular Chambers

Because the absorbers' performance is dependent on frequency, as shown in the Section 4.2.1, the dimensions of a rectangular chamber must be defined considering the smallest working frequency. NF or FF systems can be installed in rectangular chambers.

In FF ranges, the following rules apply [4, 7]:

- The minimum chamber width and height should be three-times the diameter of the minimum sphere that contains the largest antenna being tested.
- The dimension of the QZ has to include the AUT.

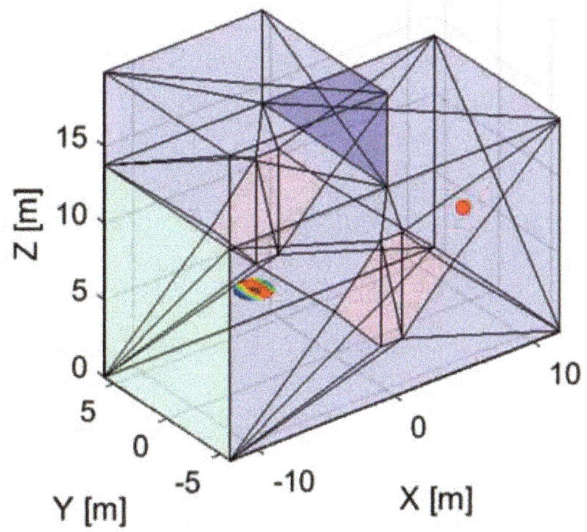

Figure 4.7 Absorber layout with different colors representing different absorbers: 24-inch represented with blue, 24-inch Chebyshev represented with pink, and 36-inch represented with green. (The red dot is the transmitting antenna, and the colored circle is the QZ.)

- The minimum length is determined by the sum of the following components:
  - A minimum spacing of 2λ between the AUT and the tips of the absorber;
  - The distance between the transmitting and receiving antenna must be $R \geq \frac{2D^2}{\lambda}$, D is the dimension of the antenna under test (see Chapter 2);
  - The thickness of the absorbers;
  - Additional space for the probe antenna.

Figure 4.8 shows the above-described configuration, expressing the dimension of the antenna and of the QZ in terms of the electrical size (i.e., in terms of wavelength).

The x value corresponding to the distance between the line-of-sight of the range to the lateral walls can be easily computed and will depend on λ, θ. The angle θ

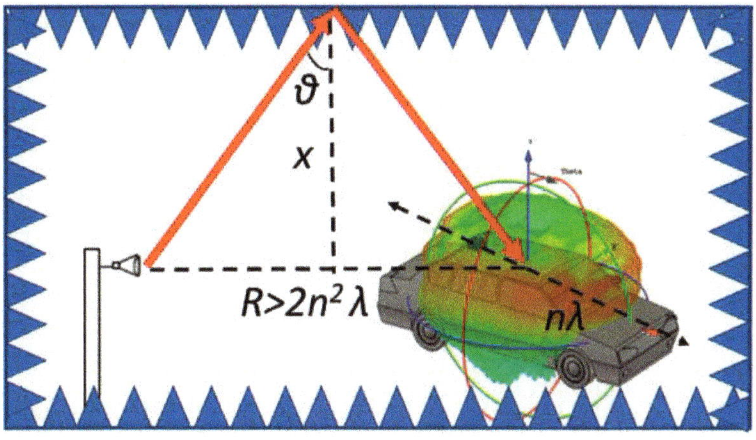

Figure 4.8 FF range dimensions.

represents the incidence angle for the absorbers placed on the side wall; therefore, it must be chosen according to the desired reflectivity, according to the behavior of the reflection versus the angle of incidence shown in Figure 4.5. It is suggested that $R/2x \leq \sqrt{3}$, resulting in recommended values for θ below 60° and values of x larger than $R/3.46$.

In NF ranges the test distance is usually shorter than the one required to meet FF conditions, as explained in Chapter 2, and a value between 3λ and 10λ is suggested [4]. For SNF systems, the following rules apply:

- The minimum chamber length is determined by the sum of the following components:
  - The encumbrance of the probe antenna and its positioner;
  - The encumbrance of the AUT intended as minimum sphere enclosing the antenna;
  - The thickness of the absorbers placed on the end walls of the chamber;
  - The test distance: 4λ;
  - The distance between the minimum sphere and the absorber tips on the corresponding end wall: 2λ;
  - An additional distance of 1m to allow the movements of the probe or the AUT.
- The minimum chamber height and width are determined by the sum of the following components:
  - The encumbrance of the AUT intended as a minimum sphere enclosing the antenna;
  - The thickness of the absorbers placed on the side walls of the chamber;
  - The distance between the minimum sphere and the absorber tips on the two side walls: 4λ;
  - The additional distance of 1m to allow the movements of the probe or the AUT.

For the angle of incidence on sidewalls θ, ceiling and floor, it is suggested that $\theta < 60°$ to achieve good performance.

For planar NF systems, the dimensions of the chamber are related to the dimensions of the scan plane. The following rules apply:

- The minimum chamber width is determined by the sum of the following components:
  - The dimension of the scan plane along the width of the chamber;
  - The distance between the scan plane and the absorber tips on the two side walls: 4λ;
  - The space for the scanner structure;
  - The thickness of the absorbers on the side walls.
- The minimum chamber length is determined by the sum of the following components:

- The dimension of the scan plane along the length of the chamber including also the probe length;
- The distance between the scan plane and the AUT;
- The length of the AUT and its positioner;
- The thickness of the absorbers on the wall behind the scan plane;
- The distance between the AUT and the absorber tips on the wall behind the AUT: 4λ;
- The thickness of the absorbers on the wall behind the AUT if there any. (Note that for high-directivity antennas—usually tested with planar field methods—no absorber lining is requested on the wall behind the AUT.)

4.2.3 Tapered Chambers

Tapered chambers [8–10] are FF ranges that do not have a traditional rectangular shape; instead, they have a tapered shape, as shown in Figure 4.9, with a taper angle below 30°. In these chambers, the reflections from the side walls are not intentionally minimized, since it is used to illuminate the QZ. Tapered chambers are suggested for applications where the working frequency is below the UHF band.

Unlike rectangular chambers, the absorbers do not cover all the walls of the chamber. The most influent absorbers for the QZ illumination are the absorbers on the end wall behind the AUT. The absorbers placed on the lateral side of the rectangular section of the tapered chamber can be half the thickness of the absorbers placed on the end wall. In particular, in such areas, the use of wedge absorbers is advisable (see Figure 4.10) in order to avoid scattering on the QZ.

The following empirical rules [7] apply:

- The width, which is equal to the height and to the length of the rectangular section, is determined by the sum of the following:
  - The dimension of the QZ;
  - The distance between the AUT and the absorber tips on the side walls: 4λ;
  - The thickness of the absorber on the end wall behind the AUT.
- The length of the tapered area should be twice the width of the rectangular section.

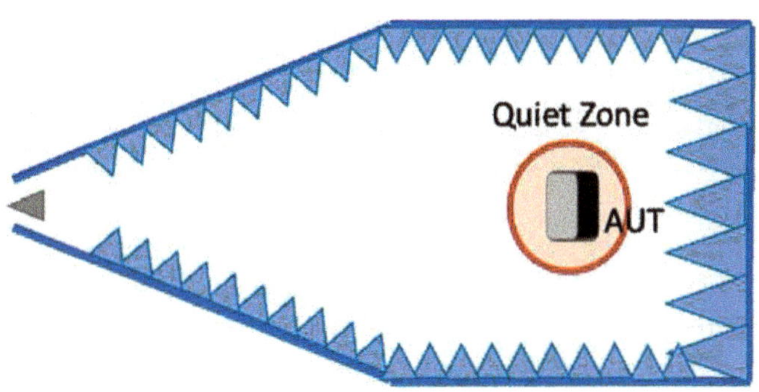

Figure 4.9 Tapered chamber.

Figure 4.10 Wedge absorbers.

4.2.4 Chambers with Conductive Floors

As mentioned in Chapter 3, automotive testing scenarios sometimes use a conductive/metallic floor generating a reflective environment, instead of a floor covered by absorbing materials, to emulate a free-space condition [11, 12].

A fully reflective ground is sometimes preferred for the following reasons:

1. It makes it possible to emulate specific driving ground conditions, like wet asphalt, and hence include this information directly into the measurements.
2. It makes it possible to reduce the downtime in the measurement setup phases, because the vehicle is simply parked in the middle of the system.
3. It makes it possible to significantly reduce the truncation errors during the NF/FF transformation by exploiting the image theory. (See details in Chapter 3.)

To properly exploit the image theory during the NF/FF transformation, the conductive floor must be assumed to be a perfect electric conductor (PEC). The electrical conductivity of the ground material should hence be chosen to be as high as possible to avoid residual errors due to the PEC assumption during data processing. Similarly, the planarity and electrical continuity of the ground should be enforced to avoid additional errors. Since a turntable is normally used to perform the scanning in azimuth (see Figure 4.2, left), the electrical continuity of the ground is particularly critical at the junction between the turntable itself and the outer part of the ground. In such places conductive contacts like metallic brushes are typically used to meet this requirement.

It is also specified that the application of the image theory to data measured under such conditions is simplified when the height of the ground floor coincides with the center of the spherical scanning system (i.e., ground floor at $Z = 0$). In fact, as described in Chapter 3, in such cases, the PEC condition can be applied by performing a simple mirroring of the measured data. An optimal design of these types of ranges should thus consider the conductive ground at $Z = 0$. Unfortunately, due to mechanical constraints, this is not always possible. Nevertheless, in such cases, the PEC condition can still be enforced, but a more computationally expensive processing must be considered.

4.3 Analysis Methods

The performance prediction of an anechoic chamber, which is extremely important for chamber suppliers, can be enabled by simulation tools. The most widespread numerical methods used for these applications are the method of moments (MoM), the finite-difference time-domain method (FDTD), and ray tracing [13]. Alternatively, suitable scaled models of the measurements' environments can be considered to predict their measurement performance.

Next, some details and some examples obtainable with the different methods are provided. In particular, Section 4.3.1 discusses full-wave models; Section 4.3.2 describes ray tracing; and Section 4.3.3 provides a particular application of ray tracing based on image theory. Finally, an example of analysis performed with the scaled model method is reported in Section 4.3.4.

4.3.1 Full-Wave Models

Full-wave models (MoM and FDTD) can yield extremely accurate results only if the modeling is well defined. In particular, the modeling of the absorber material can be a critical process, since the data sheets of the absorbers do not include the exact geometry but only the overall dimensions; in addition, concerning electromagnetic properties, only the normal reflectivity is provided. Even if all the information to model the absorber is available, this is a time-consuming process that must be added to the long computational time needed to run the simulation. The time increases with the chamber size, since many linear equations must be solved, especially in the case of a complicated absorber model. In addition to that, it must be underlined that, usually, before finding the optimal chamber configuration, several iterations are needed to consider different dispositions and different kinds of absorbers. This means that a different model is necessary for each kind of absorber and that several simulations must be launched.

Figure 4.11 shows an example of an analysis of a chamber for automotive applications with full-wave methods. Results have been obtained using the TD solver available in the commercial electromagnetic software CST [14]. The chamber was entirely covered by 60-inch absorbers. The plots show the E-field distribution at 64 MHz in different cuts, horizontal (top and bottom) and vertical.

4.3.2 Ray-Tracer Solvers

Unlike the full-wave models, ray-tracer solvers do not require a time-consuming modeling for the absorbers, since the only relevant EM property to be input is the reflectivity. The environment of the ray-tracer simulation has the shape of the room where the walls are associated to a certain reflectivity according to the absorber layout. In addition to that, the order of magnitude for the computational time of ray-tracer simulations is seconds; this can increase to minutes only in the case of multifrequency analyses. It is well-known that ray tracers cannot provide the same level of accuracy of full-wave models, especially in cases where the chamber dimensions are only a few wavelengths at the lowest operating frequency. Nonetheless, a ray tracer can give quick insight into the different models involved in the trade-off for the design of a chamber, leaving the full-wave analysis to check one or a few selected models at the end of the iteration process [15, 16].

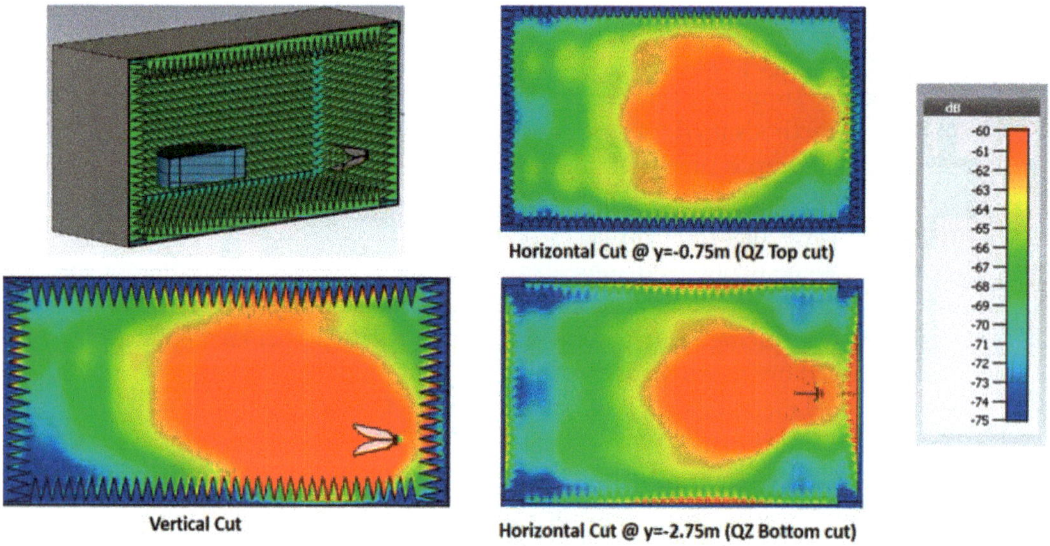

Figure 4.11 Analysis of a chamber with the full-wave method.

Figure 4.12 shows an example of an analysis of a chamber with the ray-tracer method obtained using the MVG SW Satsim [17]. Satsim is based on the combination of numerical modeling of the environment based on ray tracing (astigmatic beam tracer [18–21]) with NF measurements of the antenna. The numerical ray-tracing approach adopted is based on the computationally efficient astigmatic beam tracer method algorithm, which considers geometrical optics and the uniform theory of diffraction. It is based on a complete partition of the 3-D space using beams. A beam is conceived as a bundle of rays that originate from a caustic. Beams generated from the source interact with the environment while they propagate. Each beam is divided into smaller beams and additional beams of reflections and diffractions when partly illuminating an obstacle. The forward ray-tracing approach allows an unlimited number of diffractions and reflections from objects represented by polygonal flat surfaces with high speed and accuracy.

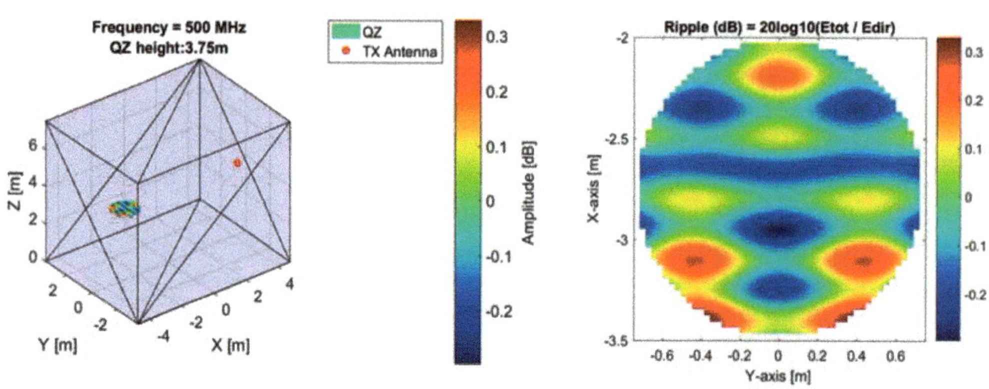

Figure 4.12 Analysis of a chamber with the ray-tracer method: (a) representation of the chamber, QZ and TX antenna, and (b) map of the ripple on the QZ. ©2021 IEEE. (Reprinted with permission.) [34].

This type of ray tracing is well-suited for parallel computation and has proven to be much faster and more accurate than conventional backward-propagating ray-tracer algorithms, since the time-consuming solution of nonlinear equations is avoided. During forward beam propagation of the environment, in this case the anechoic chamber, a beam-tree is generated, and the pertinent information is gathered in a data structure. From this data structure, the ray path is simply obtained by exploring the beam-tree from a leaf to the root. Since the beam-tree is based only on the geometrical properties of the structure, the same beam-tree is used to evaluate antennas with different characteristics, frequencies, and materials composing the environment.

A recently released version of Satsim enhances the previous one, adding some functionalities developed to ease and speed the investigation of anechoic chambers' performance. The geometry of the chamber can be easily drawn or imported from an image using the available CAD functionality. The absorbers are defined only in terms of reflectivity, and any profile layout can be easily inserted. The transmitting antenna is represented in terms of spherical wave expansion coefficients coming from a real measurement or a synthetic Gaussian beam. The illumination on the QZ can be then computed, and the contributions can be separated in terms of total, direct, and scattered fields. Once the ray-tracer computation has been performed, different options are available. The field on the QZ can be evaluated on 2-D or 3-D maps, and scattering effects due to small objects, such as a sprinkler in the chamber's fire-suppression system, can be computed. The performance of the chamber is computed in terms of peak-to-peak ripple; equivalent reflectivity; down-range and cross-range tapers and ripples (as visible in Figure 4.12) for amplitude and phase; and equivalent reflectivity computed as an interferer causing such down-range taper and peak-to-peak ripple. It is also possible to map hot spots on the walls from where the reflected contributions have a strong impact on the QZ illumination.

The classic ray-tracer formulation considers the radiation pattern of the transmit antenna, while the receiving antenna is considered as omnidirectional. A new feature has been added that also makes it possible to consider the actual radiation pattern of the receiving antenna. In this way, a FF antenna measurement can be emulated including the probe pattern effect, and the coupling between the transmitting and receiving antenna can be computed.

4.3.3 Image Theory–Based Technique

A particular implementation of the ray-tracing method that can be used for the evaluation of antenna measurement scenarios is explained in this section. The method is based on the spherical wave expansion (SWE) technique and image theory [15, 22]. It includes both the anechoic chamber performance and the range antenna/probe effect on the measurement. The SWE technique is known from the traditional SNF measurement theory. (See Chapter 3 and [23].) Image theory is a well-established concept in electromagnetics and modeling [24]. To be specific, it exploits the transmission formula introduced in Chapter 3, which is revisited in the following. This formula can be used in the accurate evaluation of the coupling expressed as (r, χ, θ, ϕ) between a transmitting AUT and a receiving probe/range antenna if the SWCs of both antennas are known:

$$w(r,\chi,\theta,\phi) = 0.5\sum_{\substack{smn \\ \sigma\mu\nu}} Q^{(3)}_{smn} e^{jm\phi} d^{n}_{\mu m}(\theta) e^{j\mu\chi} C^{sn(3)}_{\sigma\mu\nu}(kr) \tag{4.2}$$

The only approximation applied in the transmission formula is that only first-order coupling between the two antennas is considered. This is generally true for most common measurement scenarios [1]. It is also recalled that the symbols $d^{n}_{\mu m}(\theta)$ and $C^{sn(3)}_{\sigma\mu\varpi}(kr)$ are, respectively, rotation and translation operators that, together with the two complex exponentials $e^{j\mu\phi}$ and $e^{j\mu\chi}$, are used to describe the probe position/orientation in each measurement point.

Direct application of the transmission formula allows for the estimation of the probe effect at a certain distance (r) in a free-space condition. Reflections in chamber walls are simulated, as the direct coupling, with the AUT image weighted by angular-dependent absorber attenuation. In the current implementation, reflections of first order are taken into account, but there is no upper limit for the number of reflections that can be included in the method. An example of a first-order image source is depicted in Figure 4.13 in the case of a reflection from a lateral wall. A typical roll-over-azimuth system with the AUT mounted on it and a static probe is considered. The direct line-of-sight signal is computed with the transmission formula at distance R and direction $(\phi, \vartheta) = (0,0)$. The reflected signal is computed with the same formula considering the image of the AUT and evaluating the signal at R^i and direction $(\phi, \vartheta) = (0, \beta)$. It should be noted that to properly include the effect of the probe on the reflected signal the probe should be rotated by $-\beta$. This process is then repeated for each scanning angle by properly rotating the AUT and its image. To emulate a rectangular-shaped chamber, contributions from five more walls (specifically, the other side wall, the ceiling, the floor, the back wall, and the front wall) are included in a similar way.

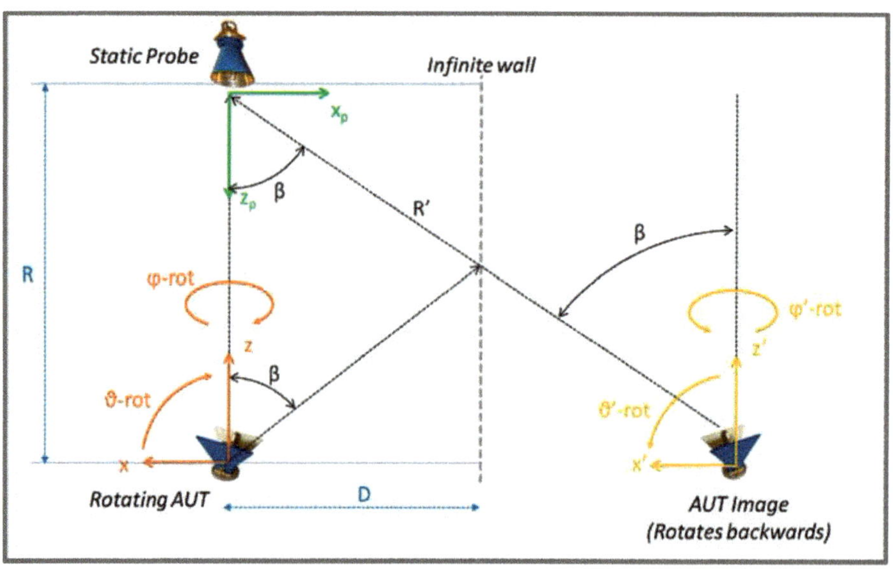

Figure 4.13 Schematic of the SWE/image-based method. Reflections in chamber walls are simulated, as the direct coupling, with the AUT image weighted by the angular-dependent absorber attenuation. ©2021 IEEE. (Reprinted with permission.) [34].

QZ probing [1] can be emulated in a similar manner by placing the measurement device [either a probe/range antenna or a plane wave generator (PWG)] in the AUT coordinate system and moving the probe in the xy-plane while the measurement device is kept fixed.

4.3.4 Measurements with the Scaled Model Method

An alternative method to analyze the performance of an anechoic chamber is to consider a scaled model of the measurement environment as done in [25–28].

The scaled model is a well-known technique that is often considered when very large objects (e.g., ships, aircrafts, and trains) cannot be tested directly. In such cases a high-fidelity time scaled model of the test object is manufactured, making sure that their dimensions, type of materials, and working frequencies are properly scaled by that N-factor [1].

In the present case, the measurement environment, being the object of the investigation, is the one that is scaled in as much detail as possible. On the other end, the test object is selected so that it provides a suitable illumination of the scaled test environment. In other words, the test object should be similar to the one of the final system, but not exactly the same as we do not know in advance the specific device to be measured.

The main advantage of the scaled-modeling technique against the simulation tools previously described is the possibility to consider a realistic measurement environment where most of the measurement-error contributions present in the final system are included. Moreover, full-wave simulation analysis can be very time-consuming when the coupling between the DUT and a realistic probe must be evaluated in several scanning points. In such cases, the scaled-modeling technique is thus preferred.

An example of the application of the scaled-modeling technique to the analysis of the measurement performance of an automotive testing environment is reported in this section.

The objective of the analysis is the extension to the 70–400 MHz band of an automotive multiprobe system like the one shown in Figure 4.14. Such a system was originally designed to work from 400 MHz to 6,000 MHz. (See Chapter 5 for more details on automotive multiprobe systems.) The focus of this analysis was to select a proper configuration for the floor that also provided a good approximation of a free-space condition and eased the positioning of the car in the center of the system. Three possible solutions were identified:

1. The use of 48-inch absorbers combined with a car-lift mechanism: these absorbers ensure low reflectivity down to 70–80 MHz [3], but, due to their size, they are difficult to move, and the vehicle needs to be raised from the floor to avoid the shadowing effect. Consequently, more time would be needed during the measurement setup phases.
2. The use of 18-inch absorbers without a car-lift mechanism: the nominal reflectivity of these absorbers is relatively high at lower frequencies (70–400 MHz), but they would be a cost-effective solution that would make it possible to reduce the downtime in the measurement setup phases.

Figure 4.14 Automotive multiprobe system with a floor covered with absorbers.

3. The use of 18-inch absorbers combined with a car-lift mechanism: this solution allows for improvements to the reflectivity level in a post-processing step where a spatial filtering technique (see details in Chapters 3 and 7 and [28, 29]) is applied. Additional time to raise the car is required as in the first solution, but the absorbers can be handled more easily as in the second solution.

The three solutions have been analyzed with the scaled model technique and are depicted in Figure 4.15.

The scaled model technique has been applied considering a scaling factor of $N = 12$, meaning that all the considered physical dimensions are 12-times smaller with respect to the actual scenario and that the frequencies are 12-times higher (i.e., the same electrical dimensions).

The considered (scaled) measurement system is the StarLab-18GHz (SL18GHz) multiprobe system shown in Figure 4.15. The SL18GHz is comprised of two interleaved probe arrays capable of performing measurements in the frequency ranges 0.4–6 GHz and 6–18 GHz, respectively. The measurement radius of the system is 45 cm.

The vehicle under test is a 1:12 scaled model of an old Morris Minor 1000 (year 1965). To generate radiation patterns like the ones expected in the final (full-scaled) scenario, three wideband patch antennas have been installed on three different positions on the car body: close to the windshield, on the rear part of the roof, and on the hood of the scaled vehicle. Measurements are conducted by feeding one patch at the time in the 1.008–6-GHz frequency band. With the considered $N = 12$ scaling factor, the performed measurements are equivalent to the ones of a full-size vehicle (real dimensions of L × W × H = 3.76 × 1.55 × 1.52m) measured in a system with a 5.4-m radius in the 84–500-MHz band as shown in Table 4.1.

As shown in top-left side of Figure 4.15, the scaled vehicle is first measured in free-space conditions over the full sphere to obtain the reference patterns.

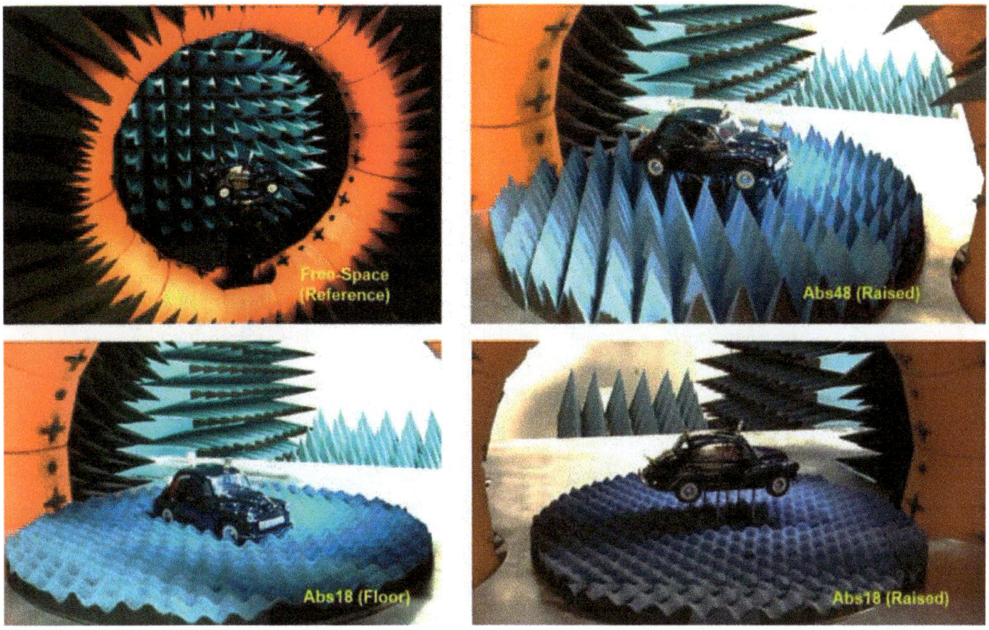

Figure 4.15 Scaled automotive measurement setups: free space scenario and modeling of the three possible solutions.

The three different solutions to be analyzed are then implemented, introducing a metallic ground inside the system. Such a ground is composed of a 75-cm diameter turntable that rotates with the antenna and of a fixed metallic part that extends outside the system. To emulate real automotive systems where the top of the car is located close to the center of the spherical scan, the metallic floor is placed 11 cm below the center of the scanner (corresponding to a position of 1.32m in the equivalent scenario). This displacement allows for measurements down to approximately $\theta = 100°$ (i.e., 10° below the horizon).

The first solution is modeled by placing 4-inch absorbers (12-times scaled version of the 48-inch absorbers) on top of the turntable and raising the vehicle 7.5 cm (equivalent to 90 cm) from the floor as shown on the top-right side of Figure 4.15.

Similarly, for the second solution, 1.5-inch absorbers (12-time scaled version of the 18-inch absorbers) are considered, and the scaled vehicle is placed directly on the floor as shown on the bottom-left side of Figure 4.15.

Finally, in the third option, the same 1.5-inch absorbers are considered but the vehicle is raised 7.5 cm from the floor. (See bottom-right side of Figure 4.15.)

It is specified that in this analysis, while the physical dimensions of the absorbers are properly scaled, their losses are not perfectly scaled as they should be according to the scale-model method [1]. In this specific case, the reflectivity of the considered scaled absorbers are slightly higher than the ones of the full-size absorbers, meaning that worst-case scenarios are considered. Nevertheless, this has been assumed to be a reasonable approximation, providing a representative emulation of the real scenario.

FF radiation patterns have been obtained applying the spherical NF/FF transformation to the measurements performed in the four scenarios shown in Figure 4.15. A vertical gain pattern comparison along the vehicle's longest dimension

Table 4.1 Measured (Scaled) and Equivalent Automotive Measurement Scenario

	Measured Scenario	*Equivalent Scenario*
Car size	31.3 × 12.9 × 12.7 cm	3.76 × 1.55 × 1.52 m
Frequency	1.008–6 GHz	84–500 MHz
Meas. radius	45 cm	5.4m
Turntable diameter	75 cm	9.0m
Floor position (Z)	–11 cm	–1.32m
Car height (H)	0 and 7.5 cm	0 and 0.9m
Absorber height	1.5- and 4-inch	18- and 48-inch

($\phi = 0°$) at 210 MHz (scaled from 2,520 MHz) is reported in Figure 4.16 for the hood-antenna. The black-dashed trace is the free-space measurement to be considered as reference, while the red, green, and blue traces are the measurements of the three scenarios under analysis. The shadowed areas indicate the unreliable FF regions associated to the scan truncation (at 100°). As expected, measurements over the scaled 48-inch absorbers (Abs48, raised) agree well with the free space especially within $|\theta| \leq 90°$. More deviations are instead observable when the car is measured over the scaled 18-inch absorber in the floor configuration (Abs18, floor). This is due to the relatively high reflectivity of the absorbers which is greater than −20 dB at this frequency. By raising the car over the scaled 18-inch absorbers (Abs18, raised) and applying a spatial filtering [28, 29], the pattern agreement with the reference is significantly improved.

Considering again the hood antenna, a similar comparison is reported in Figure 4.17 for the horizontal cut at $\theta = 85°$. Even along this cut, the gain radiation patterns from the first and third scenarios (Abs48 and Abs18, raised) agree well with the free space, while more deviations are observed with the second scenario (Abs18, floor).

The measurement scenarios under analysis and the different measured antenna positions are compared over the whole frequency band, considering the equivalent noise level (ENL) defined as

$$ENL = 20\log_{10}\left(RMSE\left| \frac{E(\theta,\ \phi) - \tilde{E}(\theta,\ \phi)}{E(\theta,\ \phi)_{MAX}} \right| \right) \tag{4.3}$$

where $E(\theta, \phi)$ is the reference gain pattern (free space), $\tilde{E}(\theta, \phi)$ is the test gain pattern, and RMSE is the root mean square error.

The ENL has been evaluated over the whole upper hemisphere. The ENL comparisons are reported in Figure 4.18 for the hood (left), rear-roof (center), and windshield antenna (right). As expected, for almost each frequency and antenna position, the lowest equivalent errors are reached when the scaled 48-inch absorbers are used (red traces). The performance of the scaled 18-inch absorbers with the car placed on the floor is worse as can be seen from the higher error levels (green traces). By raising the car over the scaled 18-inch absorbers and applying the spatial filtering, the measurement performance is significantly improved (blue

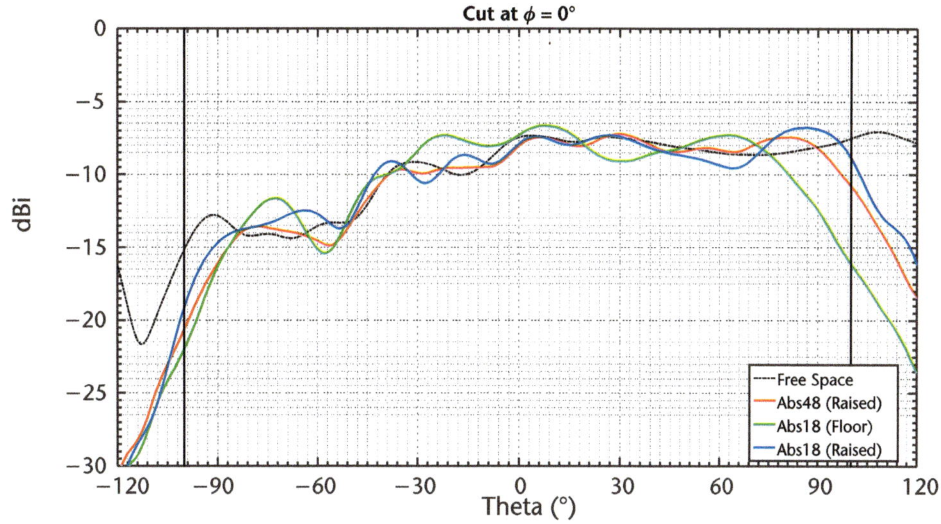

Figure 4.16 Hood antenna: comparison of vertical gain pattern cut at 210 MHz (scaled frequency).

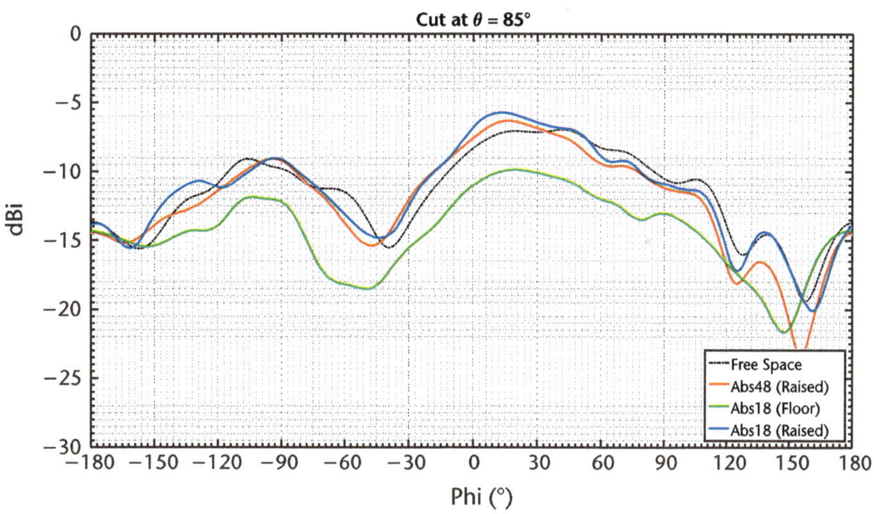

Figure 4.17 Hood antenna: comparison of horizontal gain pattern cut at 210 MHz (scaled frequency).

traces), reaching almost the same error level obtained with the larger absorbers. This interesting behavior is observed for each measured frequency and considered antenna position.

This analysis performed with scaled measurements suggests that the measurement accuracy achievable with the combination of 18-inch absorbers on the floor, the raising of the vehicle, and spatial filtering (the third solution among the ones listed above) is comparable with the one of achievable by using 48-inch absorbers (first solution).

It is hence concluded that to extend the operational bandwidth of an automotive system like the one shown in Figure 4.14 to 70–400 MHz, 18-inch absorbers could be placed on the floor if the car is raised and spatial filtering is applied in a

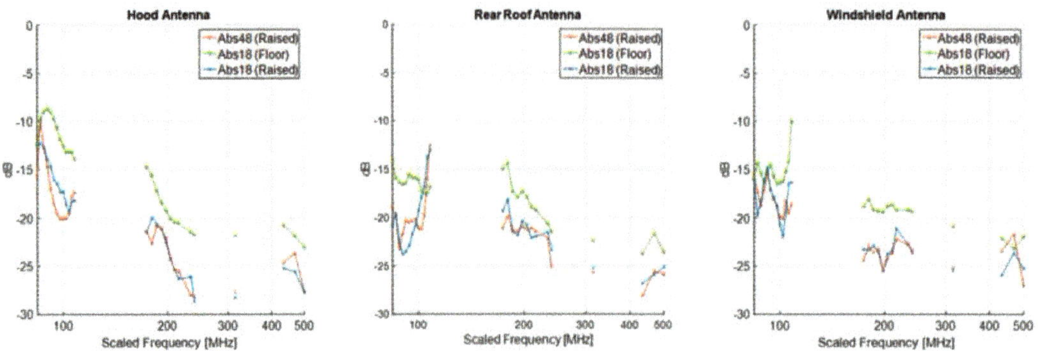

Figure 4.18 ENL comparison among the different scenarios and antennas: hood antenna (left); rear-roof antenna (center); and windshield antenna (right).

post-processing step. This is a cost-effective solution that also significantly simplifies the measurement setup phases, since such absorbers are less bulky and hence easier to handle than the 48-inch ones.

4.4 QZ Evaluation with Measurements

This section shows two different techniques for the evaluation of the QZ in anechoic chambers. According to the type of chamber and range, there are methods more appropriate than others. These methods can be found in [1, 30]. The first one is based on field probe measurements over the quiet zone, and it is very appropriate for evaluating the quality of the signal in the FF ranges. The AUT is in this volume, guaranteeing enough accuracy for antenna sizes lower than that of the QZ. The second method is based on the repetition on the measurements at different distances. If enough acquisitions are done, the comparison of the results at those distances determines the uncertainty of the system.

4.4.1 Field Probe Measurements Over QZ

This method is based on the acquisition of the electromagnetic field in a plane using, for instance, a linear slide and an azimuth turntable, or, in a more general way, a XY scanner. The acquisition must be performed for different frequency points to cover the whole frequency band; this is usually done for the following reasons:

- In the fields probing, make sure the antenna source is directed to the center of the QZ (measurement aperture);
- The reflections are evaluated, in a general case, assuming that the acquired signal is the addition of two signals, the direct and the reflected one (Figure 4.19). The ratio between the direct ray and the reflected one can be easily calculated using the following equation, in which σ is the peak-to-peak value of the ripple:

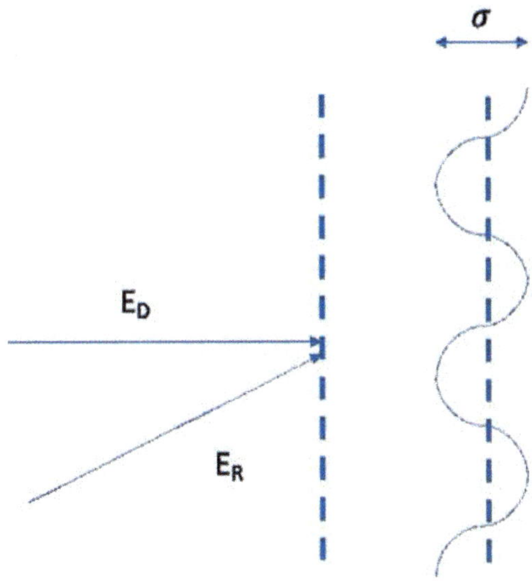

Figure 4.19 Schematic of the sum of direct ray and reflected ray.

$$\frac{|E_R|}{|E_D|}(dB) = 20log\left(\frac{10^{\sigma/20} - 1}{10^{\sigma/20} + 1}\right) \tag{4.4}$$

Figure 4.20 shows the evaluation of the QZ in the LEHA-UPM compact antenna test range. In this case, a horn is in the linear slide, placed on the roll positioner to get the samples on a plane. The acquisitions can be done for the different cuts and in both orthogonal polarizations. In this case, in order to have enough information on the QZ, the following three graphs are plotted:

- The ripple of the amplitude of the field in the quiet zone;
- The ripple of the phase in degrees;
- The co-polar to cross-polar ratio.

These values are used to evaluate the uncertainty of the range and the valid frequency band of the measurement range.

4.4.2 Antenna Pattern Comparison

The second method is based on repeating the same acquisition but slightly modifying the distance between the AUT and probe. If the variation in the difference of the distance between the direct ray and the reflected ray is larger than a wavelength, there will be enough information to have all the possible values of the interference between both rays (Figure 4.21).

This method is useful for detecting reflections coming from a wide angular range, and, if it is measured in FF, it directly provides the uncertainty of the measurement. The uncertainty contribution is measured for the specific AUT, used in

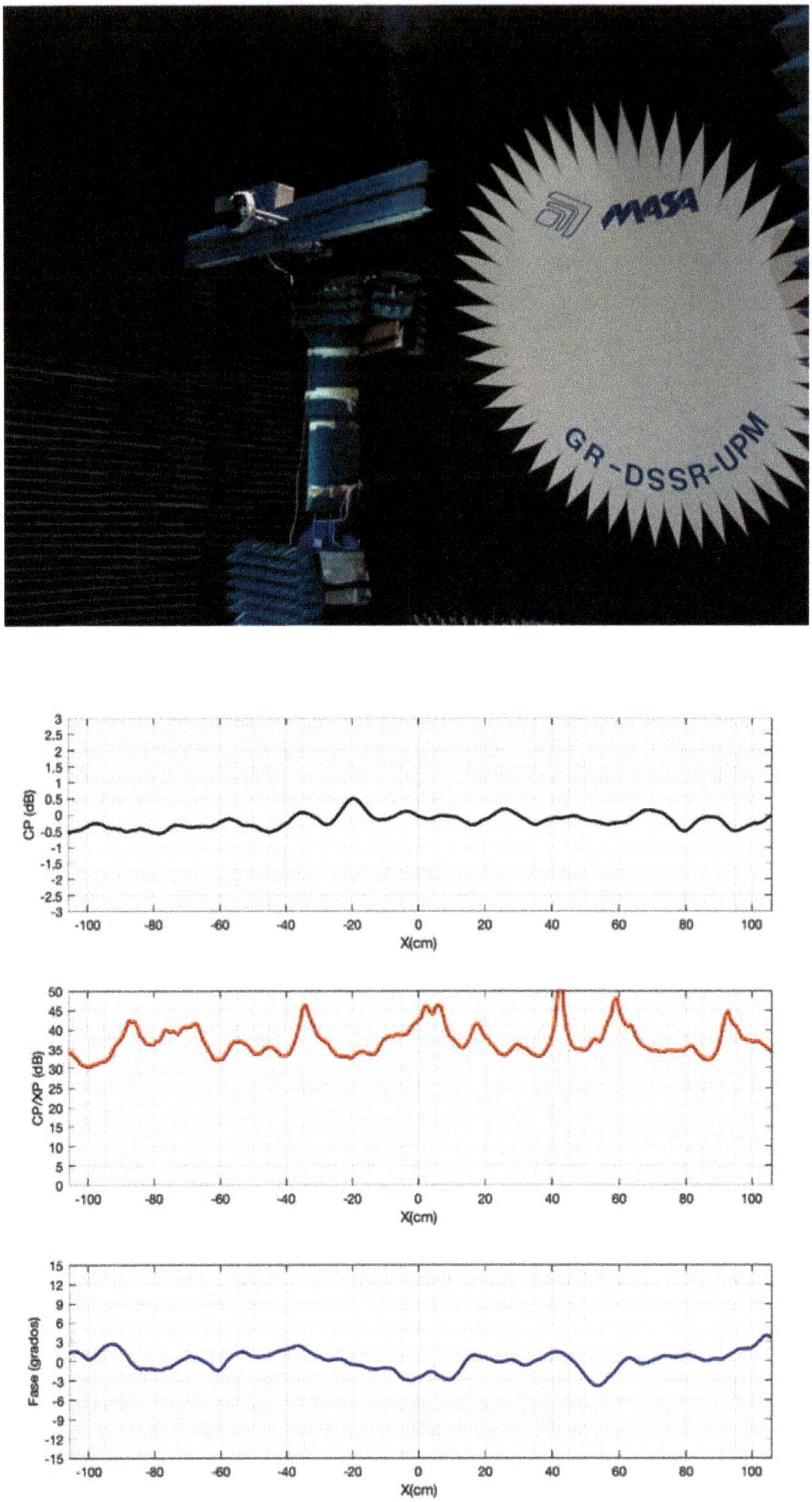

Figure 4.20 Evaluation of the QZ at UPM CATR; measurement results at 15 GHz, for the horizontal plane and H-polarization.

Figure 4.21 Setup of antenna pattern comparison.

the evaluation at different distances. As in the previous case, from the acquired ripple, it is possible to determine the amplitudes of the direct and reflected ray.

This technique is used in the spherical field system of the LEHA-UPM to evaluate the performance of the anechoic material at UHF frequencies (660 MHz). Figure 4.22 shows the setup, where a horn antenna was used as AUT. This horn antenna was slightly moved to nine different positions with a distance between AUT and probe around 3m, to have a sufficient number of relative positions between the direct and reflected rays.

Figure 4.22 shows a comparison of the different acquisitions of the co-polar field for the horizontal pattern. To compare all the patterns, they are normalized at the maximum, where the effect of the reflection is minimum. As shown, this method directly yields the uncertainty of the measurement for each frequency band and for this antenna. There are two main contributions to the chamber reflections: AUT positioner and the absorbers. In this case, absorber performance is the limiting factor on measurement performance.

These methods represent only two examples of ways to evaluate the quality of the antenna measurement system. Other methods can be found in [1, 31–33].

References

[1] IEEE Standard Test Procedures for Antennas, ANSI/IEEE Std 149-1979.

[2] Rodriguez, V., *Anechoic Range Design for Electromagnetic Measurements,* Norwood, MA: Artech House, 2019.

[3] https://www.mvg-world.com/en/products/absorbers/standard-absorbers/pyramidal-absorbers-aep-series.

[4] Rodriguez, V., "Basic Rules for Indoor Anechoic Chamber Design [Measurements Corner]," in *IEEE Antennas and Propagation Magazine,* Vol. 58, No. 6, Dec. 2016, pp. 82–93.

[5] https://www.mvg-world.com/en/products/absorbers/ferrite-tiles/ferrite-tiles.

[6] "Unique Application of Chebyshev Absorbers," https://www.mvg-world.com/en/products/absorbers/adapted-absorbers/chebyshev-applications.

[7] Stockman, H., "Communication by means of reflected power," *Proc. Inst. Radio Engineers,* Vol. 36, No. 10, 1948, pp. 1196–1204.

[8] Emerson, W. H., and H. B. Sefton, Jr., "An improved Design for Indoor Ranges," *Proceedings of the IEEE,* Vol. 53, No. 8, Aug 1965, pp. 1079–1081.

[9] King, R. W. P., et al., *Antennas in Matter: Fundamentals, Theory, and Applications*, Cambridge, MA: MIT Press, 1981.

[10] D'Abreu, G., Z. Xiong, and Z. Chen, "Chambers for the Evaluation of Vehicle Mounted Antennas," *EuCAP 2018,* 9–3, April 2018, London, United Kingdom.

[11] Mauermayer, R. A. M., and T. F. Eibert, *Spherical Field Transformation for Hemispherical Antenna Measurements Above Perfectly Conducting Ground Planes, AMTA,* Austin, TX, October 2016.

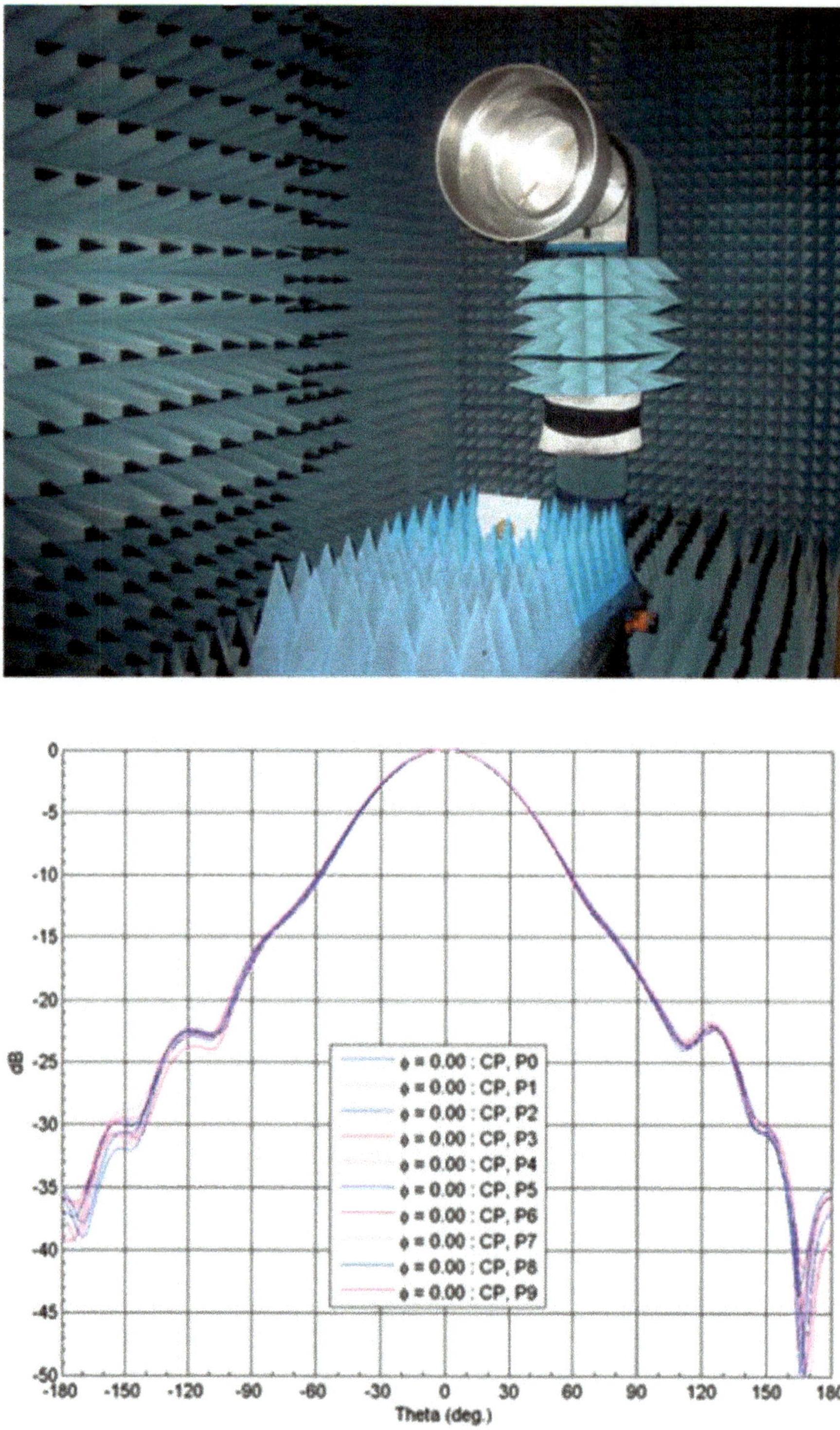

Figure 4.22 Evaluation of reflections at low frequencies in the UPM spherical range.

[12] Asghar, M. E., "Comparative Analysis of Spherical Near-Field Automotive Antenna Measurement Facilities," *2019 13th European Conference on Antennas and Propagation (EuCAP),* 2019.

[13] Campbell, G. Gampala, and C. Reddy, "Modeling and Analysis of Anechoic Chamber Using CEM Tools," *Appl. Comput. Electromagn. Soc. J,* Vol. 28, No. 9, 2013, pp. 755–765.

[14] https://www.3ds.com/products-services/simulia/products/cst-studio-suite/user-interface/ (accessed November 2021).

[15] Saporetti, M. A., "Low Computational Cost Modelling of Anechoic Chambers and Probe Effect in Antenna Measurement," *2021 15th European Conference on Antennas and Propagation (EuCAP),* 2021, pp. 1–5.

[16] Rodriguez, V., "Simple Approaches to Range Performance Estimation," *2021 15th European Conference on Antennas and Propagation (EuCAP),* 2021, pp. 1–4.

[17] https://www.mvg-world.com/en/products/antenna-measurement/software/satsim.

[18] Mourgues, S., F. Jensen, and M. Sabbadini, "Advanced CAD applications with GTD," *JINA Conference,* Nice, France, November 1992.

[19] Hussar, P. E., et al., "An Implementation of the UTD on Facetized CAD Platform Models," *IEEE Antennas and Propagation Magazine,* Vol. 42, No. 2, April 2000, pp. 100–106.

[20] Di Giampaolo, E., M. Sabbadini, and F. Bardati, "Astigmatic Beam Tracing for GTD/UTD Methods in 3-D Complex Environments," *Journal of Electromagnetic Wave and Application,* Vol. 15, No. 4, 2001, pp. 439–460.

[21] Sabbadini, M., F. Bardati, and E. Di Giampaolo, "A recursive Beam-Splitting Algorithm for Forward Ray Tracing," *USNC/URSI National Radio Science Meeting,* 203, July 11–16, 1999.

[22] Saccardi, F., et al., "Efficient Modelling of Antenna Measurements Including the Impact of the Anechoic Chamber," *APS 2021.*

[23] Hansen, J. E. (ed.), "Spherical Near-Field Antenna Measurements," London, United Kingdom: Peter Peregrinus Ltd.,1988.

[24] Balanis, C. A., *Antenna Theory: Analysis and Design*, Hoboken, NJ: John Wiley & Sons, 2016.

[25] Saccardi, F., et al., "Experimental Comparison of Absorber and Conductive Floor Automotive Near Field Antenna Measurement Systems," *International Journal of Microwave and Wireless Technologies,* 2021, pp. 1–12.

[26] Saccardi, F., et al., "Experimental Comparison of Vehicular Antenna Measurements Performed Over Different Floors," *EuCAP 2020,* March 15–20, 2020, Copenhagen, Denmark.

[27] Saccardi, F., et al., "Accurate Calibration of Truncated Spherical Near Field Systems with Different Ground Floors using the Substitution Technique," *AMTA 2019,* October 6–11, San Diego, CA.

[28] Saccardi, F., et al., "Comparative Investigation of Spatial Filtering Techniques for Ground Plane Removal in PEC-Based Automotive Measurements" *AMTA 2019,* October 6–1, San Diego, CA.

[29] https://www.mvg-world.com/en/products/antenna-measurement/software/mv-echo.

[30] Evans, G. E., *Antenna Measurement Techniques*, Norwood, MA: Artech House, 1990.

[31] University of Michigan Report 5391-1-F, February 1963 (Free space VSWR test).

[32] Zhong, C., X. Zubiao, and A. Enayati, "Limitations of the Free-Space VSWR Measurements for Chamber Validations," *38th AMTA Proceedings,* Austin, TX, October 2016, pp. 144–148.

[33] Tian, B. B., "Free Space VSWR Method for Anechoic Chamber Electromagnetic Performance Evaluation," *30th AMTA Proceedings,* Boston, MA, November 2008, pp. 524–529.

[34] Saporetti, M. A., et al., "Low Computational Cost Modeling of Anechoic Chambers and Probe Effect in Antenna Measurement," *2021 15th European Conference on Antennas and Propagation (EuCAP).*

CHAPTER 5

Implementation and Calibration of Automotive Ranges

Francesco Saccardi, Lars J. Foged, Thierry Blin, M. E. Asghar, F. Wollenschläger, C. Bornkessel, M. A. Hein, and Manuel Sierra Castañer

The innovation of new communication systems for vehicles created the need for accurate and fast antenna measurement systems. The primary resulting automotive testing solutions are presented and discussed in this chapter.

The most common configurations are based on spherical scanners, where the vehicle is placed on a turntable performing the scanning in azimuth, while the elevation scanning is performed by a moving probe or an array of probes.

The treatment of the ground floor where the vehicle is placed during the test is one of the main aspects influencing the measurement results. The most common floor scenarios are the absorbing and conductive ones. In the former, absorbing materials are used to emulate a free-space condition. In the latter, a metallic floor is considered to create a fully reflective (or PEC) ground. The pro and cons of these two types of automotive systems, along with the different gain calibration methods applicable to different scenarios, are discussed in Section 5.1. Comparative examples obtained from scaled automotive measurements performed with absorbing and conductive floors are also reported.

Beside the different treatment of the floors, automotive ranges are categorized as direct FF or NF and according to the scanning mechanism.

In direct FF ranges, the data is collected at a sufficiently large distance to meet the plane wave condition, and basically no data processing is applied. Such types of systems are introduced in Section 5.3.

As widely described in Chapter 3, in NF systems the field is sampled on a closed surface around the DUT, and a mathematical NF-to-FF transformation is applied to compute the FF. Such systems basically differ according to the scanning mechanism implemented to acquire the field around the DUT.

Single-probe systems are based on the mechanical movement of the probe to perform the elevation scanning. In most of these systems, a gantry arm is used to perform this scanning, while other systems instead make use of an arch where the probe is moved on a rail. Single-probe NF systems are described in Section 5.3. Besides describing the main implementation aspects, we focus on the use of wideband probes and associated probe pattern compensation to improve measurement efficiency and accuracy.

Other classical implementations of scanning mechanism are the multiprobe systems. In such systems, the movement of the probe is replaced by an arch where a set of probes performing an electronic scanning is installed. The main advantage of multiprobe systems is reduced measurement time. The fast electronic switching or scanning from probe to probe is significantly faster than mechanical movements and allows for very fast measurement, since only one movement (azimuth) is required. Multiprobe systems are also the only technique that enable low-frequency measurements, down in the FM range. The main drawback of such systems is the need for probe-array calibration and, in some cases, post-processing to ensure measurement performances. Multiprobe systems are detailed in Section 5.4.

Finally, Section 5.5 discusses the scaled measurement.

5.1 Absorber and Conductive Floor Systems

As discussed in Chapters 1–4, many automotive antenna systems are based on spherical NF ranges where the scanning surface is truncated. The truncation is generally close to or below the horizon, over an absorbing covered or conducting floor.

An example of an absorbing floor is shown on the right side of Figure 4.1. Absorbing floors are used to emulate a free-space environment, which, in most of the cases, can be considered a good approximation of realistic ground scenarios. Moreover, as detailed in Chapter 8, the availability of the free-space response of the vehicle-mounted antennas also enables straightforward emulations of dielectric and lossy driving grounds, such as asphalts or soils, performed with dedicated post-processing tools. The ability to measure down to FM frequencies like 50–110 MHz is another advantage of such absorbing floor-based systems.

An example of a conductive floor system is shown on the left side of Figure 4.1. In these cases, a metallic floor is used to generate a fully reflective ground, which is then assumed to be a PEC in the data processing. The main advantages of such measurement scenarios are the ease of the vehicle positioning in the system (which is usually simply parked in the center of the system) and the significant reduction of the truncation errors during the NF/FF transformation, as explained in detail in Section 3.1.2.2. On the other hand, a fully reflective ground can be considered a good approximation of a realistic ground only in a few cases (such as under wet asphalt conditions), and the emulations of generic dielectric grounds require a preprocessing step to retrieve the free-space response. As detailed in Chapter 7, such a free-space response retrieval is not a trivial task and, in many cases, does not provide accurate enough results. Moreover, because of the strong coupling with the reflective floor, the lowest frequency limit of such systems is typically around 400 MHz.

In this section, we focus on the different gain-calibration techniques for such absorbing and conductive floor systems and present some experimental and comparative measurement results. Other comparative investigation and consideration of such systems can also be found in [1–5].

5.1.1 Gain Calibration

As discussed in Chapter 2, system calibration is a fundamental step in any type of antenna measurements to test power-related quantities such as gain, efficiency, EIRP, and TRP. In some types of systems, the calibration can be more challenging than in others, and the selection of a proper calibration method is critical. In this section we focus on the (passive) gain calibration of truncated automotive antenna ranges with absorbing or conductive floors.

The primary gain calibration methods are the so-called direct-gain and gain/efficiency substitution methods.

In the case of FF measurements, the direct method is simply based on the inversion of the Friis transmission formula (5.1) by means of the gain of the AUT (G_{AUT}). (See also the description in Section 2.1.1.)

$$S_{21}^{dB} = G_{AUT}^{dB} + G_{Probe}^{dB} - L_{freespace}^{dB} - L_{system}^{dB} \tag{5.1}$$

As can be seen, beside the measured coupling between the AUT and the probe (S_{21}^{dB}) this method requires the knowledge of different quantities including the gain of the probe (G_{Probe}^{dB}), the free-space losses ($L_{freespace}^{dB}$) and the loss of system (L_{system}^{dB}). In some cases, the probe gain can be characterized with dedicated measurements. Unfortunately, this is not possible, for example, when the probe is embedded in complex structures like multiprobe arrays. (See Section 5.4 for more details.) The loss of the system includes all the RF path (e.g., cables, switches, attenuators, and amplifiers) and they should be characterized with a conducted measurement where the two ends of the RF chain are connected using a transmission line with known losses. This measurement is often not practicable in the case of large systems where the probe and the AUT are far away from each other. Alternatively, each component of the chain could be characterized independently but at the cost of a higher uncertainty due to the insertion losses introduced during the characterization of each component. Finally, the free-space losses depend on the working frequency and the separation distance between the AUT and the probe.

In the case of NF measurements, the free-space losses are compensated for during the NF/FF transformation processes. To apply the direct calibration to NF measurements, the other terms in (5.1) are then used to scale the whole transformed pattern [6, 7]. The main advantage of the direct method is that it requires only one radiating measurement, but the accurate knowledge of the probe gain and the losses of the system might often be critical.

To overcome these limitations, a reference antenna with known gain and/or efficiency can be measured to calibrate the system, implementing the so-called substitution (or transfer) method. Even though it requires an additional measurement, this second method generally achieves higher accuracy, because the uncertainty of the system calibration is basically reduced to the uncertainty related to the reference antenna.

When a reference antenna with known gain (G_{REF}^{dB}) is measured, the Friis formula can be rewritten as follows:

$$C_{Gain}^{dB} = G_{REF}^{dB} - S_{21,REF}^{dB} = L_{freespace}^{dB} + L_{system}^{dB} - G_{Probe}^{dB} \tag{5.2}$$

obtaining a gain calibration (or correction) term (C_{Gain}^{dB}) that depends only on the known gain and "raw" measurement of the reference antenna. Such a calibration term is then used to scale the actual measurement(s) of the AUT(s).

Hence, the calibration uncertainty depends on the uncertainty with which the gain of the reference antenna is known and the uncertainty of the reference measurement itself ($S_{21,REF}^{dB}$). The latter can be reduced in two ways.

First, as explained in Chapter 2, it is considered good practice to consider reference antennas with radiation characteristics as close as possible to the ones expected by the AUT. If the two devices have a similar behavior, they most likely experience the same measurement errors, which will tend to cancel each other out. For this purpose, monocones, monopoles mounted on metallic ground planes, or horns are typically used in automotive antenna measurements.

Second, the so-called efficiency substitution method can be used instead of the "conventional" gain substitution [5]. Integrating (in linear scale) the left and right side of (5.2) over the whole measurement sphere, we obtain on the left side the delta (in decibels) between the (known) real efficiency (η_{REF}^{dB}) and the "raw" measured efficiency (η_{Meas}^{dB}) of the reference antenna, while the right side does not change as all the quantities do not depend on the location on the sphere.

$$C_{\eta}^{dB} = \eta_{REF}^{dB} - \eta_{Meas}^{dB} \tag{5.3}$$

Ideally, C_{Gain}^{dB} and C_{η}^{dB} would be identical, but the efficiency calibration allows to reduce some measurement errors. For example, the ripples caused by the reflections (in both FF and NF ranges) or due to the truncation errors (in NF ranges) tend to be averaged out when the measured pattern is integrated to compute the efficiency. In other words, the efficiency, since is an integral quantity (not a local one like the gain), is less sensible to many error sources.

Of course, the accuracy of the efficiency substitution method depends on the ability of the system to be calibrated to measure the total efficiency (i.e., the efficiency associated to the radiated power on the whole sphere). Unfortunately, in a truncated absorbing floor-based system, the power radiated on the backward hemisphere is lost, and thus the total efficiency cannot be measured. In such cases, it is hence convenient to consider the UHE (or η^{UH}) defined as the integral of the gain pattern over the forward hemisphere

$$UHE = \frac{1}{4\pi} \int_{\phi=0}^{2\pi} \int_{\theta=0}^{\pi/2} G(\theta,\phi) \sin\theta d\theta d\phi \tag{5.4}$$

The calibration equation can be reformulated in terms of the UHE as follows

$$C_{UHE}^{dB} = UHE_{REF}^{dB} - UHE_{Meas}^{dB} \tag{5.5}$$

where UHE_{REF}^{dB} and UHE_{Meas}^{dB} are respectively the (known) real UHE and the measured UHE of the reference antenna, and is (again) mathematically identical to C_{Gain}^{dB} and C_{η}^{dB}.

When conductive floor-based systems are considered, the power radiated in the backward hemisphere is fully reflected by the presumed lossless ground. Hence, the resulting power on the upper hemisphere is the same as the total radiated power that the antenna would radiate in free space over the full sphere. The calibration term in this case can be obtained as the difference between the reference (total) efficiency $\left(\eta_{REF}^{dB}\right)$ of the calibration antenna and the measured hemispherical efficiency (UHE_{Meas}^{dB}) as shown in (5.6):

$$C_{\eta PEC}^{dB} = \eta_{REF}^{dB} - UHE_{Meas}^{dB} \tag{5.6}$$

Note that this latter calibration method can be used only if the ground floor does not modify the matching of the calibration antenna. This is a reasonable assumption for antennas with low back-radiation, such as horns and monopoles/monocones mounted on their own ground plane.

5.1.2 Gain Calibration Accuracy

The accuracy of the gain and efficiency substitution methods when applied to absorber and conductive floor-based measurements has been investigated.

The scenario is the 12-times scaled measurement setup discussed in Section 4.3.4 in different scenarios as shown in Figure 5.1. In this setup, two dual-ridge horn antennas, the SH800 and the SH2000, are considered as calibration devices. The SH800 (not shown in Figure 5.1) is used to check the calibration performance in the 84–500-MHz band (scaled from 1 to 6 GHz real measured frequencies), while the SH2000 is used for the 500–1,500-MHz band (scaled from 6 to 18 GHz real measured frequencies).

The calibration antennas are first measured in free-space conditions over the full sphere, as shown on the left of Figure 5.1. As shown in Figure 5.1's second and third photographs, two absorbing floor scenarios are considered. The larger absorbers emulate 48-inch absorbers (Abs48), while the smaller absorbers scale approximately to 18-inch absorbers (Abs18). In both absorbing floor configurations, elevation data has been collected down to approximately 10° below the horizon.

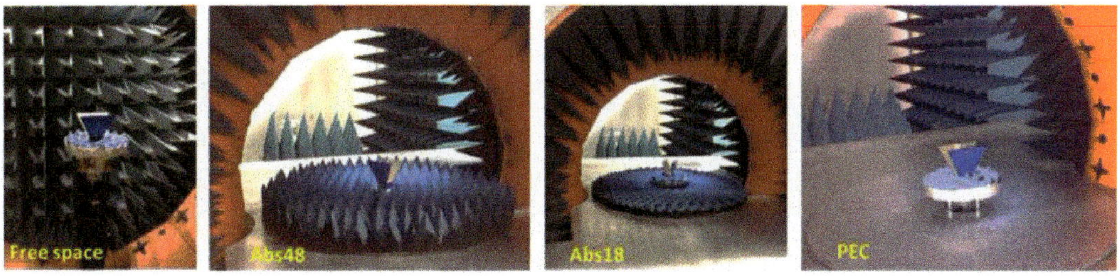

Figure 5.1 SH2000 antenna during calibration in the different scenarios: free-space, 48-inch (Abs48) and 18-inch (Abs18) scaled absorbers, and metallic floor (PEC).

A conductive floor-based measurement scenario is also considered as shown on the right side of Figure 5.1.

A comparison of the measured E-plane pattern cut at 500 MHz (SH2000 scaled from 6 GHz) in the four scenarios is shown in Figure 5.2. The patterns are not normalized, and their levels are thus associated to the losses of the system. The blue-dashed trace is the pattern measured in free-space over the full-sphere (the baseline). The orange and green traces are the patterns measured respectively over the scaled 48-inch and 18-inch absorbers. The effect of the truncation is evident at angles larger than 90°, where the pattern is strongly attenuated. Moreover, some ripples caused by the abrupt discontinuity introduced by the zero-padding are present in the proximity of the truncation angle (i.e., θ = 100°). The black trace is the pattern measured over the conductive floor. Such a pattern is, of course, defined only in the forward hemisphere and is affected by strong ripples caused by the interaction with the reflective floor.

The gain and efficiency (or UHE) substitution methods described to this point have been applied to the three different floor scenarios and compared to the free-space configuration, assumed to be the reference. Results from this calibration analysis are summarized in Figure 5.3, where the calibration errors (i.e., the delta between the calibration term obtained in the free-space condition and in the absorbers or conductive floor conditions) are shown for the gain (left) and efficiency (right) calibration methods. The green traces are calibration errors for the 18-inch absorbers. The red and black traces are calibration errors for the 48-inch absorbers (Abs48) and PEC floor (PEC), respectively.

As expected, the variations are much smaller when the efficiency (or UHE) calibration is applied (< 0.2 dB at almost any frequency point and scenario); the larger calibration errors for the PEC-based measurements at 500 and 600 MHz are associated to other measurement error of the implemented measurement setup, not to the calibration method itself. The efficiency is in fact an integral quantity and errors due to spurious radiations in the measurement setup tend to be averaged out.

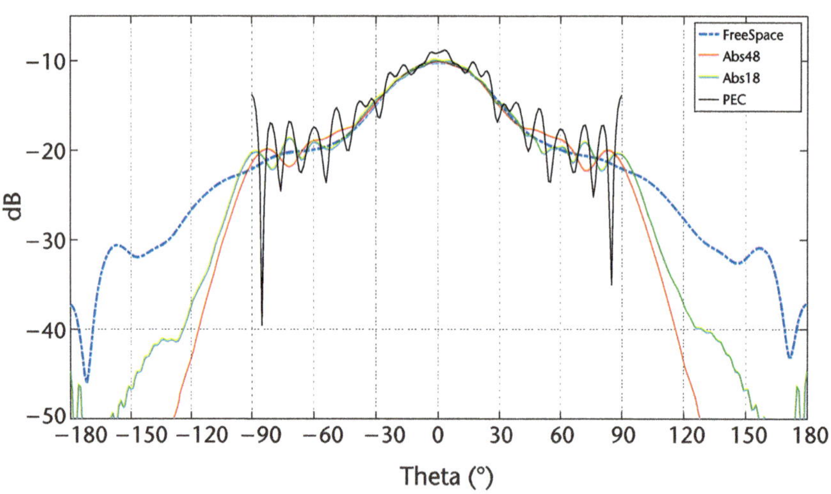

Figure 5.2 Measured E-plane pattern cut of the SH2000 calibration antenna at 6 GHz (representative of 500 MHz) in free space and different floor scenarios. ©2019 IEEE. (Reprinted with permission.) [5].

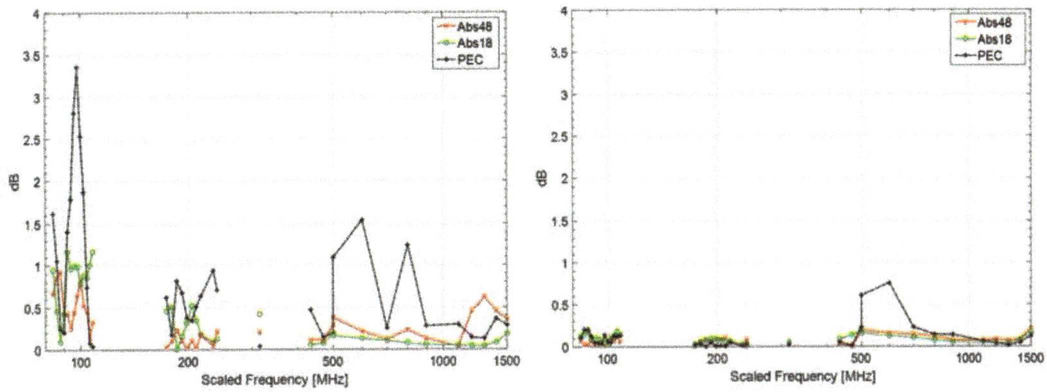

Figure 5.3 Calibration error comparison: gain calibration (left) and efficiency calibration (right). ©2019 IEEE. (Reprinted with permission.) [5].

On the other hand, the gain is a pointwise quantity and hence is more subject to measurement errors.

5.1.3 Measurement Examples

Three of the scaled automotive measurement configurations presented in Section 5.1.2, free space, scaled 48-inch absorbers and PEC, are now considered in the measurement of the scaled car model, shown in Figure 5.4. The vehicle is the same 12-times scaled model considered in Section 4.3.4 with three patch antennas mounted on the hood, the rear part of the roof, and the windshield.

Here, the aim of these scaled measurements is to investigate the differences among automotive testing when performed under free space (or absorber floor) and PEC (conductive floor) conditions.

Figure 5.5 compares the vertical ($\phi = 0°$, cut along the car largest dimension) and horizontal ($\theta = 85°$, cut 5° above the horizon) NF/FF transformed gain pattern cuts of the hood antenna at 1,200 MHz (where the real measured frequency is 14.4 GHz). The dashed-blue traces are associated with the free-space measurement, while the orange ones relate to the measurement over the scaled 48-inch absorbers. As expected, the agreement between these two measurements is good at angles corresponding the untruncated region (i.e., $\theta = [-100°, 100°]$), meaning that the absorbers are able to emulate the wanted free-space condition. The black

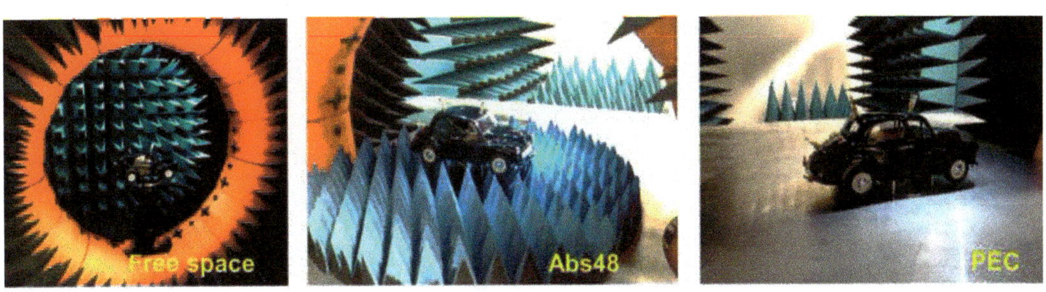

Figure 5.4 Scaled vehicle measured in free-space conditions, over absorbing floor (Abs48) and over conductive floor (PEC).

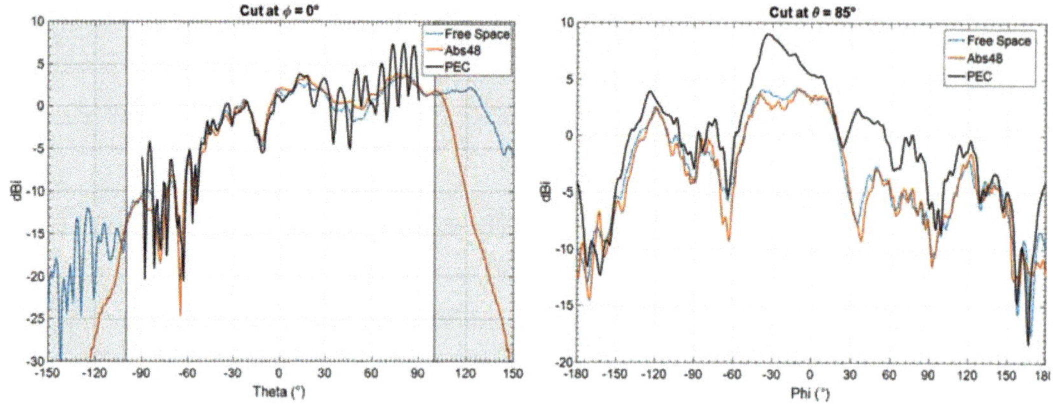

Figure 5.5 Hood antenna: vertical (left) and horizontal (right) gain pattern comparison at 1,200 MHz (scaled frequency).

traces are the conductive floor-based measurements. The interaction with the fully reflective floor is evident on both cuts. From the vertical cut, it can be seen that the constructive and destructive interference create a significant ripple. Such behavior is the result of the considered PEC boundary condition and is fully in line with expectations.

It is also interesting to analyze the effect of the different floor configurations on figures of merit like the partial radiated powers (PRP) defined in Section 2.2.3, namely the upper-hemisphere radiated power (UHRP), the near-75 degrees PRP (N75PRP), and the near-horizon PRP (NHPRP).

It should be noted that in active or OTA testing, the above PRP is measured and usually expressed in decibels relative to one milliwatt. (See Chapter 6 for more details.) When passive measurements are performed, using a VNA, the obtained PRPs are normalized to the input power, and thus the corresponding partial efficiencies are found instead (expressed in decibels). The different floor scenarios will affect the partial efficiencies and the PRP in the same way; hence, without loss of generality, we refer to the latter, even though the former have been measured.

Figure 5.6 shows the computed UHRP for all antenna positions and scaled frequencies. The measured UHRP in the absorber-based scenario agrees very well

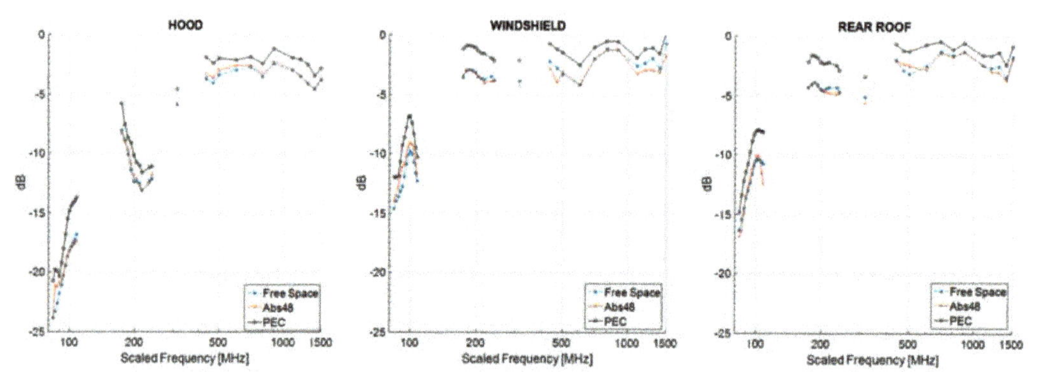

Figure 5.6 UHRP for the different antennas and measurement scenarios.

with the "true" free-space condition for all frequencies and antennas. This indicates the correct emulation of the wanted boundary condition by the absorber covered floor. As expected, the conductive floor scenario overestimates the UHRP; in fact, the power radiated toward the floor is reradiated in the upper hemisphere. A similar behavior can be observed by looking at the computed N75PRP shown in Figure 5.7.

The trend is different when the NHPRP are computed as shown in Figure 5.8. The NHPRP is computed integrating the measured powers in the θ = [60°–120°] angular range [8]. Computing the NHPRP from a conductive floor-based measurement is equivalent to integrating from θ_{min} = 60° to θ_{max} = 90° (because there is no field beyond 90°), but, due to the presence of the PEC, the power that would be radiated in the θ = [90°–120°] range is radiated back in the θ = [60°–90°] range. Hence, the NHPRP parameter should lend itself to a better comparison between the PEC-based and free-space case. Indeed, the NHPRP calculated from the conductive floor measurements is closer to the free-space condition, while the NHPRP is underestimated in the absorber floor scenario. This is strictly related to the truncation of the spherical scanning surface, which, in this case, was at θ = 100° (i.e., 10° below the horizon), while the NHPRP requires integration up to 120°. This means that part of the integration comprises the radiated power that is lost in the

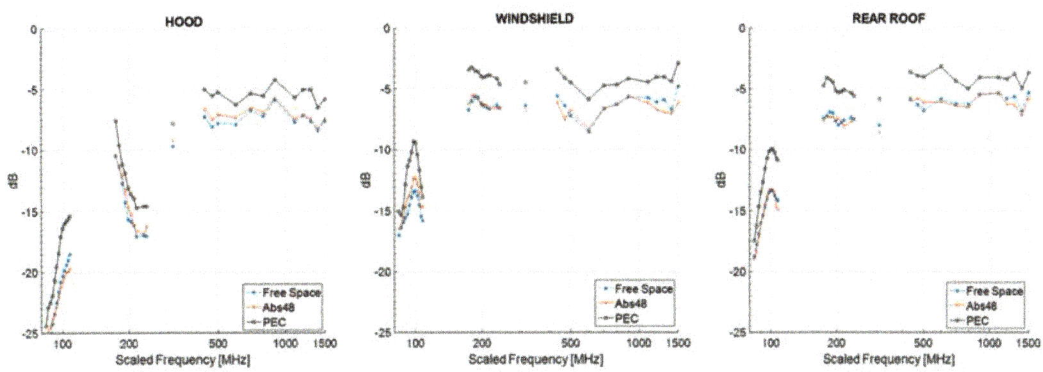

Figure 5.7 N75PRP for the different antennas and measurement scenarios.

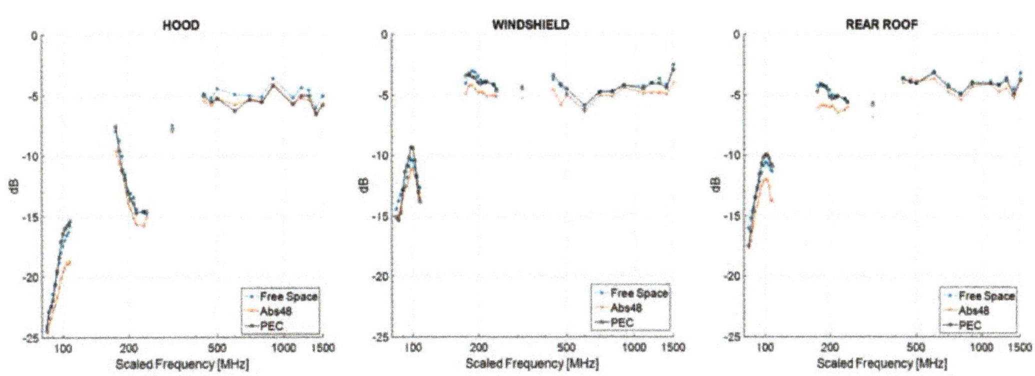

Figure 5.8 NHPRP for the different antennas and measurement scenarios.

truncated region of the absorber-covered floor scenario. It should be noted that the NHPRP is often used to ease the comparison between conductive and absorber floor-based systems [8].

5.2 FF Ranges

In direct FF ranges, a sufficiently large measurement distance between the DUT and the probe (or range antenna) is needed. Basically, no data processing is applied. As described in Chapter 2, a well-known and widely used criteria to meet the FF condition (or plane wave approximation) is to consider at least a measurement distance of $d = 2D^2/\lambda$, where D is the maximum diameter of the DUT.

As already pointed out in Chapter 2, in the case of automotive measurements, considering the whole vehicle size to evaluate the range distance could lead to very large distances that, due to budget and/or space limitations, cannot be realized in an indoor anechoic environment. For these reasons, direct FF automotive ranges are often implemented outdoors.

A sketch of a typical implementation of a direct FF range is shown in Figure 5.9. Systems like that consist of a rotating turntable where the vehicle under test is placed and a probe tower installed at distance from the center of turntable. The probe can usually be moved vertically to measure azimuth cuts from the horizon to a few degrees in elevation. The maximum measurable elevation angle depends on the measurement distance and the height of the probe tower. Since the distance cannot be too short, because of FF criteria to be met, the maximum elevation angle typically does not exceed 5°. Outdoor FF systems often use RF transparent radomes, where the vehicles under test are placed inside to ensure protection from outdoor weather conditions. An example of this type of direct FF systems is shown on the left side of Figure 5.10.

To enable direct FF measurements at higher elevation angles, gantry arm systems are sometimes used. As described in Section 5.3, gantry arms are typically used for spherical/hemispherical NF measurements to sample the 3-D field around the vehicle under test. If the gantry arm is large enough, they can also be used for direct 3-D FF measurements. The 11-m gantry arm inside the radome shown in Figure 5.10 (right) is an example of these types of systems [9].

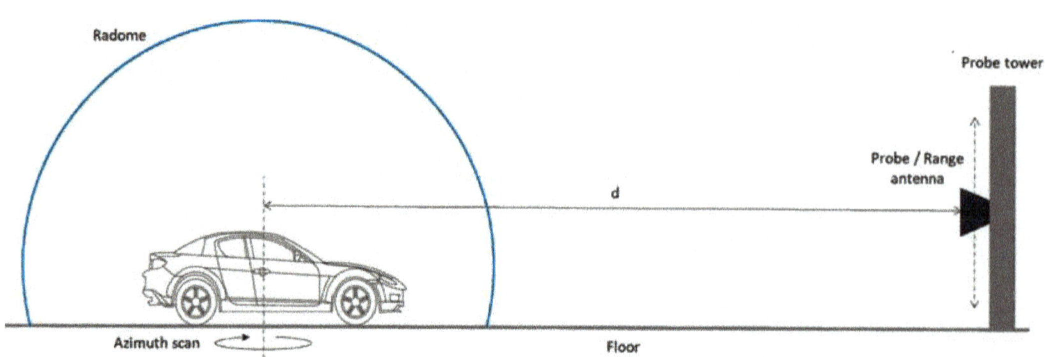

Figure 5.9 Sketch of a direct FF system for azimuth cuts. A radome is often used in the case of outdoor installations.

Figure 5.10 (a) Example of a FF outdoor system for direct azimuth cuts, and (b) example of an 11-m-high system for direct hemispherical FF measurements. (*Source:* [9]. Reprinted with permission.)

Outdoor FF systems are sometimes the preferred measurement solution at low frequencies like 60–100 MHz. At such frequencies, common cars (e.g., D = 5m) are electrically small, and the whole car structure contributes to the radiation. A measurement distance of at least 15m must be considered to meet the proper plane wave condition, and the indoor solution is often the only practicable one.

Direct FF indoor ranges are an attractive solution because they make it possible to perform the measurements in a controlled, repeatable, and anechoic environment. As described in Section 4.4, the performance of FF indoor ranges is usually characterized by evaluating the field variations in terms of ripple and taper of the within the quiet zone. Indoor FF systems work as schematized in Figure 5.9, but instead of a radome, both the turntable and the probe tower are installed inside a relatively large anechoic chamber where the measurement distances are on the order of 10–15m.

As previously noted, the main limitation of indoor FF ranges is the measurement distance, which in many cases is insufficient. The errors caused by the reduced range distance are usually accounted for in the measurement uncertainty budgets by estimating or measuring the magnitude of such errors for specific test cases and frequencies. While it might seem that the measurement uncertainty due to the reduced distance would increase dramatically with the frequency, this is not always the case. The reason is that when the cars become electrically large (but the antenna is electrically small), only a small portion of the car structure around the antenna contributes to the radiation, keeping the error due to the distance relatively low. (See Section 2.3 for a detailed discussion and example.)

An interesting approach to compensate for the errors due to the limited measurement distance is a single-cut NF/FF transformation. With such an approach, the measured cuts are no longer considered direct FF, but a special and approximate NF/FF transformation is applied to reach a better plane wave condition in post-processing. Different techniques have been proposed in the literature for this purpose [10]. The most common and accurate one is based on the cylindrical wave expansion (CWE) of the measured cuts [11, 12]. Like the SWE, the CWE is typically involved in cylindrical NF measurements to perform NF/FF transformation [13]. In this case the CWE is simply applied considering each single azimuth cut a cylinder of zero-height, making it possible to better approximate the FF condition. Due to the limited information associated to the height of the cylinder, this approach cannot be considered an exact transformation technique. Nevertheless, it has been shown that, if the measurement distance is not too short, the CWE-based single-cut transformation can provide sufficiently accurate results [11, 12].

5.3 Single-Probe NF Systems

Many NF automotive systems are based on spherical geometry where mechanical scanning is performed along two axes: the elevation one (or θ-scanning) and the azimuth one (or ϕ-scanning).

Most implementations are based on a gantry arm and spinning turntable, as shown in Figure 5.11, where the measurement probe is installed at the edge of the gantry while the vehicle under test is placed on the turntable. The elevation scanning is performed by rotating the gantry about its axis, and azimuth scanning is achieved by the rotating turntable. A full 360° azimuth rotation is normally performed, while the extent of the elevation scanning depends on the implementation. To reduce chamber interaction, the gantry arm size is generally minimized, and the structure is usually covered by absorbing materials as shown in the two examples in Figure 5.11. The measurement radius, measured as the distance from the center of the scanning sphere to the probe, is usually in the range of 4–10m.

A variation of these systems can be found in the arch-rail systems where the elevation scan is performed by moving a probe antenna with a rail-based mechanism along an arch. As in the gantry arm system, the azimuth scan is performed by rotating the vehicle. Rail-based systems are usually less common than gantry arm systems, because more errors due to the bending of the cable are experienced during the scanning. A single-probe arch-rail system is shown in Figure 5.12.

As described in Section 5.2, the turntables of such systems are sometimes exploited to perform direct azimuth FF cut measurements. For example, the system on the left side of Figure 5.11 is installed inside a radome to protect from the environment, and a probe tower placed outside can be used for direct azimuth FF tests. Similarly, the system on the right side of Figure 5.11 also has a probe tower for azimuth FF measurements. In such a case, however, a large anechoic chamber hosts

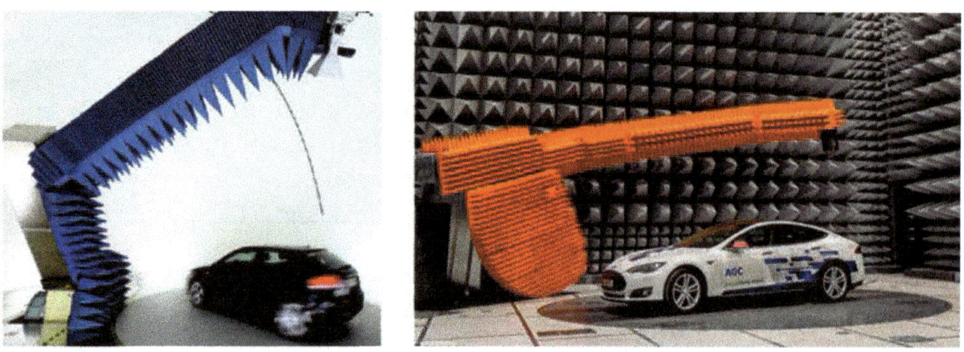

Figure 5.11 Example of a gantry arm system installed inside a radome (left) and a gantry arm inside an anechoic chamber (right). (Image courtesy of Pulsaart by AGC.)

Figure 5.12 Examples of single-probe arch-rail systems.

the entire FF and gantry arm system. Moreover, when the FF measurements are to be performed, the gantry arm can be hidden inside the floor.

Single-probe automotive systems often make use of conductive floors, but absorbing floor solutions can be also found. When a conductive floor is considered, the elevation scanning axis (θ-axis) usually lies on the same floor plane (as shown on the left side of Figure 5.13). In other words, the conductive floor is at the Z=0 position of the measurement coordinate system. This simplifies the NF/FF transformation, as pointed out in Section 3.1.2.2. In this case, the elevation scanning is performed from the zenith, $\theta = 0°$, to close to the conductive floor close to $\theta = 90°$. However, $\theta = 90°$ cannot readily be measured because of the physical encumbrance of the probe. Hence it must be extrapolated, as described in Section 3.1.2.2.

When absorber-covered floors are considered, as sketched in Figure 5.13 (center), the elevation scanning axis is usually above the floor, so that a larger spherical coverage can be measured. The larger angular coverage can be up to 10°–20° with respect to the center of the scanning sphere. Modern gantry arm systems, like the one shown on the right side of Figure 5.11, allow the use of both conductive and absorber floors. In such cases, absorbers are manually moved to switch from one configuration to the other, and a vertical scanning axis is used to move the gantry up/down, increasing/decreasing the measurable portion of the sphere below the center of the scanning sphere.

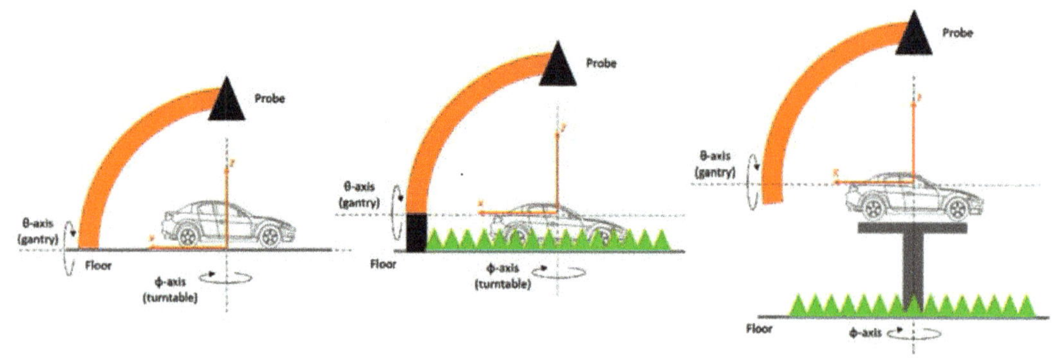

Figure 5.13 Sketches of three different implementations of a gantry arm system.

Other recent gantry arm systems make it possible to measure a much larger sphere and, in some cases, almost the full sphere. In such situations, a mechanism to raise the vehicle is needed, as shown on the right side of Figure 5.13.

The main drawback of a single-probe system, such as the gantry arms system, is the measurement time. In fact, unlike the multiprobe systems described in Section 5.3.1, a mechanical scanning is performed along both the elevation and azimuth axes.

To improve the measurement time, one of the two scanning axes, usually the azimuth, is often performed in continuous movement, also called "on-the-fly," while the other axis is stepped. In on-the-fly measurements, the data acquisition is performed continuously with a time-trigger mechanism. The speed of the motor and the trigger time must be synchronized with the sweeping time of the RF measurement equipment, often a VNA. The sweeping time depends on frequency-related parameters, including the measurement bandwidth, the density of the sampling points, and the IF bandwidth. The effect of the noninstantaneous frequency data acquisition leads to a spreading of the spatial samples over the measurement sphere. The speed of the motor must hence be set so that the data spreading is minimized to achieve a negligible effect on the final transformed data. Nevertheless, the effect of spreading, if quantified accurately, can be compensated for during the NF/FF transformation. In most cases it can be achieved by a simple phase adjustment of the FFT operation.

The on-the-fly acquisition scheme makes it possible to significantly reduce the measurement time (e.g., usually by a factor of 2 or more). To further reduce measurement time, the acquisition along both scanning axes is sometimes performed simultaneously, leading to a spiral scanning scheme [14, 15]. This approach is very convenient from the measurement time point of view, but it complicates the system architecture and the data processing stage as described in Section 3.3.

The selection of dual-polarized and/or wideband probes also significantly contributes to a reduction of the measurement time. Dual-polarized probes allow for the simultaneous acquisition of the two orthogonal field components needed for the NF/FF transformation; hence they are often used to halve the acquisition time.

Probes with large bandwidths, such as 15:1, make it possible to avoid otherwise frequent manual probe changes during multiservice campaigns. These strongly reduce the downtime between measurements. The drawback of wideband probes is that full-probe correction is often needed in the NF/FF transformation to compensate for the complex and frequency-dependent probe radiation. (See Section 3.1.3.2 for more details.)

Practical examples of full-probe corrected automotive measurements performed with wideband probes are shown in Section 5.3.1. Similar examples obtained with wideband and dual-polarized probes are reported in Section 5.3.2.

5.3.1 Use of Wideband Probes

In this section, the measurement results from a validation campaign regarding the upgrade of an automotive gantry arm system are reported [16]. The system is the conductive floor-based gantry arm shown in Figure 5.14. Such systems originally made use of five to six half-octave open-ended waveguides (OEWs) to cover the 0.6–6.0-GHz bandwidth. The challenge of the system upgrade was to replace all the OEWs with a single wideband probe antenna able to cover the same bandwidth.

The considered wideband probe is the SH600 dual-ridge horn [17] shown in Figure 5.14. As also shown in Figure 5.14, the validation is performed considering biconic antennas as test objects placed at a large distance with respect to the center of the scanning sphere. Measurements have been performed dividing the whole bandwidth into five subbands where baseline/reference results are obtained using five OEWs as probes. The same measurements are then repeated using only the SH600 as the probe and processed with the full-probe-compensated NF/FF described in Section 3.1.3.2. The accuracy of the measurements performed with the new probe are evaluated by comparing them with the ones taken with the OEWs.

The full-wave simulated main pattern cuts at different frequencies of the SH600 are shown on the left side of Figure 5.15. The radiation pattern significantly changes with frequency hence, a proper probe compensation (PC) should be applied to equalize the measurement results over frequency. On the right side of Figure 5.15, the azimuthal modal content over the considered bandwidth is shown; note that only odd azimuthal spherical modes are radiated as even modes are well below –50 dB. The order modal content of the SH600 is significantly high, especially at

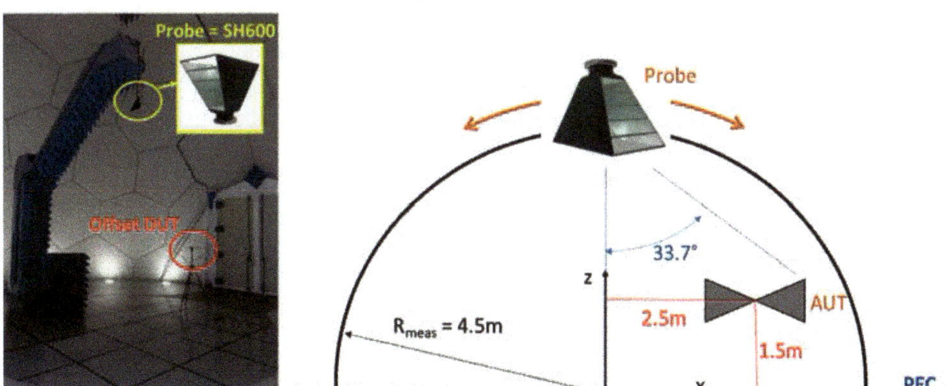

Figure 5.14 Photo of the gantry arm and measurement layout with the wideband probe (SH600) at Hirschmann Mobility (TE connectivity). ©2018 MVG. (Reprinted with permission.) [16].

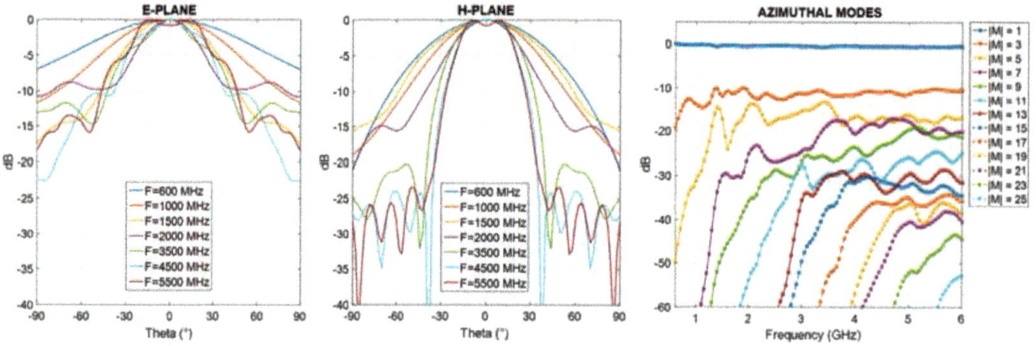

Figure 5.15 Normalized patterns (E-plane and H-plane) and spherical azimuthal modes of the SH600 probe. ©2018 MVG. (Reprinted with permission.) [16].

higher frequencies, up to $|\mu| = 23$, considering a threshold of –50 dB. To maintain an acceptable measurement accuracy the effect of such higher-order modes should be compensated for.

Two biconic antennas are considered as test objects (AUTs) to cover the whole 0.6–6-GHz frequency range. To reproduce a general vehicle measurement, the test antennas have been placed far away from the center of the system as shown in Figure 5.14. The geometrical center of the AUT is approximately at [x, y, z] = [2.5, 0, 1.5]m, generating a very large AUT-probe view angle of more than +/–33°, when considering the 4.5-m measurement radius and hence enhancing the impact of the nonideal probe.

Hemispherical SNF acquisitions above the conductive floor have been conducted, considering for each subband, the minimum sampling rate defined by the AUT minimum sphere as discussed in Chapter 3, which is ~3m. Both vertical and horizontal orientations of the AUT are considered in the validation campaign, but, for the sake of brevity, only results relative to the horizontal ones are reported here. The complete results can be found in [16].

Due to the conductive floor, each measurement has been processed imposing the PEC boundary condition, as described in Chapter 3. Different levels of PCs are then applied considering the full-wave-simulated model of the probe.

Gain pattern comparisons at 1,900 MHz and at 5,500 MHz are shown in Figure 5.16 for the elevation cut at $\phi = 90°$ (plots on left) and for the averaged azimuth cut (obtained averaging all the values in the $\theta = [60°–90°]$ range (plots on the right). In the plots, solid traces are the co-polar field components, while the dotted traces are the cx-polar ones. Blue traces are the reference pattern obtained with the OEW probe and full PC. Green and orange traces are relative to the measurement with the SH600 probe processed with first-order PC (FOPC) and full PC, respectively. When the FOPC is applied, the deviations with respect to the reference are nonnegligible. Instead, the agreement of the two full-probe-compensated measurements performed with the different probes is excellent for both the co-polar and the cx-polar.

To quantify the correlation of the measurements performed with the different probes, the ENL defined by (5.7),

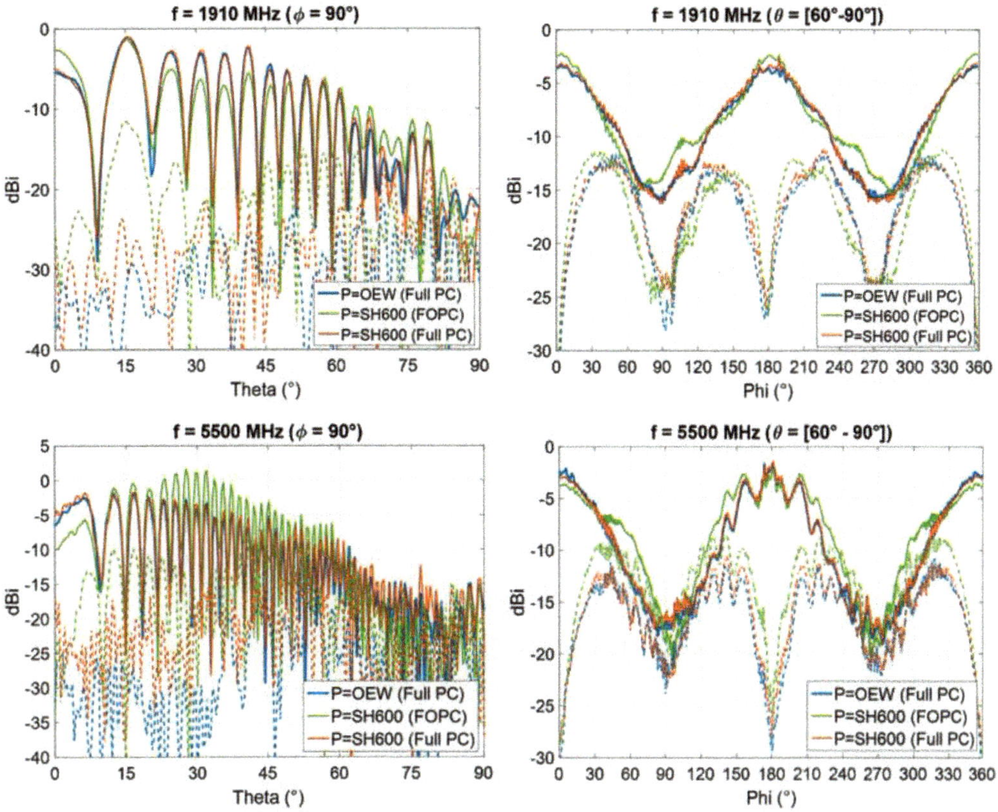

Figure 5.16 (a) Gain pattern comparison at 1,910 MHz and 5,500 MHz: elevation cut at ϕ = 90° and (b) averaged azimuth cut in the θ = [60°–90°] range.

$$ENL = 20\log_{10}\left(RMSE\left|\frac{E(\theta,\phi)-\tilde{E}(\theta,\phi)}{E(\theta,\phi)_{MAX}}\right|\right) \tag{5.7}$$

where $E(\theta, \phi)$ is the reference pattern, and $\tilde{E}(\theta, \phi)$ is the test pattern, evaluated for each measured frequency and independently for the co-polar and cx-polar pattern, as shown in Figure 5.17. In this case, the measurements with the SH600 NF/FF transformed without PC (dashed-black traces), with FOPC (dotted-green traces), and with full PC (solid-orange traces) are compared to the reference one. The lowest ENLs are obtained when the full PC is applied, highlighting the achieved improvements with respect to the processing without PC and with FOPC.

A comparison in terms of UHRP, or UHE, is also performed and reported in Figure 5.18. In this graph the difference between the UHRP obtained from the SH600 measurement processed with the different levels of PC and the referenced UHRP by the OEW measurements is shown. The FOPC is able to reduce the large differences obtained without PC from several decibels to a maximum of ~1.7 dB. By applying the full PC, the differences are further reduced. It should be noted that the differences represent measured errors from both OEW and wideband probe measurements.

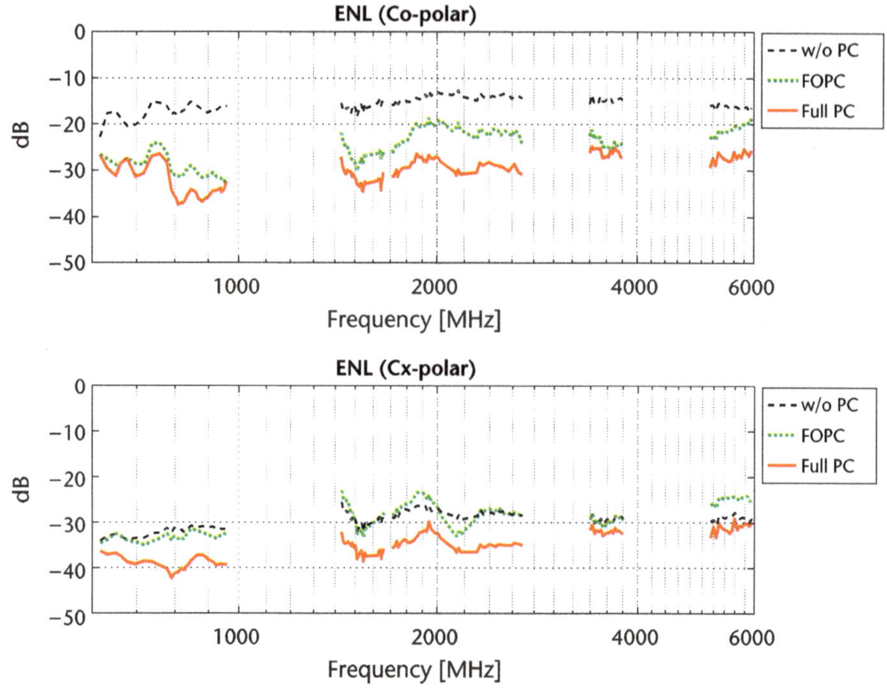

Figure 5.17 ENL comparison for each measured frequency in the case of horizontally oriented antennas.

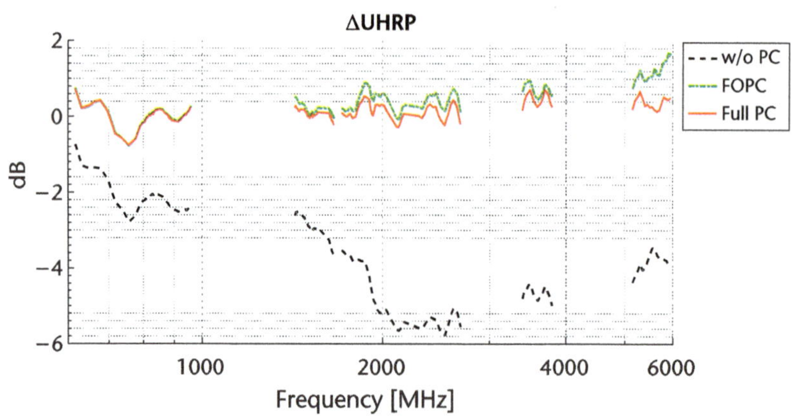

Figure 5.18 UHRP differences with respect to the measurements with the OEW probes.

5.3.2 Use of Wideband and Dual-Polarized Probes

Results from a validation campaign of another gantry arm system are reported in this section [18, 19]. Such automotive systems use just two wideband and dual-polarized probes to cover the 0.45–18-GHz frequency band. The probes are the two quad-ridge horns shown in Figure 5.19: the QH400 for the 0.4–6-GHz band, and the QH1400 for the 6–18-GHz band [20]. The dual-polarized nature of these probes makes it possible to halve the measurement time but introduces additional challenges with respect to the scenario described in Section 5.3.1. In fact, in this case we will see that for some figures of merit, it is beneficial to consider the individual patterns of the probe ports in the PC process.

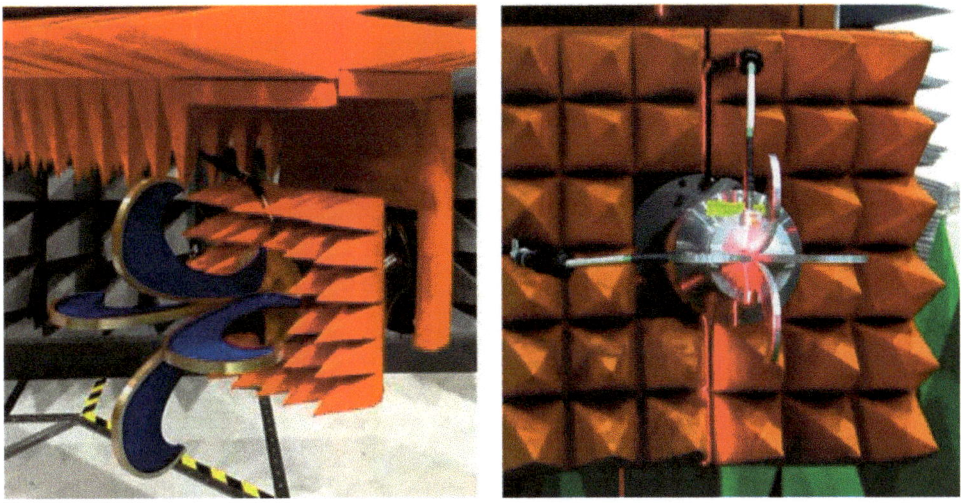

Figure 5.19 QH400 (left) and QH1400 (right) wideband dual-polarized probes installed on the gantry arm of the SNF facility at Pulsaart (by AGC). ©2021 IEEE. (Reprinted with permission.) [18].

As illustrated in Figure 5.20, the validation is performed in free-space conditions (i.e., absorbers on the floor) by measuring antennas in three different positions. The two probe ports are exploited to simultaneously measure both orthogonal field components. The on-axis response of the two probe ports has previously been calibrated in amplitude and phase to properly compensate for the port imbalance. The different antenna positions generate different probe effects resulting in different AUT pattern distortion if no PC is applied. Instead, considering a proper PC, the measured AUT pattern should be the same.

Next, monocone antennas mounted on a 1-m-diameter ground plane are considered as AUTs to cover the whole frequency band measured by the QH400. The considered antenna positions are summarized in Table 5.1. Considering a 4.5-m measurement radius, the resulting AUT-probe angles (α) are 6.5° in the case of onset configuration and 10° and 24° when the offsets are respectively 30cm and 150m. For the QH1400 measurements, a dual-ridge horn, SH2000 is considered

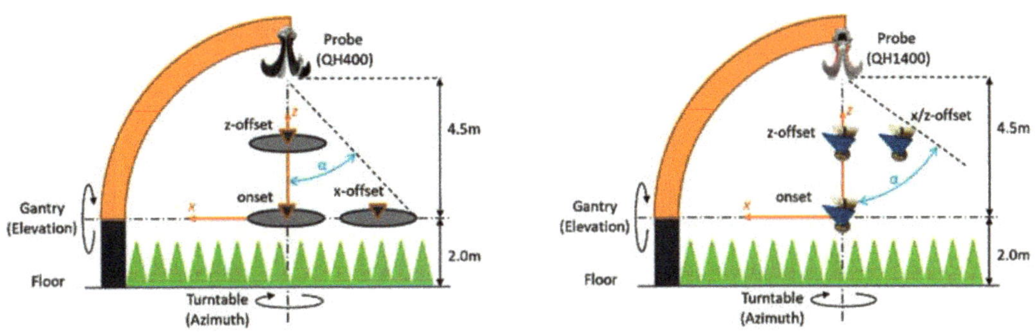

Figure 5.20 Sketch of the measurement setups with the different AUT positions: QH400 probe measuring monocone antennas (left) and QH1400 probe measuring the SH2000 horn (right). (Reprinted with permission.) [18].

Table 5.1 Details of the Measurement Configurations

Probe	*Antenna*	*Frequency*	*Antenna Positions*
QH400	SMC220+GP100	0.45 – 0.7 GHz	– [x, y, z] = [0, 0, 0] m
			– [x, y, z] = [0, 0, +1.5] m
			– [x, y, z] = [–1.5, 0, 0] m
QH400	SMC700+GP100	0.7 – 2.2 GHz	– [x, y, z] = [0, 0, 0] m
			– [x, y, z] = [0, 0, +1.5] m
			– [x, y, z] = [–1.5, 0, 0] m
QH400	SMC2200+GP100	2.2 – 6.0 GHz	– [x, y, z] = [0, 0, 0] m
			– [x, y, z] = [0, 0, +0.3] m
			– [x, y, z] = [–0.3, 0, 0] m
QH1400	SH2000	6.0 – 18 GHz	– [x, y, z] = [0, 0, 0] m
			– [x, y, z] = [0, 0, +0.3] m
			– [x, y, z] = [–0.3, 0, +0.3] m

in the whole 6–18-GHz band as indicated in Table 5.1. In this case the resulting AUT-probe angles are 1° for the onset configuration and 4° and 6° for the z-offset and x/z-offset configuration, respectively.

Full-wave simulated co-polar and cx-polar radiation patterns (E-plane cut) of the QH400 probe are shown in Figure 5.21 for three frequencies and both ports. The complexity of the probe pattern, the cx-polar level, and the differences among the probe ports increases with the frequencies, suggesting that a full PC accounting for both probe ports is needed.

The azimuthal modal content of the of the QH400 (χ-port, obtained from the full-wave simulation of the probe) for the whole frequency band is shown in Figure 5.22. The presence of the higher-order modes (e.g., modes $|m|>1$) is evident, and it increases with the frequency. It should also be noted that, unlike the SH600 probe considered in the example in Section 5.3.1, the QH400 exhibits both odd and even modes. The even modes are usually more critical than the odd ones because they made the radiation pattern asymmetric, and, in turn, they increase the level of distortion of the measured NF [21].

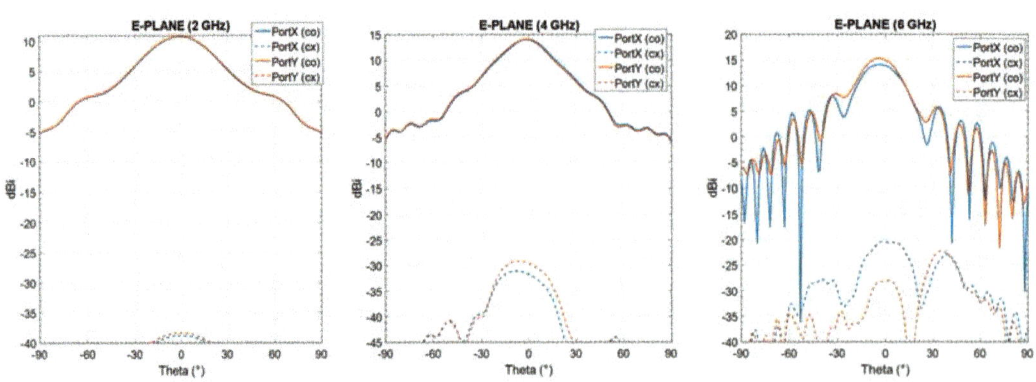

Figure 5.21 Directivity radiation pattern (E-plane) of the two ports of the QH400 probe at different frequencies.

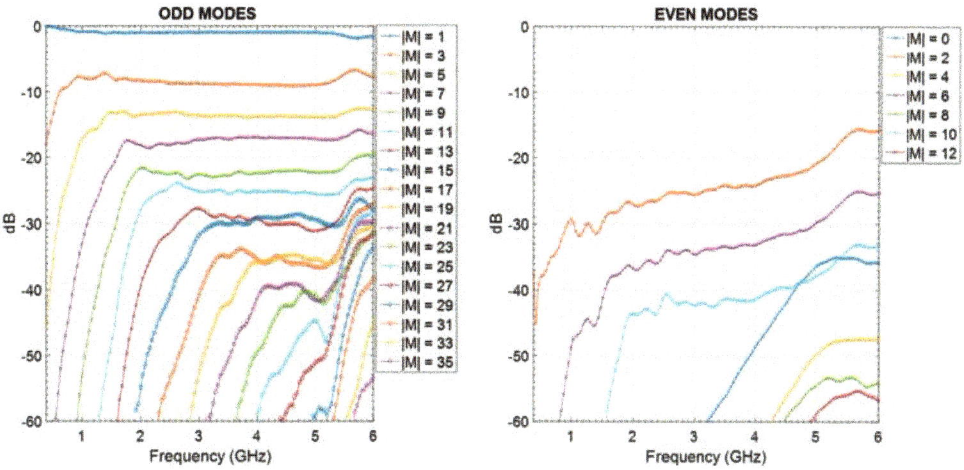

Figure 5.22 Azimuth spherical modal content versus frequency of the x-port of the QH400. ©2021 IEEE. (Reprinted with permission.) [18].

The gain radiation pattern cut of the monocone antenna measured with the QH400 at 2 GHz is shown in Figure 5.23. The comparison shows the different measurement configurations and the impact of the different levels of PC. Solid and dashed traces are the co-polar and cx-polar patterns, respectively. Burgundy, cyan, and yellow traces are the patterns obtained from measurements performed in the onset configuration and with the z- and x-offset, respectively.

The plot on the top-left of Figure 5.23 shows the pattern comparison when PC is not applied. As expected, the differences are evident both in the co-polar and cx-polar patterns. A peak-to-peak (P2P) difference of almost 2 dB can be observed at the maximum of the measured radiation pattern. Moreover, the cx-polar patterns obtained with the two offset AUT configurations deviate significantly from the expected/nominal one, which, for this type of antenna, should have a random-like shape and be around 35 dB from the peak.

The patterns obtained with the FOPC are shown in the plot on the top-right side of Figure 5.23. The correlation of the co-polar patterns is improved significantly and the P2P variation at the peak of the pattern is now about 0.5 dB. Instead, it can be observed that the FOPC is not able to improve the cx-polar.

The pattern comparison on the bottom-left of Figure 5.23 is obtained with the single-pol full PC, assuming that the probe ports are identical. The correlation of the co-polar patterns is further improved so that the P2P difference at the peak of the pattern is less than 0.2 dB, but neither in this case, the cx-polar patterns is improved.

Finally, on the bottom-right side of Figure 5.23, the pattern comparison obtained with the dual-pol full PC are shown. The co-polar patterns are basically identical to those obtained with the single-port full PC, but the cx-polar pattern patterns are now improved significantly, obtaining the expected random-like shape with an average level lower than 35 dB from the peak.

The correlation of the AUT patterns measured in the different configurations has been evaluated with the combined ENL. Such a metric is obtained by first

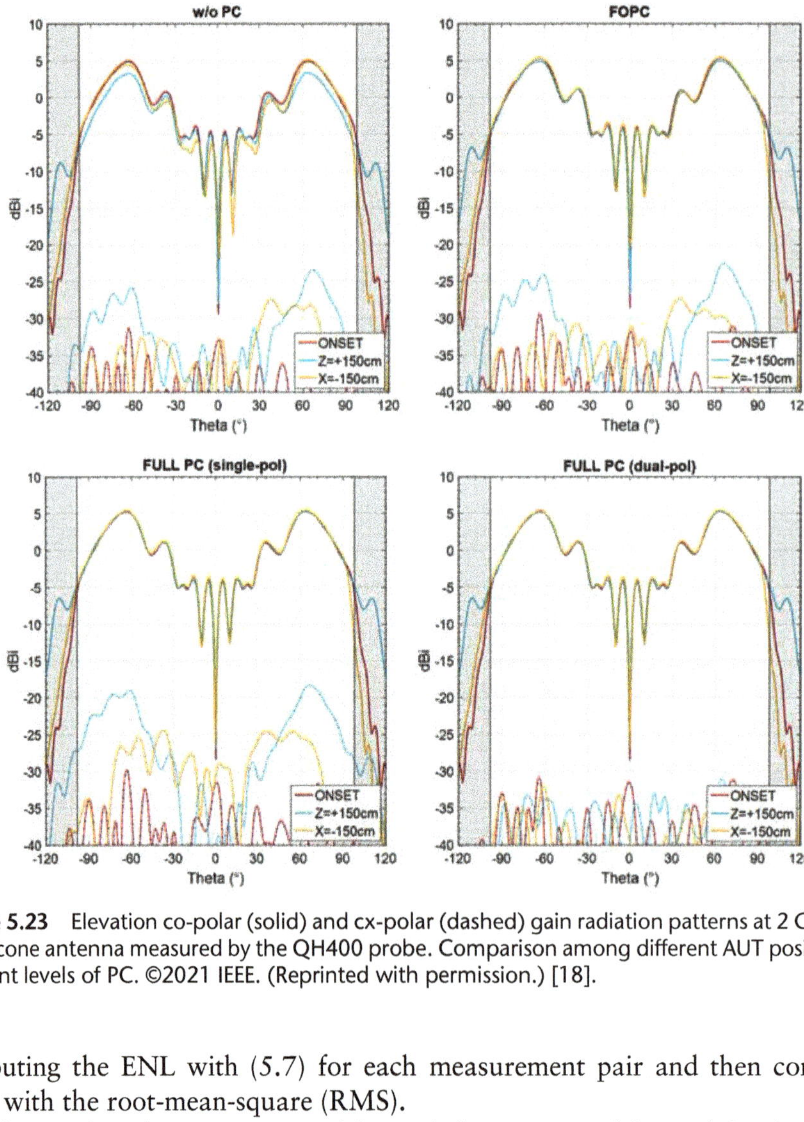

Figure 5.23 Elevation co-polar (solid) and cx-polar (dashed) gain radiation patterns at 2 GHz of the monocone antenna measured by the QH400 probe. Comparison among different AUT positions and different levels of PC. ©2021 IEEE. (Reprinted with permission.) [18].

computing the ENL with (5.7) for each measurement pair and then combining them with the root-mean-square (RMS).

The combined ENLs computed for each frequency and for each level of PC are shown in Figure 5.24, for both the co-polar and cx-polar patterns from the measurements with the QH400. The areas shadowed with different colors are associated with the three different bands and antennas considered. As concerns the co-polar, when the PC is not applied at all (blue traces), the combined ENLs increase with the frequency, reaching a maximum level of approximately −20 dB. This frequency dependence is in line with the increasing level of higher-order modes of the QH400 probe. The drop of the ENL at 2.2 GHz is due to the smaller offset considered with the AUT working at the last frequency band. The FOPC (black traces) has a positive effect, especially in the second half of the middle band where the improvement of the ENL is on the order of 10 dB. As expected, the lowest ENLs are obtained with the two full PC techniques (orange and green traces), which approximately return the same co-polar results. The improvements obtained with the full PC are particularly evident at the very end of the band of the probe (5.5–6.0 GHz), where indeed the higher-order modes of the probe are even more pronounced.

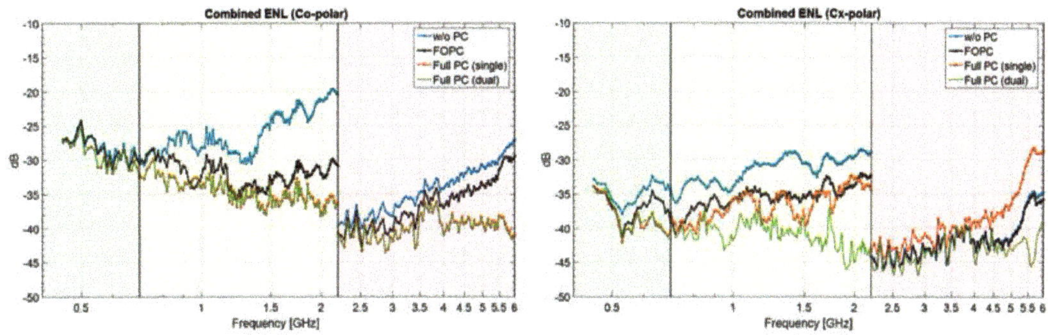

Figure 5.24 Measurements of the monocone antennas with the QH400 probe: combined co-polar (left) and cx-polar (right) ENL versus frequency obtained with different levels of PC. ©2021 IEEE. (Reprinted with permission.) [18].

Concerning the cx-polar, in the first two bands, gradual improvements of the combined ENL are obtained when the different levels of PC with growing complexity are applied. In the last band the single-pol PC shows worse performance than the FOPC and the NF/FF without PC. This is due to differences between the two ports of the probes that are not negligible at higher frequencies as seen in Figure 5.21. It is observed that the dual-port full PC can provide excellent cx-polar results at all the considered frequencies. The achieved ENLs vary from –35 dB to –45 dB, respectively, at lower and higher frequencies. Considering a cx-polar level of –20 dB from the peak, such equivalent errors would result in uncertainties ranging from approximately +/–1.5 dB to +/–0.5 dB.

For simplicity, only results in terms of combined ENL are shown here for the validation measurements in the 6–18-GHz range performed with the QH1400 as the probe. (Interested readers can refer to [19] for more details.) The comparison shown in Figure 5.25 is like the one previously shown for the QH400 probe and a similar trend in the results can also be observed. It should be noted that the achieved ENLs are in this case lower than approximately −40 dB even if PC is not applied. Due to the considered relatively small AUT-probe angles, the effect introduced by the probe could be considered negligible in this case. Nevertheless, the pattern correlation gradually increases when the different levels of PC with growing complexity are applied.

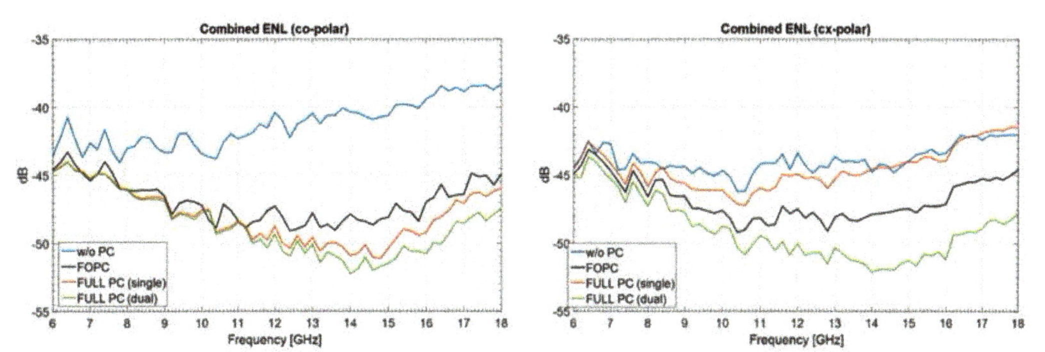

Figure 5.25 Measurement of the SH2000 horn with the QH1400 probe: (a) combined co-polar and (b) cx-polar ENLs versus the frequency obtained with different level of PC.

5.4 Multiprobe NF Systems

In automotive MP systems, like the ones shown in Figures 5.26–5.28, the mechanical movement of the gantry arm needed for the elevation scanning is replaced by an array of probes performing a much faster electronic scanning. As in the previously presented gantry arm systems, the azimuth scanning is performed by a spinning turntable where the vehicle under test is placed.

The primary objective of a MP system is to drastically reduce the test time. Depending on the measurement configuration, automotive MP systems generally allow for up to 10 times faster measurement than a single-probe system based on the conventional physical movement of the probe.

The design challenge is to achieve the speed increase without sacrificing the measurement accuracy of the system. The reduced testing time has several advantages including the decrease in the cost of the measurement campaigns. By reducing the cost of the tests, more complete characterizations of the vehicles under test within a product development cycle can be performed. Hence, better products can be designed, if the development takes advantage of more design iterations involving performance tests. The reduced testing time also reduces drift in the measurement equipment or test environment and hence improves accuracy.

Automotive MP systems have evolved over the years to cover the frequency band from 50 to 70 MHz to 18 GHz (up to 50 GHz in nonautomotive MP systems). This has been accomplished by integrating different arrays of probes covering different subbandwidths in the same system.

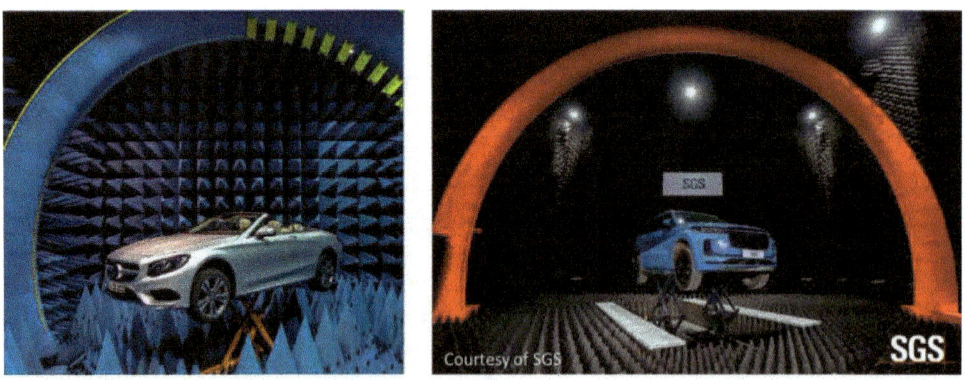

Figure 5.26 MP automotive system at TU Ilmenau (left) and at SGS (right).

Figure 5.27 Example of a MP automotive system with an automatized car-feeding mechanism. The system is installed at Renault Aubevoye, France. ©2012 IEEE. (Reprinted with permission.) [27].

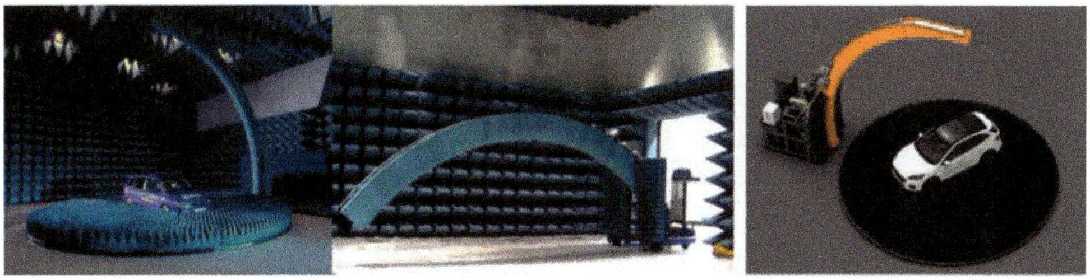

Figure 5.28 Example of a MP automotive system with movable arch. (The left pictures are from the GM system at Milford, MI.) ©2012 IEEE. Reprinted with permission.) [27].

The measurement radius (i.e., distance from the center of the scanning sphere to the probe) ranges from 4 to 6m. Such distances are usually considered "near field" distances; hence in the case of passive measurements, the same NF/FF transformations considered with single-probe systems are used with the MP systems. As discussed in Chapter 6, MP systems are also used for active/OTA testing. In such cases, due to the absence of phase information, the collected data is used directly as FF performance data, or special compensation techniques are used (e.g., quasi-NF/FF transformation through parallax compensation). (See Chapter 6 for details.)

The main design challenges of MP systems are detailed in Sections 5.4.1 and 5.4.2.

5.4.1 Probe-Related Design Challenges

The main design challenges of MP systems are related to the probe and the probe array where they are installed. These include the bandwidth of the probe, the switching and mutual coupling among the probes, the calibration/equalization of the probe array, and the probe pattern compensation.

The probe array elements are wideband with a frequency range up to 15:1. The elements are specifically designed and optimized for probe array applications. They are dual-polarized and aligned according to vertical and horizontal polarizations, or in some cases, to +/–45° slant polarizations. The probes are completely reciprocal and can be used both in receive and transmit modes. Their assembly also houses a circuit board containing the control electronics for the scanning of the probe. The probe array elements are specifically designed to be embedded in multilayer conformal absorbers. The shape of the conformal absorbers is optimized to reduce the mutual coupling among the probes and to keep the reflectivity of the probe array (including scattering and reflections from the support structure and cabling) at minimum levels.

The probes are connected to the RF system through a combiner network. The individual probes can be interrogated by on/off switching or by using the advanced modulated scattering technique (A-MST) [22]. In an A-MST system, a low-frequency modulation is applied to each probe to perform multiplexing. The measured field information, in terms of amplitude and phase at each probe position, is extracted by coherent detection. For measurements of active systems, the switching network approach is generally preferred.

The probes of any MP system must be properly calibrated to equalize their response in amplitude and phase. The probe array calibration is a mandatory step needed to compensate for minor differences and imperfections of the probes and the array beamforming network. In fact, due to the mechanical tolerances of the elements and the electrical errors of the feeding networks, some differences in the radiating performances of the probe are expected. The probe array calibration procedure is based on the measurement of the complex on-axis response of each dual-polarized probe. This is normally performed by pointing a linearly polarized auxiliary antenna with high polarization purity (e.g., on-axis cx-polar better than –45 dB) toward the probes. For each probe, the two ports are measured with the auxiliary antenna in two orthogonal orientations (four measurements for each probe). From each of these measurements, a 2 × 2 correction matrix is computed. Such matrixes are used to compensate for the on-axis cx-polar and the port amplitude/phase unbalance of the each dual-polarized probe. Ultimately, a correction factor is computed for each probe to equalize their amplitude and phase responses. After calibration, the typically measured probe uniformity is better than ±0.05 dB in amplitude and ±1° in phase.

The probe array calibration is sufficient to achieve accurate measurement results without resorting to probe pattern correction. As described in Section 3.1.3.2, accurate nonprobe-compensated SNF measurements can be performed when the probe directivity is relatively low and/or when DUT-probe view angle is sufficiently small. These conditions are usually met with MP systems like the ones shown in this chapter. At higher frequencies, pattern correction techniques like the ones described in Chapter 3 should be considered in some cases to further improve the measurement accuracy. Such techniques can be applied in a similar manner as described in Sections 5.3.1 and 5.3.2 in the case of single-probe systems. The additional challenge here is characterization of the specific pattern of each individual probe. In [23, 24] it has been shown that assuming a sufficient high correlation among the probes of the MP system, the pattern correction can be accurately applied considering the "average" probe pattern, obtained from measurements of few probes of the same production batch. A good probe pattern correction is ensured in the MP system shown in this chapter; hence, when needed, this pattern correction approach is applied.

5.4.2 Sampling in MP Systems

Meeting the spatial sampling requirements, or Nyquist criteria as described in Chapter 3, can be another challenge for a MP system with fixed-probe array spacing, particularly when electrically large DUTs are to be measured. In automotive MP systems the azimuthal or ϕ sampling is performed with a rotating turntable where the vehicle is placed. The ϕ-sampling is hence not problematic because its density can be chosen arbitrarily.

The elevation or θ sampling is performed by the probe array where the probes are installed on fixed and evenly spaced positions. Depending on the frequencies, the probe spacing usually ranges from 5° to 1°. In many testing scenarios these sampling steps are sufficient to include the whole vehicle in the equivalent sampling sphere, but in other scenarios they may not be sufficient. For example, according to the sampling requirements based on the DUT minimum sphere as discussed in Chapter 3, $\Delta\theta = 1°$ is sufficient to measure the entire whole 5-m vehicle up to ~3

GHz. Beyond this frequency, part of the vehicle will be outside the equivalent sampling sphere.

To overcome this limitation, a limited mechanical elevation movement of less than a probe spacing can be considered. This enables unlimited increase of the sampling density in elevation. This operation is commonly called arch oversampling or simply oversampling and is usually a default feature of smaller MP systems for applications other than automotive. Due to size and complexity, this feature was not available in early automotive systems. Recently, oversampling features have been introduced in advanced systems as an option.

An alternative solution to the oversampling is to use the translated spherical wave expansion (TSWE) technique in the NFFC transformation as described in Section 3.1.3.3. Practical examples of the latter are reported in Section 5.4.5.

5.4.3 Different Implementations of Automotive MP Systems

As shown in Figures 5.26–5.28, multiprobe automotive systems usually implement a partial spherical scanning surface terminated to a floor covered by absorbers to emulate a free-space condition. The ground floor is usually conductive; hence, if needed, the systems can also be operated in "PEC mode" by simply removing the absorbers. The free space is normally the default operational mode; this impacts the design and the geometry of the system. As shown in Figure 5.26, the center of the measuring sphere and thus the reference coordinate system is placed above the ground floor at a ~1.2–1.5m distance. Such configurations center the roof of most vehicles in the measurement system, so close to $Z = 0$, and thus minimize the minimum sphere and reduce the sampling requirements. From this position it is possible to measure scanning angles below the horizon, thereby increasing the measured portion of the sphere and thus reducing truncation errors. (See Chapter 3 for details.) The portion of the sphere measured below the horizon usually ranges from 10–30° (i.e., $\theta_{max} = 100°$ to $\theta_{max} = 120°$). As discussed in Chapter 9, the availability of measurements at angles below the horizon is essential to emulate the vehicle's performance on arbitrary driving grounds.

It should be noted that when such systems are operated in PEC mode, using the conducting floor below the vehicle, the floor is in a position well below the coordinate system center, with $Z < 0$ instead of $Z = 0$, hence a TSWE-based NF/FF transformation must be used as described in Section 3.1.3.5.

The most common automotive MP systems solutions are similar to the systems shown in Figure 5.26. The system on the left is composed by two truncated arches. One arch is equipped with low-frequency probes, covering 70–400 MHz with 5° spacing using 22 probes. The other arch has higher-frequency probes covering 400 MHz to 6/10GHz with 1° spacing using 111 probes. The maximum elevation scanning angle for such a system is $\theta_{max} = 107.5°$ and $\theta_{max} = 110°$, respectively, for the low- and high-frequency arch.

Such systems are equipped with a raising mechanism needed to optimize the z-position of the vehicle. This is particularly needed at low frequencies, between 70 and 400 MHz to avoid the shadowing effect of the large absorbers. Usually 48-inch pyramidal absorbers are used at these frequencies.

Automotive MP systems may differ depending on the car-feeding mechanism. For example, in systems like the ones shown in Figure 5.26, the vehicle is driven to the center of the turntable, by first removing the absorbers and then placing them around the vehicle. To minimize such manual operations, automated solutions like the one shown in Figure 5.27 are also available; in this case the vehicle is placed on a platform outside the anechoic chamber, which is then automatically moved inside, and positioned in the center of measurement system.

Other implementations of MP systems are based on a movable arch like the ones shown in Figure 5.27. Such types of systems are useful when the antenna measurement environment is shared with another facility like EMC. In this way the MP array can be lowered and moved out the facility when antenna measurements are not to be performed.

5.4.4 Low-Frequency Measurements

As discussed in Section 5.4.3, MP systems as shown in Figure 5.26 (left) and in Figure 5.27 can measure frequencies down to 50/70 MHz as the low-frequency arch/probes operating at 50–400 MHz. It is well-known that antenna measurements at such low frequencies are very challenging because of the equivalent electrical sizes of the anechoic chambers, which becomes smaller, and the limited reflectivity provided by the absorbers. Moreover, as described in Section 3.1.2.1, truncation errors, introduced during the NF/FF transformation process, can become nonnegligible at these frequencies.

In MP systems the 48-inch pyramidal absorbers provide −15-dB reflectivity (normal incidence). The RF interaction with the probe array is minimized by using conformal absorbers whose shape is properly optimized. To cope with the truncation errors, special filtering techniques like the iterative modal filtering (IMF) and the equivalent currents (EQCs) techniques introduced in Chapter 3 can be considered, as detailed in Chapter 7.

In this section we present a measurement example of a low-frequency monocone antenna mounted on a 4-m GP. The antenna is the SMC70+GP400 shown on the top-left of Figure 5.29. As discussed in Chapter 2, such devices are normally used as the gain/efficiency/UHE calibration standard in automotive systems. The antenna has been measured in a 12-m diameter MP system able to measure down to 17.5° below the horizon (θ_{max} = 107.5°). The system is installed in a 17 × 15 × 11m anechoic chamber with 48-inch absorbers on the floor.

Standard data processing based on the SWE-based NF/FF transformation is applied to the measurement. The measured peak gain is shown on the right side of Figure 5.29 (green trace). Since the SMC70+GP400 is a reference calibration antenna, the gain in this case is obtained by enforcing the known UHE of the antenna (substitution method). The red trace shown in the same plot is the gain of the 10-times scaled version of the antenna (SMC700-GP400) measured in a smaller MP system able to perform a full spherical acquisition. As described in Chapter 2, scaled antennas like this one are commonly used to characterize the full scaled reference devices. It should be noted that the two measurements have been normalized so that they have the same UHE. In this case, the difference between the two gain curves is thus only due to the differences in the measured pattern. Such a difference is approximately +/–1 dB at maximum and is usually considered good for such low frequencies.

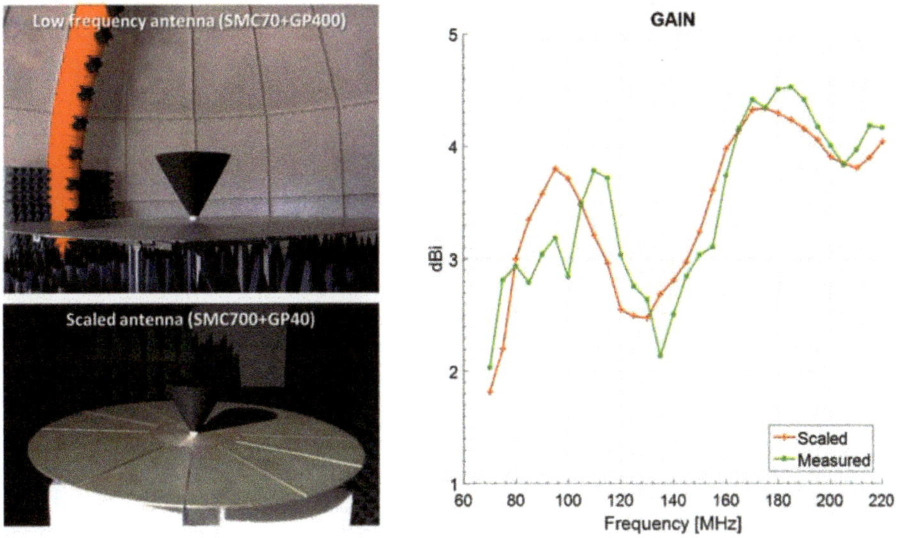

Figure 5.29 Measurement of a low-frequency monocone antenna with a 4-m ground plane (SMC70+GP400) and comparison with a 10-times scaled version of the antenna (SMC700+GP40).

To better investigate the difference among these measurements, the radiation patterns at 70, 100, 150, and 200 MHz are compared in Figure 5.30. The patterns of the full-size antenna measured in the MP automotive system (green traces) show some ripples, which are due to the residual reflections and truncation errors. Nevertheless, the agreement with the pattern measurement of the scaled antenna can be considered very good given the expected error sources at such low frequencies and the possible differences between the two antennas due to the manufacturing. The full-wave simulated model of the SMC70+GP400 is also included in comparison (blue traces), and it also agrees well with the two measurements.

It should also be noted that when such reference antennas are used to calibrate the systems, the UHE is considered, which, as seen in Section 5.1.2, makes it possible to average out the ripples caused by the measurement errors, thereby improving the accuracy of the calibration. Moreover, it is recalled that measurement accuracy can be improved by using advanced post-processing filtering techniques, which are discussed with some examples in Chapter 7.

5.4.5 High-Frequency Measurements

In this section practical examples of high frequency ~5.9-GHz measurements, performed in the MP system installed in the Virtual Road Simulation and Test Area (VISTA) facility at the Technical University of Ilmenau, are reported [25]. The VISTA MP system is shown on the left side of Figure 5.26. No arch-oversampling capabilities are available in this system, and the probe spacing of the high-frequency arch is $\Delta\theta = 1^\circ$ in elevation. Such a sampling can provide a ESS ~54λ diameter. This is sufficient to fully enclose a common passenger car up to ~3 GHz, where the ESS is about 5.4m. At higher frequencies like 6 GHz, the ESS become ~2.7m, leaving out parts of the vehicle.

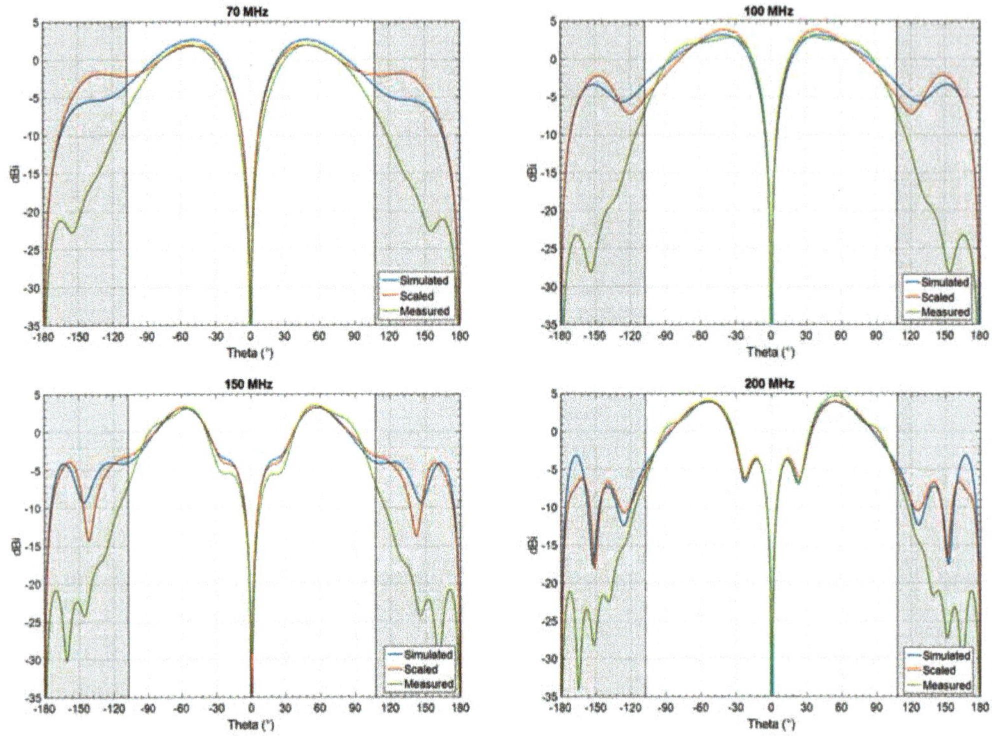

Figure 5.30 Measured gain radiation patterns of a low-frequency (70–220-MHz) monocone antenna (SMC70) on a 4-m ground plane compared with the full-wave simulation and measurement of a 10-times scaled model of the antenna.

As described in Chapter 3, in measurement scenarios like this one, the local measurement approach implemented by the TSWE-based NF/FF transformation can be used to measure offset mounted antennas on vehicles. With TSWE, the reference coordinate system can be positioned conveniently on the antennas and thus be included in the ESS.

Section 5.3.1 considers different test cases of vehicle-mounted antennas, as shown in Figure 5.31. Two vehicle-to-vehicle (V2V) communication antennas exploiting the intelligent transportation system (ITS) band at 5.9 GHz are measured. The antennas are embedded in the same module and have been installed on the rear edge of a station wagon and on the center of sedan mock-up as shown on the left and right side of Figure 5.31, respectively.

The Cartesian positions of the antennas when installed on the two cars and the total offset (i.e., Euclidean distance) with respect to the center of the measurement sphere are reported in Table 5.2. The radius of the ESS at 5.9 GHz when the conventional SWE is applied is also reported in Table 5.2. As also depicted in Figure 5.32, the ESS centered in the measurement system is not sufficient to include the antenna installed on the station wagon. Incorrect measurement results are hence expected from this setup with only conventional SWE-based processing. Considering the TSWE instead, the ESS can be translated to the antenna position properly applying the local measurement approach.

The TSWE has been applied to both setups considering the antenna's position as the expansion center. The achieved ESSs (obtained with [3.27], where the 4-m

Figure 5.31 Measurement setup at the TU Ilmenau (VISTA facility); V2V 5.9 GHz antenna installed on the rear edge of the roof of a station wagon (left) and on the center of the roof of a sedan mock-up (right). ©2019 Ilmenau. (Reprinted with permission.) [25].

measurement radius of the system is considered) are reported in the last column of Table 5.2. It should be noted that the ESSs obtained with TSWE displacement are smaller than the centered ESSs obtained with SWE. This effect was explained in detail in Section 3.1.3.3.

The NF data of the measured configurations has been NF/FF transformed with both the SWE and TSWE techniques. The radiation pattern results are reported in Figure 5.33. For each case, two vertical cuts associated with the lateral and longitudinal directions of the cars are shown. For both ITS antennas installed on the sedan mock-up (patterns on the left side of Figure 5.33), the SWE- and TSWE-based processing are in good agreement. Indeed, with this antenna position the conventional SWE can still be applied because, as depicted in Table 5.2, the antenna falls within the ESS volume. It should be noted that the TSWE results show a less pronounced ripple, which can be attributed to the intrinsic modal filtering caused by the smaller ESS obtained with the TSWE technique.

The station wagon setup exhibits significant quantitative and qualitative differences between the patterns generated by the two expansion techniques. (See the patterns on the right side of Figure 5.33.) This is completely in line with expectations because the antenna is located significantly outside the ESS provided by the conventional SWE technique. It is hence confirmed that in such a case the TSWE should be used to achieve sufficiently accurate results.

To further analyze the achieved results, the patterns of the vehicle-mounted antennas obtained with the TSWE technique are compared to a third setup obtained by measuring the same antennas on a simple metallic, flat ground plane. In the latter, the conventional SWE-based transformation is applied because the antenna

Table 5.2 Position of the Antennas and Radius of the ESSs at 5.9 GHz

	Antenna Position (cm)				*ESS Radius (cm)*	
Setup	*x*	*y*	*z*	*Total offset (cm)*	*SWE*	*TSWE*
Station wagon	+4.0	−175.9	+75.4	191.4	137.4	71.4
Sedan	−3.3	−31.1	+92.5	97.6	137.4	103.6

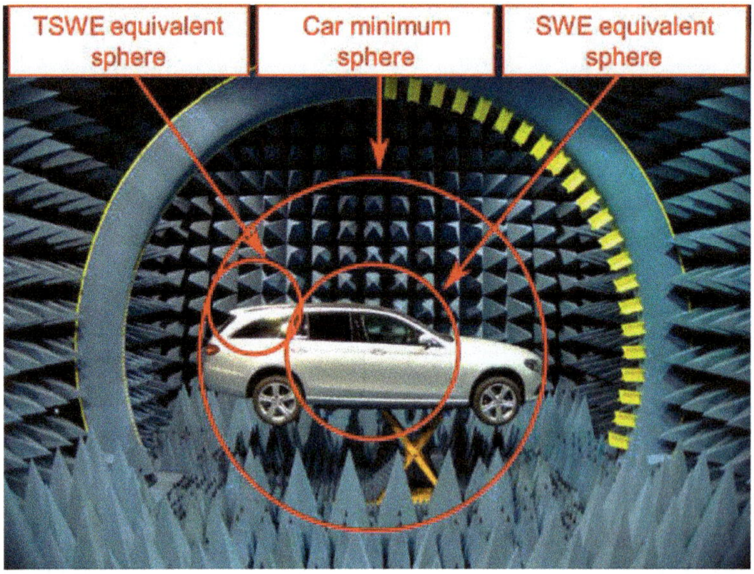

Figure 5.32 Illustration of the equivalent sampling sphere for the station wagon with the 5.9-GHz V2V antennas installed on the rear edge of the roof. ©2019 Ilmenau. (Reprinted with permission.) [25].

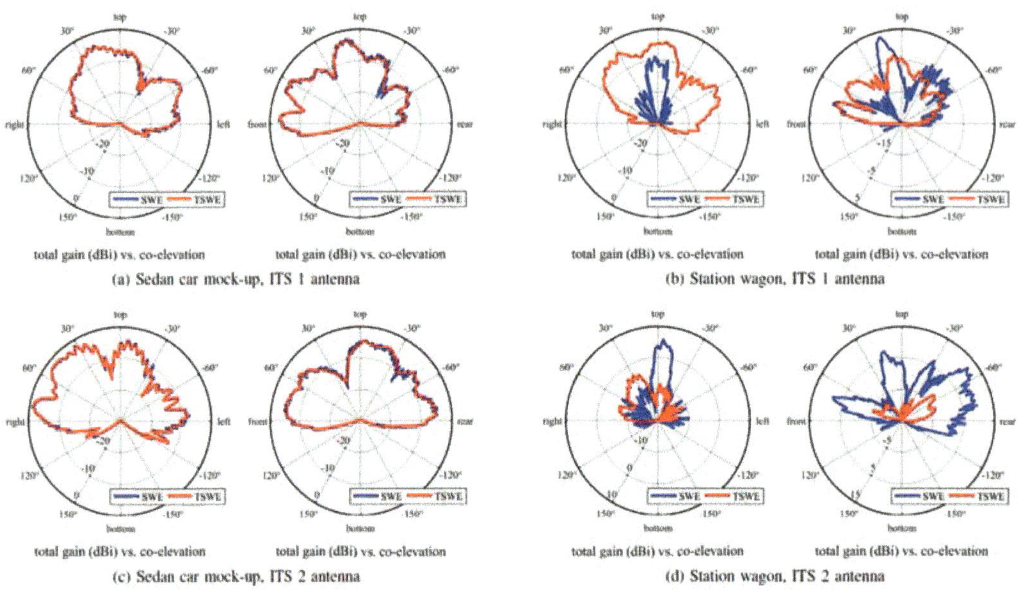

Figure 5.33 Elevation pattern comparison obtained with the SWE and TSWE NF/FF transformation both ITS antennas. For all four setups, the lateral cut (left panel) and longitudinal cut (right panel) are shown. ©2019 Ilmenau. (Reprinted with permission.) [25].

is well-centered in the system. The pattern comparisons are shown in Figure 5.34, again considering lateral and longitudinal vertical cuts.

For the lateral cuts (left panel) a good agreement between the patterns was found. The longitudinal cuts of both antennas show instead more deviations. In

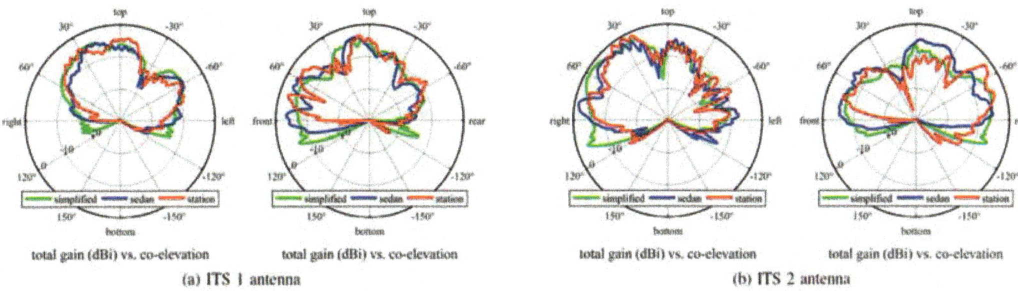

Figure 5.34 Elevation pattern comparison of both ITS antennas installed on the station wagon (red), on the sedan car (blue), and on a simple metallic ground plane (green): lateral vertical cuts (left panel) and longitudinal vertical cut (right panel). ©2019 Ilmenau. (Reprinted with permission.) [25].

particular, with the station wagon setup, a tilt toward the rear caused by a declining roof toward the edge is observed. Additionally, the roof structure of both cars differs since the sedan car mockup has a solid metal roof, while for the station wagon, large areas are covered by glass or plastic. This last comparison showed that accurate analysis, such as the impact of the car structure on the radiation performance of the antennas, can be performed at high frequency in MP systems by using TSWE-based processing.

5.5 Scaled Measurements

Even if important progress in antenna measurements in complex environments has been achieved (as shown in this book), the measurements of antennas installed on physically large structures could be difficult and/or impractical in some cases. In these situations, the scale-modeling technique, reducing the antenna and structure size, is a useful alternative, as explained in [4]. Scale modeling can be applied to antennas placed in complex environments such as big trucks, buses, ships, aircrafts, satellites, or other structures that can have an important effect on the properties of the antennas. In those situations, the measurement of the antenna is performed in a setup that emulates as much as possible the antenna's real operating environment.

The accuracy of such measurements strongly depends on the quality of the scaled models. The technique has been used often in the aerospace and defense domains to measure antennas on physically large objects. The method has also been used, but less frequently, in the automotive domain. Two scaled measurement examples are shown in Figure 5.35, namely a ship model measured over a conductive floor to emulate the effect of the sea and a scaled locomotive measured in a multiprobe free-space system [26].

While the radiation pattern is usually the antenna characteristic of most concern in scaled model measurements, other antenna properties of interest can also be reliably reproduced. If the exact scaling procedure is implemented, all fields are reproduced exactly both externally and within the feed line. Thus, power gain, directivity, radiation efficiency, input impedance, mutual impedance, boresight error, and generally all properties dependent only on field ratios are preserved. However, certain antenna characteristics, such as the power level for high-voltage breakdown

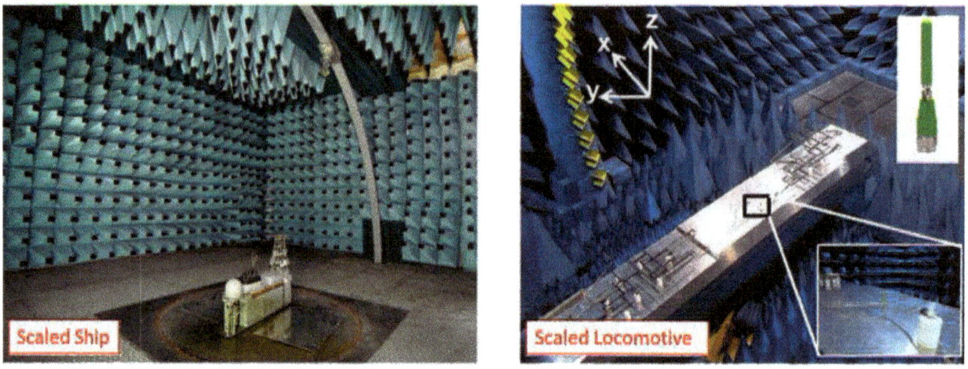

Figure 5.35 Examples of scaled model measurements.

or the noise temperature cannot be scaled because of the frequency-dependent nature of the mechanisms involved.

Scale modeling is based on the application of Maxwell's equation: Basically, the electromagnetic performance of a system depends on the dimensions in terms of wavelengths. The important dimension parameter is the electrical size, that is the physical dimension divided by the wavelength. This means that if the frequency is multiplied by a factor N, and the dimensions are divided by the same factor N, the electromagnetic behavior will be consistent. As an example, an antenna working at 80 MHz in a 3-m-length car has the same performance of a 10 times smaller antenna working at 800 MHz in a 30-cm-scaled automotive.

Therefore, geometrical scaling is performed where the model size is reduced from full scale model. When model is scaled down the following conditions should be followed:

- Geometrical dimensions of the model shall be $1/n$ times those of the full-scale model.
- Operating frequency and conductivity of the materials used in the model shall be n times those of the full-scale model.
- Permittivity and permeability of the materials used in the model shall have the same values at the scaled frequency as at the original frequency.

Table 5.3 summarizes the relations among quantities for the real-size model and the scaled model. As can be observed, the electrical parameters of the antenna remain the same, without modifying the permittivity or permeability of the materials.

The most problematic parameter is the change in the conductivity of the materials, since it must be multiplied by the factor N of the scaled model. Although it is possible, in some situations, to find materials that fit this scaling, in practical situations, the antennas and vehicles are metallic, with high conductivity. Therefore, the model can be constructed with metals of as high conductivity as possible, and, for antennas that are not highly resonant, the error will usually be small if good conductors such as copper or aluminum are used to simulate the metal parts.

Table 5.3 Relationship Among Quantities in Scaled Measurements

Parameter	*Real Model*	*Scaled Model*
Length (linear dimension)	L	L/n
Frequency	f	f*n
Dielectric permittivity	ε	ε
Magnetic permeability	μ	μ
Conductivity	σ	σ *n
Input impedance	Z	Z
Isolation between antennas	Z	Z
Antenna gain	G	G

References

[1] Saccardi, F., et al., "Experimental Comparison of Absorber and Conductive Floor Automotive Near Field Antenna Measurement Systems," *International Journal of Microwave and Wireless Technologies,* 2021.

[2] Ehtisham Asghar, M., et al., "Comparative Analysis of Spherical Near-Field Automotive Antenna Measurement Facilities," EuCAP 2019, Krakow, Poland, 31 March–5 April 2019.

[3] D'Abreu, G., Z. Xiong, and Z. Chen, "Chambers for the Evaluation of Vehicle Mounted Antennas," EuCAP 2018, 9–13, April 2018 London, United Kingdom.

[4] ANSI/IEEE Std 149-1979 "IEEE Standard Test Procedures for Antennas."

[5] Saccardi, F., et al., "Accurate Calibration of Truncated Spherical Near Field Systems with Different Ground Floors using the Substitution Technique," *AMTA 2019,* October 6–11, San Diego, CA.

[6] Newell, A. C., "Error Analysis Techniques for Planar Near-Field Measurements," IEEE Trans. Antennas Prop., Vol. 36, June 1988, pp. 754–768.

[7] IEEE Std 1720-2012 "Recommended Practice for Near-Field Antenna Measurements."

[8] 5GAA VATM (Vehicular Antenna Test Method) Technical report version 1.0, April 2021.

[9] Betjes, P. N., D. Pototzki, D. Janse van Rensburg. "A Hemi-spherical Near-Field System for Automotive Antenna Testing," *AMTA 27th Annual Meeting & Symposium,* Newport, RI, 2005.

[10] Cornelius, R., et al., "A Comparison of Different Methods for Fast Single-Cut Near-to-Far-Field Transformation," *in IEEE Antennas and Propagation Magazine,* Vol. 56, No. 2, April 2014, pp. 252–261.

[11] Salmerón-Ruiz, T., et al., "A Fast Single Cut Spherical Near-Field-to-Far-Field Transformation Using Cylindrical Modes," *EuCAP,* The Hague, The Netherlands, 6–11 April 2014.

[12] Sierra-Castañer, M., and F. Saccardi, and L. J. Foged "Gain Antenna Measurement using Single Cut Near Field Measurements," *AMTA 2016,* Austin, TX, October 30–November 4, 2016.

[13] Leach, W. M., and D. T. Paris, "Probe Compensated Near-Field Measurements on a Cylinder," *IEEE Transactions on Antennas and Propagation,* Vol. 21, No. 4, Jul. 1973, pp. 435–445.

[14] D'Agostino, F., et al., "Experimental Assessment of an Effective Near-Field–Far-Field Transformation with Spherical Spiral Scanning for Quasi-Planar Antennas," in *IEEE Antennas and Wireless Propagation Letters,* Vol. 12, 2013, pp. 670–673.

[15] Mauermayer, R. A. M., and T. F. Eibert, "Spherical Field Transformation Above Perfectly Electrically Conducting Ground Planes," in *IEEE Transactions on Antennas and Propagation*, Vol. 66, No. 3, March 2018, pp. 1465–1478.

[16] Saccardi, F., et al., "Using Standard Wideband Antennas as Probe in Spherical Near Field Measurements with Full Probe Correction: Experimental Validation," *AMTA 2018,* 4–9 November 2018, Williamsburg, VA.

[17] https://www.mvg-world.com/en/products/antennas/measurement-probes-and-feeds/dual-polarized-probes (accessed January 2022).

[18] Saccardi, F., et al., "Experimental Validation of Full Probe Correction Technique using Wideband and Dual-Polarized Probes in Spherical NF Antenna Measurements," *AMTA 2021,* Daytona Beach, FL, October 24–29, 2021.

[19] Saccardi, F., et al., "Investigation of Different Levels of Probe Compensation in Spherical Near Field Measurements Performed with Wideband and Dual-Polarized Probes," *EuCAP 2022,* Madrid, 27 March–1 April 2022.

[20] https://www.mvg-world.com/en/products/antennas/measurement-probes-and-feeds/open-boundary-quad-ridge-horns (accessed January 2022).

[21] Laitinen, T. A., "Double ϕ-Step θ-Scanning Technique for Spherical Near-Field Antenna Measurements," *IEEE Transactions on Antennas and Propagation,* Vol. 56, June 2008, pp. 1633–1639.

[22] Bolomey, J. C., and F. Gardiol, *Engineering Applications of the Modulated Scatterer Technique,* Norwood MA: Artech House, 2001.

[23] Saccardi, F., et al., "Fast Multi-Probe Planar Near Field Measurements With Full Probe Compensation," *EuCAP 2021* (virtual conference), 22–26 March 2021.

[24] Saccardi, F., et al., "Analysis of Probe Compensation Techniques for Fast Multi-Probe Planar Near Field Measurements," *AMTA 2021,* Daytona Beach, FL, October 24–29, 2021.

[25] Wollenschläger, F., et al., "Spherical Wave Expansion Applied to the Measured Radiation Patterns of Automotive Antennas in the Installed State in the GHz Range," *2019 13th European Conference on Antennas and Propagation (EuCAP)*, 2019, pp. 1–5.

[26] Asghar, M. E., et al., "Simulation and Measurement of Locomotive Antennas in the Installed State Based on a Scaled Mockup," *2021 Antenna Measurement Techniques Association Symposium (AMTA)*, 2021, pp. 1–6.

[27] Noren, P., L. J. Foged, and P. Garreau, "State of the Art Spherical Near-Field Antenna Test Systems for Full Vehicle Testing," *2012 6th European Conference on Antennas and Propagation (EUCAP).*

CHAPTER 6

OTA Measurements

Alessandro Scannavini, Kim Rutkowski, Mathieu Mercier, Nicolas Gross, Francesco Saccardi, and Lars J. Foged

This chapter introduces OTA measurement techniques applied to automotive antennas. The test setup and its calibration and common test parameters are introduced, and differences and analogies with concepts introduced in Chapter 2 are explained. Further, different OTA measurement methods are presented and compared to investigate accuracies using practical examples. Finally, an introduction to MIMO-OTA is given, and its relationship with virtual drive testing (discussed in Chapter 8) is explained.

With the growing number of communication technologies installed in modern vehicles, significant efforts are being directed toward standardized testing of these devices. The first draft technical report (TR) of the 5G Automotive Association (5GAA), on vehicular antenna test methodology (VATM) has been completed [1]. The document outlines the methodology to evaluate the installed performance of vehicular antennas (i.e., when fully integrated) in the vehicle. The 5GAA effort is closely related to 3GPP recommendations on general device testing [2]. In a similar manner, the CTIA is investigating methods for cellular testing of automotive antennas [3]. As the antenna and antenna system become integral parts of the vehicle, full vehicle measurements have become crucial to verify performance. The objective of standardization is to provide flexible test procedures [1, 4–6]. 5GAA's TR does not restrict the use of existing measurement systems but reports on different measurement configurations and how to achieve comparable results. Examples of such flexibilities in the automotive test domain include the use of different ground materials in the test environment; such materials can be either free-space, absorbing, or reflective, and both NF and FF testing methodologies using multiple or single probes are permitted.

6.1 OTA Test Setup

OTA testing differs from traditional passive antenna testing in the absence of a direct connectorized link to the unknown DUT. To perform the measurement, an external device capable of setting up a communication and control link with the unknown DUT (which could be a wireless/mobile communication terminal or GNSS device) is needed. This slightly complicates OTA testing with respect to traditional antenna testing.

The differences in passive and active OTA measurement configurations are illustrated in Figure 6.1. In passive antenna measurements, a network analyzer or equivalent equipment directly provides the amplitude and phase relationship between power into and out of the passive DUT. In the OTA testing scenario, a radio communication tester/system simulator or equivalent is used to manage the communication link with the active DUT. It emulates the technology used by mobile networks, satellites, road side units (RSUs), and vehicles, among other examples. For radiated power measurements, spectrum analyzers or power meters may be used.

6.2 OTA Testing Parameters

OTA testing of wireless devices involves the determination of transmitter, and receiver characteristics such as power, sensitivity, and, in some cases, data throughput. To evaluate the field performance of the DUT, the entire RF system must be tested to determine the interaction between the transceiver, antenna, and the structure in which it is integrated. When assessing the performance of the DUT as a system, the testing metrics are related to system performance. The traditional DUT testing parameters for both active and passive antenna testing are discussed in Chapter 2. A complete list of test metrics commonly measured using conventional OTA systems can be found in CTIA and 3GPP wireless measurement procedures [7–10].

The OTA testing of automotive wireless devices is only slightly more complicated than traditional antenna testing [1]. The testing requires external RF equipment, capable of establishing a communication and control link with the DUT. For wireless communication testing, this is often a base-station emulator. The external RF equipment regulates the communication and maintains the test mode of the device.

As discussed in Chapter 2, by measuring two spatial power quantities, usually in spherical coordinates, such as EIRP and EIS, both transmitter and receiver

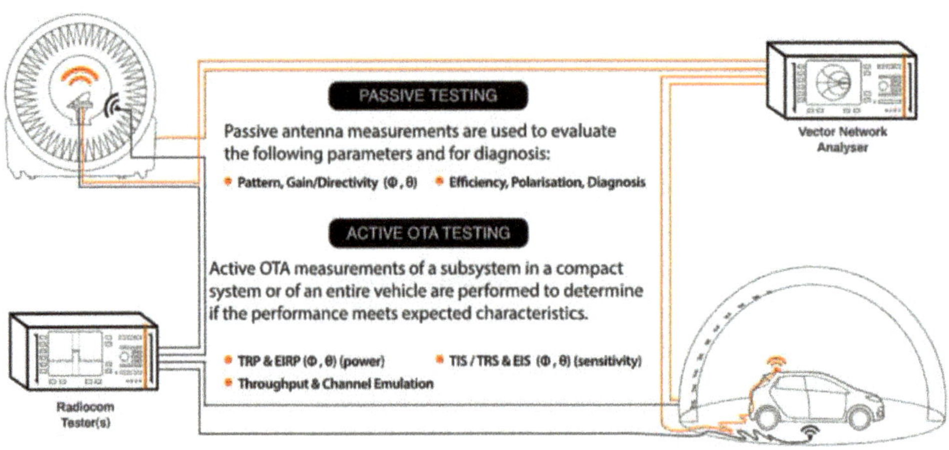

Figure 6.1 Simplified measurement setup for passive and OTA device testing.

properties of the DUT can be characterized respectively. In this chapter we concentrate on the determination of the general OTA power and sensitivity performance parameters. We also cover system parameters such as throughput including channel emulation. Other, more specialized, power-based parameters can be tested in a similar manner using different RF equipment.

6.3 OTA System Calibration

OTA testing of devices involves the characterization of the radiated power and receive sensitivity parameters. Power parameters can be measured directly in a gain calibrated system. A common method to calibrate the system is by the gain-substitution or gain-transfer technique [4–6], using a well-defined gain standard, as shown in Figure 6.2. Further details can be found in Chapter 2. This calibration is often called OTA system calibration.

The purpose of OTA system calibration is to enable measurements of the isotropic power "into" or "out of" the unknown DUT. After calibration, the probe or any other reference port used in the calibration, becomes a virtual connectorized port of the unknown DUT. OTA measurements can now be performed as the pertinent power parameters into and out of the unknown device can be determined directly at the probe port(s). As all parameters are relative to isotropic radiation. Following the discussion and conventions used in Chapter 2, the coupling $S_{21\,REF}^{dB}$ between the calibration or reference antenna (REF) and the probe/reference port can be measured using a standard network analyzer.

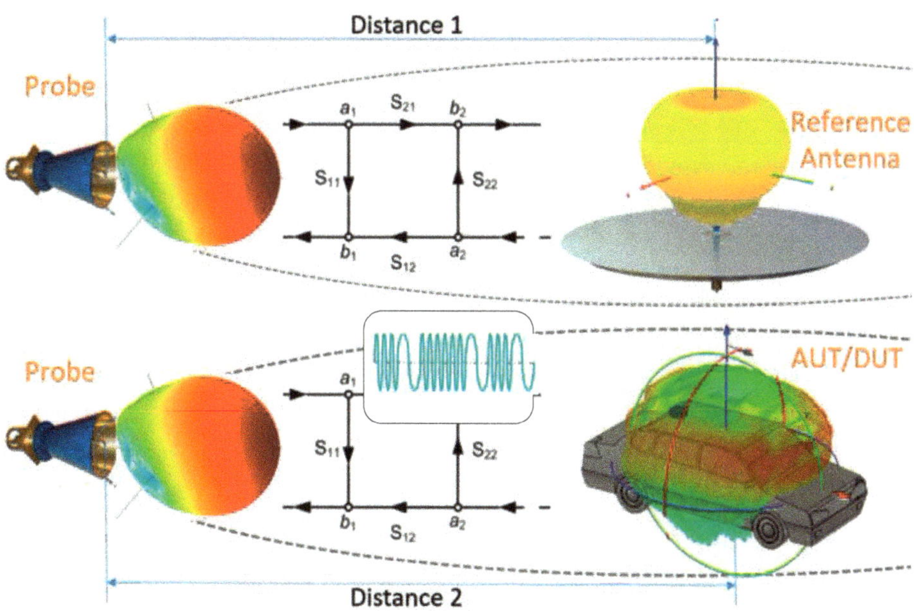

Figure 6.2 OTA system calibration using the gain-substitution or gain-transfer technique. A reference antenna with well-known characteristics is used to calibrate the system. The probe becomes a virtual connectorized port of the unknown DUT.

$$S_{21\,REF}^{dB} = G_{Probe}^{dB} + G_{REF}^{dB} - L_{freespace,REF}^{dB} \tag{6.1}$$

The known gain of the reference antenna G_{REF}^{dB} can be used to eliminate the probe gain G_{Probe}^{dB} and the free-space losses $L_{freespace,REF}^{dB}$ for the finite measurement distance in the given measurement scenario. From (6.1), we can derive the OTA *system calibration* term, S_{Cal}^{dB}. This term also gives the relationship between, respectively, power in/out and power out/in of the probe and an isotropic radiator as discussed in Chapter 2.

$$S_{CAL}^{dB} = S_{21REF}^{dB} - G_{REF}^{dB} = Power_{Probe}^{dB} - Power_{DUT\ Isotropic}^{dB} \tag{6.2}$$

Equation (6.2) is valid for both transmitting and receiving isotropic radiators. Since $Power_{DUT\ Isotropic}^{dB}$ is the isotropic radiated power by the DUT, it follows that the EIRP can be determined as:

$$EIRP^{dB} = Power_{DUT\ Isotropic}^{dB} = Power_{Probe}^{dB} - S_{CAL}^{dB} \tag{6.3}$$

Analogously, it follows that the EIS of the measured DUT can be determined in an OTA calibrated setup as:

$$EIS^{dB} = EIS_{Probe}^{dB} - S_{CAL}^{dB} \tag{6.4}$$

Equations (6.3) and (6.4) make it possible to measure the power-related performance parameters of an unknown DUT. However, the assumption in the derivation is that both the DUT and probe are in the FF. Section 6.4 examines different methods that enable such measurements.

6.4 OTA Measurement Methods

There are several different methods available for OTA radiated power and sensitivity measurements [1]. We will now briefly introduce each method. In OTA testing, the DUT consists of an antenna mounted on a vehicle. The DUT is located in the measurements setup that can be a multiprobe system [11], as illustrated in Figure 6.3, or a single-probe system. The different implementations of automotive measurement systems are discussed in Chapter 5. The system makes it possible to sample a sufficient spherical surface around the DUT. The measurement distance R, from the center of the measurement system to the probe(s), is assumed to be insufficient to qualify as classical FF distance but sufficient to avoid reactive coupling and minimize multiple coupling effects at $>1\lambda$ distance between the antenna and the probe(s), at the measurement frequency.

6.4.1 NF to FF Transform Using Phase Recovery

There are many generally applicable phaseless or phase recovery measurement techniques suggested in the literature and tested to different extents. These techniques have been developed for a variety of different applications, and not all of these are

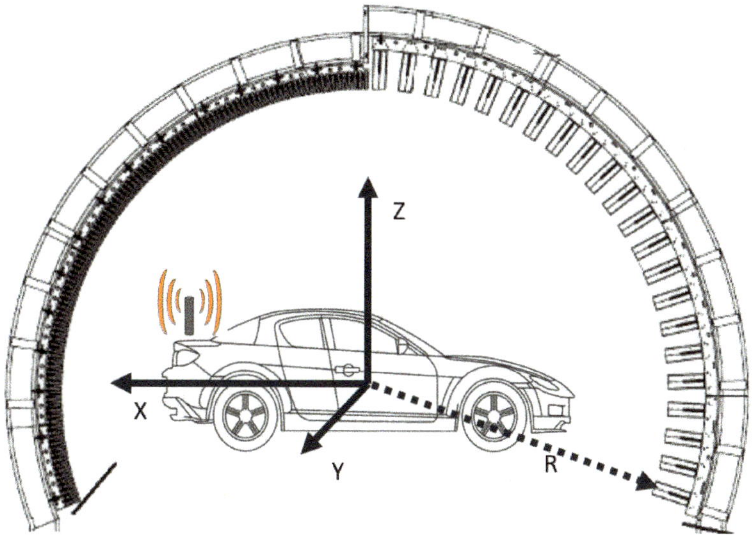

Figure 6.3 OTA automotive measurement scenario. The coupling between the probe(s) and the DUT is measured in different spatial directions.

particularly well-suited for OTA measurement applications and automotive measurements. A good overview of the different techniques can be found in [12–14]. The most common phase recovery technique applied on automotive antenna measurements is the indirect holography technique where the relative phase of the measured signal in each NF point is determined with respect to a phase-stable reference using correlation techniques [15–17]. As only relative phase is needed for NFFF transformation as discussed in Chapter 3, this knowledge can be used to determine both the FF directivity pattern $D_{DUT}(\theta,\phi)$ and the radiated or received power. From this knowledge the DUT radiated power and sensitivity profile can be determined. General EIRP mesurement techniques are also discussed in [18, 19].

In some cases, it is more convenient to use the determined FF directivity pattern $D_{DUT}(\theta,\phi)$ by phase recovery successively in a two-step OTA measurement procedure as explained in Section 6.4.2.

6.4.2 Separate Measurements with Conducted RF

EIRP is defined as the product of the antenna gain and the net input power to the antenna as discussed in Chapter 2. It follows that the power-related OTA figures of merit, such as EIRP, can be determined directly if the antenna and RF system can be measured separately. In this approach the antenna gain is $G_{DUT}(\theta, \phi)$ measured as a passive device or in a different active mode using phase recovery or similar techniques. The transmit power delivered by the RF system to the antenna or the receive sensitivity is measured as a conducted measurement directly at the RF system level. From the separately measured parameters, the system performance, including TRP, TIS/TRS, and $EIRP(\theta,\phi)/EIS(\theta,\phi)$ can be determined.

When using this approach, the power stability of the RF system must be evaluated to ensure that the measured power performance parameters are the same in conducted measurements and when loaded by the antenna. In some cases, the different loading of the amplifier may cause inaccuracies in the results [15]. Another consideration is that conducted sensitivity measurements do not capture *desense.* This is a degradation in DUT sensitivity due to noise sources, which can be generated by the RF system and coupled to the antenna when the device is assembled.

6.4.3 Two-Step Measurement or Combinational Method

In the two-step measurement process, the easier passive measurement becomes an active measurement by applying single-point active measurement as the tie point in a pattern normalization. This method is sometimes also called the combinational method [1]. It is assumed that the finite measurement distance R is too close to give an accurate estimate of the desired DUT FF power parameters. The antenna FF pattern and thus gain $G_{DUT}(\theta,\phi)$ of the DUT is initially obtained in a passive measurement or by using phase recovery techniques. Using standard spherical NFFF techniques, as explained in Chapter 3, the difference between the radiated field at the finite measurement distance R, $G_{DUT}(R,\theta,\phi)$ and the true FF $G_{DUT}(\theta,\phi)$ can be used as a correction factor to the radiated power measurements performed at the finite measurement distance R [15].

A single active measurement at the point (R, θ, ϕ) is sufficient, as a tie point, to determine the power offset for the full pattern. However, it is generally advisable to use the K offset point in different directions and average the results to improve on measurement uncertainty. Using this approach, the $EIRP(\theta,\phi)$ can be determined from:

$$EIRP(\theta,\phi) = G_{DUT}(\theta,\phi)\frac{1}{K}\sum_{1}^{K}\frac{Power_{DUT}(R,\theta_k,\phi_k)}{G_{DUT}(R,\theta_k,\phi_k)} \tag{6.5}$$

In an analogous way, the DUT sensitivity can be determined from:

$$EIS(\theta,\phi) = \frac{1}{G_{DUT}(\theta,\phi)}\frac{1}{K}\sum_{1}^{K}Power_{DUT}(R,\theta_k,\phi_k)G_{DUT}(R,\theta_k,\phi_k) \tag{6.6}$$

In [15], the two-step/combinational method discussed in Section 6.4.3 and the separate measurements with conducted RF described in Section 6.4.2 are compared. The comparison is performed on an active antenna system (AAS) at LTE frequencies; other applications are discussed in [16, 17]. The AAS is emulated using a mobile phone with the LTE protocol connected to the eight-element passive array. The phase recovery was performed in a 10-MHz bandwidth around the 1,940-MHz center frequency. The reported deviation on measured peak EIRP between the two methods is ~0.5 dB. The reported deviation on measured peak EIS is ~1 dB. Such differences are considered consistent with the associated uncertainties of both methods.

6.4.4 Direct OTA Measurements

In the direct OTA measurement method, the power or sensitivity profile of the wireless device is measured directly in a standard OTA calibrated setup at a finite measurement distance R. The finite distance leads to measurement errors if insufficient to guarantee that the full DUT, consisting of vehicle and device, is in the FF condition. The measurement error, on full or partial integral OTA metrics such as TR, and TIS/TRS can in some cases be deemed acceptable. This is particularly true for low-directivity antennas on vehicles. The error must be evaluated on a case-by-case basis, as it is strongly DUT- and scenario-dependent but is often on the order of +/−1 dB [9]. The error is further investigated in Section 6.6.

6.4.5 NF to Quasi-FF Transform Through Parallax

The NF to quasi-FF transform differs with the direct OTA measurement in that it uses knowledge of the antenna position with respect to the measurement coordinate system to improve accuracy. Knowledge of the antenna position can be used to compensate for the angular and distance variation with DUT rotation. This approach is very useful in the case of direct FF measurements performed at reduced distances, where often it is difficult to access the phase information of the DUT, and hence it is not possible to apply the conventional and rigorous NFFF transformation technique.

Classical *parallax* occurs in antenna measurements when the DUT is located off the center of rotation of the measurement system. The offset causes the resulting angle of the DUT as viewed from the probe/range antenna to be different than the angle to which the positioner is commanded. Parallax in antenna measurements cause distortions in the measured pattern, particularly when the ratio of the offset distance of the effective DUT phase location from the center of rotation during measurements is significant in comparison to the measurement distance. In classical parallax compensation, the knowledge of the DUT phase center position relative to the rotation center allows for the determination of an appropriate mapping from the DUT rotation angles to the actual angles as viewed from the probe/range antenna [20]. A further elaboration of the parallax is to use the information of the effective DUT phase location to compensate both the angular distortion and compensate the variation in measurement distance with DUT rotation angles [21]. This compensation is often referred to as NF to quasi-FF transform through parallax [22].

In automotive measurements the antenna is often located in an offset position on the vehicle with respect to the center of rotation during measurement. The main source of radiation from a vehicle-mounted antenna is generally the antenna itself and an area of a few λ in the immediate vicinity of the antenna. The radiation contribution is of secondary importance to the overall pattern of the antenna and vehicle unless there are areas of significant reflections or diffractions in close proximity to the antenna.

It is therefore advantageous to use the information of the antenna position to compensate for the angular and distance variation with the DUT rotation angle using the classical parallax formulation [20]. Considering the measurement geometry in Figure 6.4, the new angles θ', ϕ' and the measurements distance can be determined from the known position of the antenna in the system X_o, Y_o, Z_o. These can be used in place of the regular angles ϑ and distance R' given by the measurement geometry. Interpolation to a regular grid can then be performed after the coordinate change.

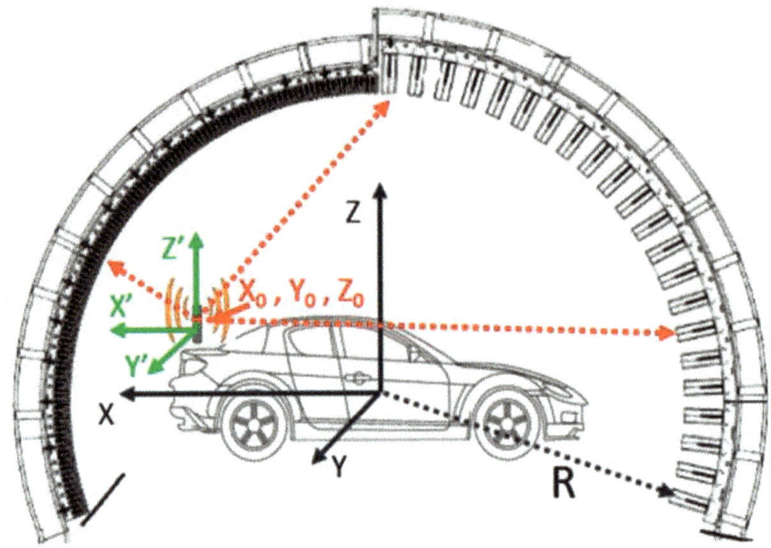

Figure 6.4 Parallax compensation. The distance and the resulting angles can be compensated for with knowledge of the position of the antenna within the measurements system.

$$R' = \sqrt{(x - x_0)^2 + (y - y_0)^2 + (z - z_0)^2} \tag{6.7}$$

$$\vartheta' = atan2\left(\sqrt{(x - x_0)^2 + (y - y_0)^2}, (z - z_0)\right) \tag{6.8}$$

$$\phi' = atan2\left((y - y_0), (x - x_0)\right) \tag{6.9}$$

The NF to quasi-FF transform is completely general and can be implemented in different measurement ranges [23–26].

6.4.6 TRP/TIS Testing of Devices on a Limited Ground Plane

While standardization efforts in organizations such as 5GAA and 3GPP support the methods outlined above [1], other organizations, such as the CTIA, has been looking into alternative evaluation methods for automotive antenna OTA testing, using a conductive ground plane with a 0.5-m diameter [3]. Given that the integral power metrics TRP and TIS are of interest, a small easily manageable ground plane size with straight edges has been chosen. The ground plane size was chosen to be large enough to ensure that the automotive antenna has roughly the same impedance matching as it would when placed on a vehicle. The limited testing metrology is illustrated in Figure 6.5 using the MVG StarLab18GHz (left) and a commercial automotive antenna from Calearo [27]. The limited ground plane interacts with the DUT. It is well-known that for a dipole-style antenna, the ground plane will enhance radiation around the 30–40° angle of elevation from the ground. This effect can also be observed in Figure 6.5 (right) where the blue trace is the infinite ground

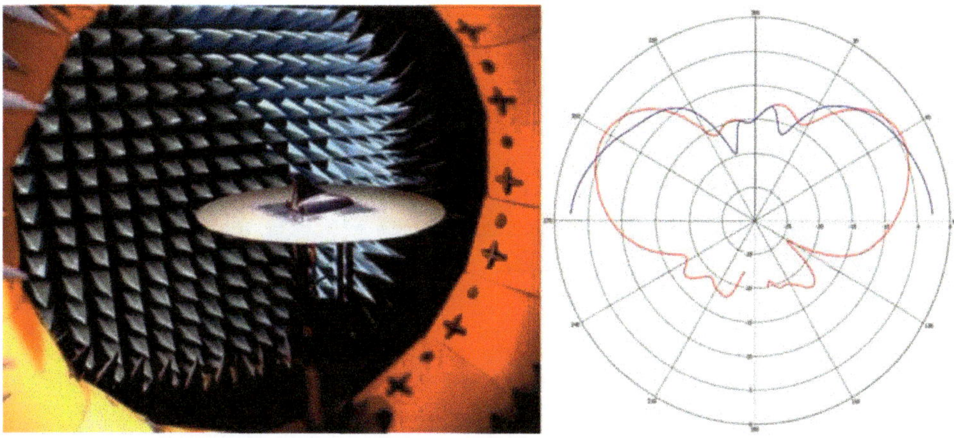

Figure 6.5 Limited ground plane evaluating method for automotive OTA device testing using the MVG StarLab-18-GHz system (left). Elevation pattern of a device on a 0.5-m ground plane (red) and on an infinite ground plane (blue) obtained in post-processing (right).

plane radiation by the device, and the red trace is the measured performance on the 0.5-m ground plane. At lower frequencies, measurement errors due to coupling with the cables and support structure must be carefully evaluated.

6.5 OTA Sampling

The total radiated power of a wireless device can be determined rapidly in a standard multiprobe system as discussed in Chapter 5. To determine integral power parameters such as TRP and TIS/TRS, the classical NF sampling requirement discussed in Chapter 3 can be relaxed. To better understand this, the sampling requirements for radiated power in a typical automotive application is examined on a 12:1 scaled vehicle as shown in Figure 6.6. The scaled vehicle and measurement is discussed in detail in Chapter 4. The measurement is performed in a spherical NF multiprobe system, MVG StarLab18GHz with a measurement radius of 45 cm [11]. The measurements of the scaled vehicle in the StarLab18GHz system, in the frequency range 1.1–18 GHz, correspond to testing the full-sized vehicle in a standard automotive multiprobe system of radius 5.4m in the frequency range 91.7–1,500 MHz.

The sampling investigation is also a comparison of the two techniques "separate measurements with conducted RF" discussed in Section 6.4.2 and the "direct OTA measurements" approach discussed in Section 6.4.4. A network analyzer was used in the radiated measurement (also called passive measurement) of the antenna in the "separate measurements with conducted RF" approach, and the RF conducted part was measured with a spectrum analyzer. The "direct OTA measurements" (or active measurement) was performed using a spectrum analyzer directly. Both measurements were conducted in the same OTA-calibrated system using the same calibration as shown in Figure 6.6.

The passive measurement enabled the measurement of the antenna gain and the evaluation of the antenna efficiency by pattern integration. The power delivered to the antenna was determined using a spectrum analyzer. This made it possible to

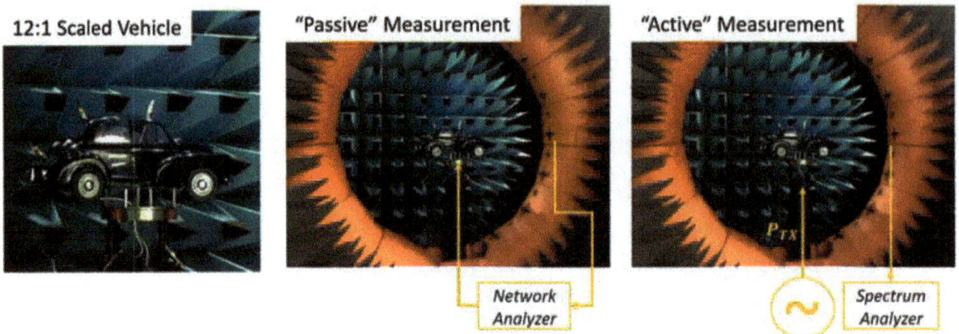

Figure 6.6 Measurements of radiated power performance of 12:1 scaled vehicle in active and passive measurement configurations using the MVG StarLab18GHz multiprobe system.

determine TRP as power delivered to the antenna less the antenna losses. In the active measurement, TRP was evaluated directly by pattern integration. The measured values for the two methods are reported in Table 6.1. Differences, observed on the order of 0.5 dB, correspond well to the expected measurement errors encountered in standard efficiency, power, and TRP measurements and, thus, to the overall expected difference between the two methods.

The $EIRP(\theta,\phi)$ variations with different sampling grids and resulting TRP are reported in Figure 6.7. The TRP difference is approximately 0.3 dB for fine sampling resolutions 2.25° and the coarser 15°. Such a difference is considered reasonable for this type of application; hence, it can be concluded that for such devices and at these frequencies a 9°/15° sampling resolution is sufficient to measure the TRP. This resolution is much less than that required for the classical NF sampling discussed in Chapter 3. It should be noted that TRP or TIS/TRS measurements at V2X application frequencies would require finer sampling.

Given that TRP and TRS/TIS are both integrated power measurements, it can be assumed that the accuracy related to the angular sampling step would apply to both performance metrics. As sensitivity measurements are more time-consuming and have a larger acceptable measurement uncertainty, the sampling step for TRS/TIS is often relaxed to about twice the step of TRP.

6.6 OTA Measurement Examples

The same DUT, vehicle, and antenna have been measured using different OTA measurement techniques as described in Sections 6.4.2–6.4.4. A comparison of the techniques makes it possible to examine factors such as measurement time and the

Table 6.1 Measured TRP of Scaled Vehicle at 750 MHz (12:1 Scaled Frequency) Using Passive and Active Measurement Setup

Frequency	*Measured PTX*	*Measured Efficiency*	*TRP (PTX – Eff) (Passive)*	*Measured TRP (Active)*	*TRP*
750 MHz	23.8 dBm	−2.0 dB	21.8 dBm	21.3 dBm	0.5 dBm

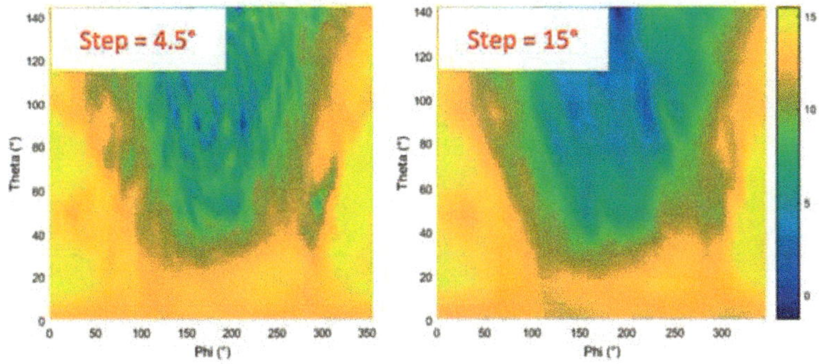

Sampling (°)	TRP (dBm)
2.25	21.3
4.5	21.2
9.0	21.0
15.0	21.0

Figure 6.7 EIRP (θ, ϕ) variations with different sampling grids and resulting TRP for 12:1 scaled vehicle active measurement configuration using MVG StarLab18GHz multiprobe system.

achievable accuracies associated with the different methods [28–32]. Accordingly, the following summarizes the outcome of each of these techniques:

1. In Section 6.4.2, the passive measurements were performed in a fine 1-degree grid in θ, ϕ. The measurements were down-sampled to θ and ϕ of 5 and 15 degrees for the power and sensitivity evaluation to match the sampling applied in the OTA measurements.
2. In Section 6.4.3, the OTA measurements were obtained from the direct-OTA measurements using four points in the TRP and TRS/TIS evaluation. The measurement distance was 4m.
3. In Section 6.4.4, the measurement distance was 4m.

The passive and OTA radiated power and sensitivity measurements on the reference vehicle were performed in the automotive antenna measurement site VISTA at the Thuringian Center of Innovation in Mobility at TU Ilmenau [32]. The system is a MVG SG3000 spherical multiprobe system as described in Chapter 5. The system is equipped with 22/111 dual-polarized probes covering the range 70–6,000 MHz on a range length/radius of 4m.

The measurements were performed on a Mercedes Benz E-Class mock-up as shown in Figure 6.8. The DUT antennas are monopole-like antennas mounted in two positions. Position 1 is in the center of the vehicle, and position 2 is the position typically used to mount shark fin antennas on the rear part of the roof. For the OTA measurements, the antennas were connected to an LTE module.

For each measurement techniques the angular step in θ and ϕ was 5 degrees for radiated power measurements and 15 degrees for sensitivity measurements. The angular scan in both cases is $\theta = 0$ to 110° and full-ϕ. The system was calibrated as described in Chapter 2 using the gain transfer method considering efficiency as the reference rather than the gain. The measurements were performed in the LTE bands 3 and 20, which correspond to the following transmit and receive frequencies:

- Tx: 1,710; 1,745; and 1,785 MHz (band 3) and 834, 847, and 860 MHz (band 20);
- Rx: 1,845 MHz (band 3) and 806 MHz (band 20).

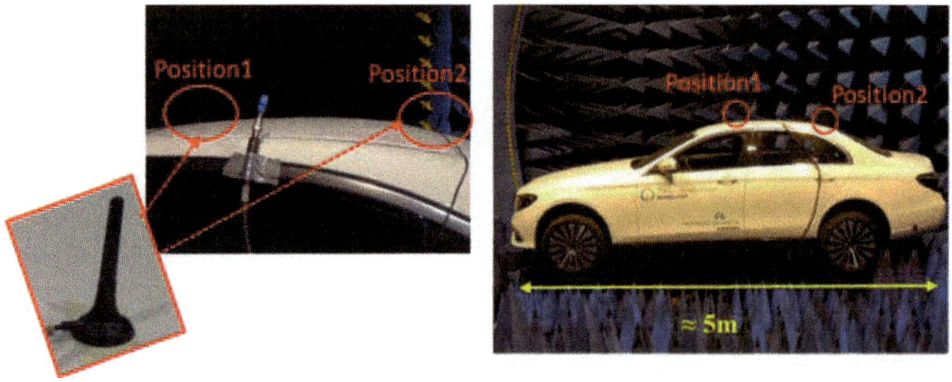

Figure 6.8 DUT test configuration. The antennas are placed on the roof of the vehicle. The vehicle is positioned in the measurement system.

For each measurement method, OTA performance metrics such as PRP/PIS in the upper hemisphere, near 75 degrees, and near horizon are evaluated for comparison as defined in Chapter 2.

The differences in measured EIRP with the three different OTA measurement techniques are reported in Table 6.2. The measured EIRPs are well-aligned for all three methods and within the expected measurement uncertainty for each step of the measurement process. As can be expected, the results for the "separate measurements with conducted RF" are slightly worse than the other two techniques. This could be due to the intrinsic uncertainty of the output power when the RF system is loaded by the antenna versus during conducted measurement. It also seems that the "direct OTA" technique has a small bias toward lower EIRP than the others. The differences for antenna position 1 and position 2 are minor. The worst-case EIRP differences occur at band 3 lower edge of 1,710 MHz. A preliminary observation is that EIRP measurements can be performed within +/−0.5 dB uncertainty depending on the chosen method.

The differences in measured sensitivity TRS/TIS with three different OTA measurement techniques are reported in Table 6.3. The measured sensitivities are well-aligned for all three methods and within the expected measurement uncertainty for each step of the measurement process. The measured sensitivity by the "separate measurements with conducted RF" are in good agreement with the "direct OTA" measurements. The "two-step or combinational method" sensitivities are biased lower and thus slightly worse than the other two techniques.

One important observation made when performing the conducted power and sensitivity measurements for the "separate measurements with conducted RF" method is that if the RF-module is not sufficiently stable, there is the potential for erroneous bias in the final result. This could also occur with the "two-step or combinational method" if the number of measured OTA radiated power or sensitivity points are insufficient to leverage potential fluctuation in the RF-module or measurement setup. If the measurement is not fully representative of the average performance of the RF-module, the overall metric will be skewed and not representative of the end user performance.

Table 6.2 Differences (in Decibels) on Measured EIRP with Different OTA Measurement Techniques

		Frequency [MHz]	Δ EIRP Between "Separate Measurements with Conducted RF" and "Direct OTA" [dB] NHPRP	N75PRP	UHRP	Δ EIRP Between "Two-Step or Combinational Method" and "Direct OTA" [dB] NHPRP	N75PRP	UHRP
Position1	*Band 3*	1,710	–1.75	–1.93	–1.95	0.41	0.21	0.19
		1,745	–0.98	–1.17	–1.22	0.21	0.00	-0.05
		1,785	–0.05	–0.25	–0.31	0.40	0.18	0.12
	Band 20	834	1.20	1.17	1.27	0.30	0.24	0.33
		847	0.96	0.87	0.93	0.49	0.36	0.41
		860	0.65	0.56	0.62	0.34	0.21	0.28
Position2	*Band3*	1,710	–1.42	–1.49	–1.48	0.58	0.49	0.50
		1,745	–0.91	–0.99	–0.95	0.50	0.39	0.43
		1,785	–0.14	–0.22	–0.22	0.43	0.34	0.33

Table 6.3 Differences (in Decibels) of Measured Sensitivity TRS/TIS with Different OTA Measurement Techniques

		Frequency (MHz)	Δ TRS (dB) between "Separate Measurements with Conducted RF" and "Direct OTA" [dB] NHPIS	N75PIS	UHIS	Δ TRS (dB) between "Two-Step or Combinational Method" and "Direct OTA" [dB] NHPIS	N75PIS	UHIS
Position1	*Band20*	806	−0.18	−0.03	0.04	−1.55	−1.38	−1.31
Position2	*Band3*	1,845	0.85	0.91	1.10	0.37	0.43	0.63
	Band20	806	−0.77	−0.79	−0.88	−1.95	−1.95	−2.04

The sensitivity TRS/TIS testing time can be rather long. It is therefore important to find the right compromise between measurement duration/number of points and desired accuracy. For the "direct OTA" method where the sensitivity was measured in each direction with a 5° sampling in θ and ϕ required five to six hours measurement time for each channel (rooftop vehicle mounted antenna < 2 GHz). With 15° sampling, the measurement time can be reduced to ~40 min and further to ~15 min with 30° sampling in θ and ϕ. A compromise must be found between desired accuracy and sampling/measurement time as discussed in Section 6.5. This compromise must be evaluated on a case-by-case basis. In the specific example above the differences betwen integral power parameters such as TRP/TRS/TIS between

5° and 15° sampling grid was 0.5 dB. A 2-dB difference was observed between the 15° and the 30° sampling grids.

Using the "two-step or combinational method" versus the "direct OTA" method reduces the EIRP testing substantially. The test time for "direct OTA" is ~10 times longer, when four OTA radiated power or sensitivity points are used to normalize the passive antenna pattern. The "two-step or combinational method" testing can thus be performed in ~35 min.

The comparison of the "separate measurements with conducted RF" to the "direct OTA" showed good correlation, proving that a shorter-range system, such as the system in Ilmenau, can be used for direct OTA measurements with good accuracy. The difference between the direct OTA and combinational method are negligible, providing good confidence that the measurement time can be reduced significantly using this two-step process.

The measurement times discussed above are indicative. Measurement time is highly influenced by the speed of the turntable, which is often compromised or slower for safety reasons or because it has other integrated measurement features. The use of distributed/parallel acquisition schemes is one way to speed up measurement times.

6.7 MIMO OTA Testing

One of the challenging aspects of vehicle-mounted communication systems is multiple-input-multiple-output over-the-air (MIMO OTA) testing. The testing objective is to determine system-level parameters such as data throughput with received power levels at the device in a realistic emulated complex scenario according to a standardized channel model.

6.7.1 Introduction

Channel models consist of both RF and spatial contributions. These can be reproduced by a channel emulator and a probe array, respectively. The RF contribution is a complex signal with elements such as modulation, polarization, temporal delay, and doppler profile associated with the complex scenario. The spatial contribution is represented by plane waves with certain angular spread. The contributions of a general channel model are illustrated in Figure 6.9. The channel model can be thought of as a time sequence of scenarios experienced by the DUT while moving in the modeled scenario. The output of the testing is the data throughput as a function of the available isotropic power level at the DUT position.

There are two general ways of MIMO OTA testing. The direct MIMO OTA testing is similar to a "hardware-in-the-loop" scenario in which the device is tested in a multiprobe anechoic chamber (MPAC), reproducing the realistic channel model at the device position. This is the most flexible and realistic test environment for MIMO OTA measurements. Alternatively, there is a two-step approach in which the antenna and RF system contribution are separated. Both methods are illustrated in Section 6.7.2 [33].

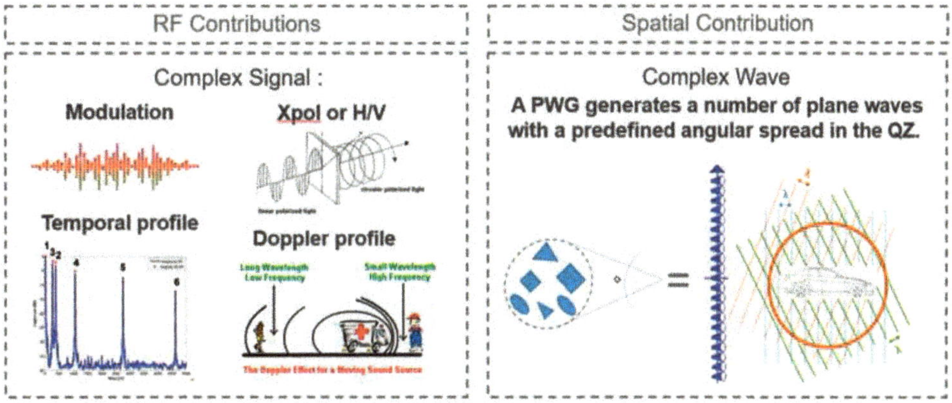

Figure 6.9 Illustration of the different mechanisms of a general channel model. The various contributions can be grouped into RF and spatial components.

6.7.2 Direct MIMO OTA Testing

The direct emulation of the complex scenario is a standard test of LTE-enabled devices in personal communication [34–36]. MIMO OTA testing requires a shielded anechoic enclosure, a circular array of probes around the DUT that simulates a complex multipath environment at the DUT location in a repeatable way by using a radio communication tester and a channel emulator. The most common spatial channel models (SCMEs) are the urbanmicro (Umi) and urbanmacro (Uma). Such models are approved by 3GPP as reference channel models for MIMO OTA testing [33]. A standard system for wireless communication device testing is shown in Figure 6.10. In this environment, the *N*-probes are positioned around the test zone; each probe is weighted in amplitude and phase to generate approximated plane waves in the test zone [34, 36]. A handheld mobile device can be carried in many different positions depending on the user. Likewise, a vehicle can be oriented in different directions depending on the travel direction. Consequently, the orientation of a terminal is considered to be unknown with respect to the channel model. The reported throughput is therefore often the averaged value in azimuth.

The antenna impact on MIMO system performance, in terms of throughput, can be illustrated by the MIMO 2 × 2 reference antenna concept reported in [37]. Three antenna pairs, representing the full range of "good," "nominal," and "bad" MIMO 2 × 2 antenna performance were conceived with low gain imbalance ($\Delta G \cong 0$dB), each covering three LTE bands (2, 7, and 13). These antenna pairs are described as follows.

- *Good*: Low correlation coefficient ($\rho \leq 0.1$), high efficiency (> 90%);
- *Nominal*: Moderate correlation coefficient ($\rho \leq 0.5$), moderate efficiency (≥ 50%);
- *Bad*: Poor correlation coefficient ($p \geq 0.9$), moderate-to-poor efficiency (≤ 50%).

To isolate antenna and modem, the antennas were designed attaching the MIMO 2 × 2 external antennas to a RF-shielded enclosure, where the DUT and its

Figure 6.10 Standard MIMO OTA testing of LTE-enabled devices in a MPAC.

RF connections are located as shown in Figure 6.11. The conducted testing shows that, as characteristic for devices, the modem is cable of full data throughput when the power level at the input is −110 dBm. As can be expected, the required power level at the device position in a realistic MIMO OTA testing scenario is higher than this value to achieve full data throughput as shown in Figure 6.9. It can be observed that the difference between a "good" and "bad" antenna design is 8 dB in this specific testing scenario. It can further be observed that the Uma SCME model is generally a more difficult environment than the Umi model; in this specific testing the difference is about 5 dB.

The implementation details are well-understood for small-form factor devices such as cellular handsets, tablets, or laptops. Direct MIMO OTA testing can be performed with a small number of probes, often ~8–16 [34, 35]. Unfortunately, the number of antennas required to generate plane wave conditions in the test zone grows with the electrical size of the test object and thus test zone. The possible setup for such a system including its automotive equivalent is shown in Figure 6.12.

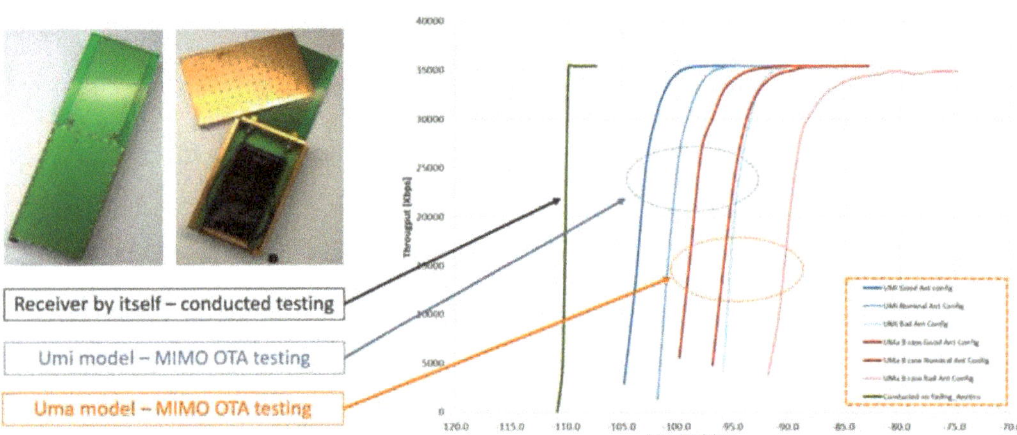

Figure 6.11 Example of MIMO OTA testing of the standard MIMO 2 × 2 reference antenna concept. Conducted throughput versus MIMO OTA testing with good, nominal, and bad antenna using the Umi and Uma spatial channel models.

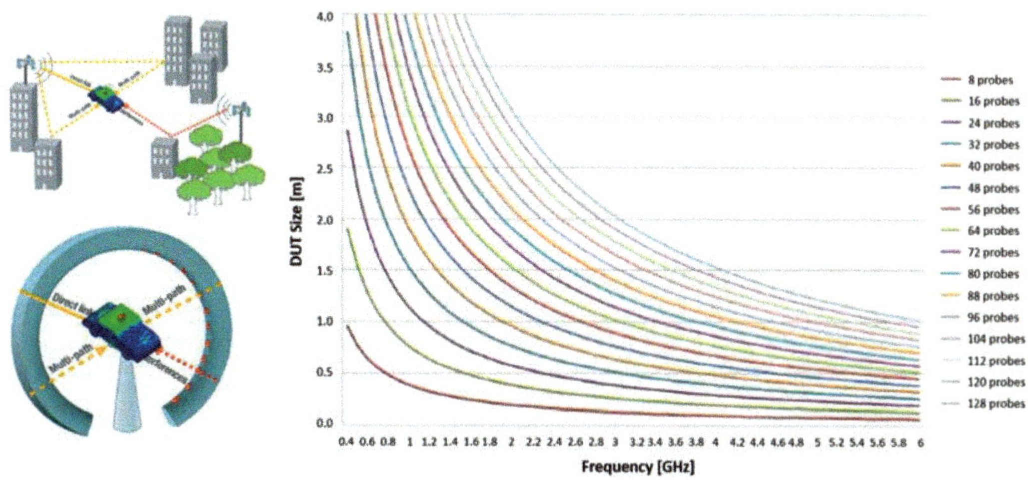

Figure 6.12 (a) Direct MIMO OTA testing, representing a realistic communication scenario using a channel emulator and a multiprobe system, and (b) number of probes needed for full ring testing with device size and frequency.

To construct the larger test area needed in full vehicle testing and to reproduce the target channel model more accurately, the number of channel emulators and probes used in the test system increases, which results in a significant increase in the cost of the test system.

The challenge presented by testing a MIMO system on a vehicle is the physical size of the article under test (i.e., car, bus, or van); even the smallest car is several orders of magnitude larger than a handheld cellular device. This means that existing MIMO OTA test methods applicable to small-form factor devices may not necessarily be fully scalable to MIMO OTA testing in an automotive environment.

One of the objectives of automotive testing and MIMO OTA testing is to optimize vehicular antenna placement. A possible solution to achieve this scope is to perform the MIMO OTA testing with a limited number of probes a suggested in [38–40]. The advantage of this methodology is that it enables throughput testing in a realistic environment using existing technology and within a reasonable cost. The method allows for the measurement of vehicle-mounted MIMO systems and makes it possible to compare the relative performance of different antenna array configurations on a test vehicle. A possible implementation of this approach inside a traditional automotive test chamber is shown in Figure 6.13 [38].

The drawback of this approach is that the number of probes and channel emulators is not sufficient to guarantee reproducible results when performed in a different test setup. And although the RF content will be realistically replicated, the spatial content of the channel models will be distorted by the insufficient number of probes and the phase curvature of the probes as the plane-wave approximation in the test zone cannot be achieved. Consequently, each test environment will have its unique channel model due to the limitations in the setup.

As the direct MIMO OTA testing with an insufficient number of probes has system-dependent realized channel models, the method is not suited for conformance testing. Thus, community-approved test procedures and plans have not been developed. Some reproducibility may be achieved if the relative positions and direction

Figure 6.13 Possible implementation of a MPAC MIMO OTA automotive test solution with a reduced number of probes. ©2019 IEEE. (Reprinted with permission.) [41].

of the MIMO antennas and the MPAC probes are recorded and reproduced. Such procedures can be implemented in a verification and R&D testing scenario but are unlikely to be agreed on and implemented for conformance testing scenarios.

6.7.3 Two-Stage Methods

The basic concept of the two-stage MIMO OTA test method is to divide the OTA test into two stages. The first stage is to acquire the embedded antenna patterns of the DUT, $G_{DUT}(\theta,\phi)_k$ with k =1 to the number of antennas in the system. This implies the gain pattern of each antenna with all other k-1 antennas terminated in 50-Ω loads. The second stage is to convolve the device antenna patterns with the desired spatial channel model (SCME). This will generate the spatial stimulus of the downlink signal that combined with the RF part of the SCME enable to perform the throughput measurements in a conducted or radiated connection with the DUT using a channel emulator [38, 39]. Another possibility is to use the measured patterns $G_{DUT}(\theta,\phi)_k$ in a simulation of the SCME and receiver system.

The two-stage MIMO OTA testing is thus very similar to the combinational method, OTA testing discussed in Sections 6.4.2 and 6.4.3. Unless already available through design, it is better to avoid enforcing a physical and/or connectorized separation of the antennas and RF system/modem since the behavior of the separated components can be different due to changes in the reference impedance. If a physical separation is already available or necessary to perform, the correct functionality of both the RF system/modem and antennas as separate components must be evaluated.

In order to accurately measure the embedded antenna pattern of an inseparable device, the DUT needs to support both amplitude and relative phase measurements of the antennas. This is not currently a standardized feature of most devices, although it is not dissimilar to other standardized device test features, such as loopback, to enable receiver BER measurements.

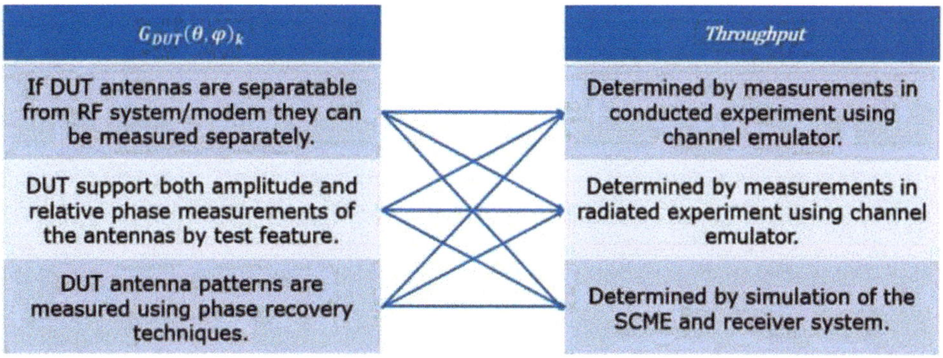

Figure 6.14 Possible solutions for $G_{DUT}(\theta, \phi)_k$ measurements of DUT and successive determination of throughput.

The gain $G_{DUT}(\theta,\phi)_k$ can also be measured in a different active mode using phase recovery or similar techniques. This requires that each antenna can be accessed separately with all other antennas turned off in a test feature of the device. The most common phase-recovery technique applied on automotive antenna measurements is the indirect holography techniques where the relative phase of the measured signal in each NF point is determined with respect to a phase-stable reference using correlation techniques [15–17]. Only relative phase information is needed in the NFFF transformation to determine $G_{DUT}(\theta,\phi)_k$ as discussed in Chapter 3.

The two-stage method has many advantages. The embedded antenna pattern measurements can be performed in a traditional automotive anechoic system, as each pattern is measured separately in a passive gain calibrated setup. The method is invariant to the complexity of the SCME that can be a 2-D, or even a partial or full 3-D model as the spatial antenna information is readily available from the antenna measurements. Once the embedded antenna patterns have been determined, the MIMO OTA performance can be evaluated in different scenarios by changing the applied SCME. The channel models are highly accurate due to being implemented electronically, and they are also fully flexible. The method is therefore scalable to higher-complexity MIMO OTA testing. The combinations of possible solutions for $G_{DUT}(\theta,\phi)_k$ measurements and successive determination of throughput are illustrated in Figure 6.14.

The availability of the measured or otherwise determined embedded gain patterns of the DUT antennas open up a range of interesting post-processing possibilities, taking advantage of powerful numerical modeling techniques [38]. These methods, known as virtual drive techniques, are discussed in Chapter 8.

References

[1] 5G Automotive Association (5GAA); DRAFT TR P-210024 "Working Group Evaluation, test beds and pilots Vehicular Antenna Test Methodology."

[2] www.3gpp.org [online].

[3] www.ctia.org [online].

[4] IEEE Standard Definitions of Terms for Antennas, 2013.

[5] IEEE Standard Test Procedures for Antennas, ANSI/IEEE Std 149-1979.

[6] IEEE Recommended Practice for Near-Field Antenna Measurements. IEEE Std 1720-2012.

[7] CTIA OTA Test Plan v 3.9.4 [online]. Available from: www.ctia.org [Accessed September 2021].

[8] 3GPP TR 25.914 V16.0.0 – Measurement of radio performances for UMTS terminals in speech mode [online]. July 2020. Available from: https://portal.3gpp.org/desktopmodules/Specifications/SpecificationDetails.aspx?specificationId=1343 [Accessed March 2021].

[9] 3GPP TS 34.114 V12.2.0 – User Equipment (UE)/Mobile Station (MS) Over-the-Air (OTA) antenna performance - Conformance testing [online]. October 2016. Available from: https://portal.3gpp.org/desktopmodules/Specifications/SpecificationDetails.aspx?specificationId=2361 [Accessed March 2021].

[10] 3GPP TS 37.114 V16.0.0 – User-Equipment (UE) and Mobile Station (MS) GSM, UTRA, and E-UTRA Over The Air performance requirements. July 2020. Available from: https://portal.3gpp.org/desktopmodules/Specifications/SpecificationDetails.aspx?specificationId=3034 [Accessed March 2021].

[11] Foged L. J., and A. Scannavini, "Efficient Testing of Wireless Devices from 800 MHz to 18 GHz," *Radio Engineering Magazine*, December 2009, 18(4), pp. 460–466.

[12] Breinbjerg, O., and J. Alvarez Fernandez, "Phaseless Near-Field Antenna Measurement Techniques—An Overview," *Proc. 38th Antenna Measurement Techniques Association (AMTA),* Austin, TX, November 2016.

[13] Castaner, M. S., and L. J. Foged, "Post-processing Techniques in Antenna Measurement, Institution of Engineering and Technology," *SciTech Publishing, Electromagnetic Waves series,* The Institution of Engineering and Technology 2019, London, United Kingdom.

[14] Hong Loh, T., "Metrology for 5G and Emerging Wireless Technologies," SciTech Publishing, *Electromagnetic Waves series,* The Institution of Engineering and Technology 2021, London, United Kingdom.

[15] Foged, L. J., et al., "Accurate Measurement of Transmit and Receive Performance of AAS Antennas in a Multiprobe Spherical NF System," *2015 IEEE International Symposium on Antennas and Propagation & USNC/URSI National Radio Science Meeting,* July 2015.

[16] Huawei. Huawei Launches the World's First SG178 BTS Antenna Testing System [online]. 2016. Available from: https://www.huawei.com/en/news/ 2016/8/shouge-sg178-jizhan-xitong [Accessed March 2021].

[17] Nokia. Nokia Annual Report on Form 20-F 2020 [online]. 2020. Available from: https://www.nokia.com/about-us/news/releases/2021/03/04/nokia-hasfiled-its-annual-report-on-form-20-f-for-2020-and-published-its-nokia-in2020-annual-report/ [Accessed March 2021].

[18] Hess, D. W., "Measurement of EIRP and Antenna Response for Active Antennas with Spherical Near-Field Scanning," *AMTA Conference 2003,* pp. 465–470.

[19] Boomstra A., and T. Lyes, "High Power Antenna Measurements in a Near-Field Facility: A Practical Approach," *AMTA Symposium 2004.*

[20] Aubin, J., C. Arnold, and T. Johnson, "Parallax Corrections in Antenna Measurements," *AMTA Conference*, Newport, RI, 2005.

[21] Sheffield, L., "Near-Field to Quasi-Far-Field Transform Through Parallax," in *Proc. 33th Antenna Measurement Techniques Association (AMTA),* 2008, Boston, MA.

[22] Saccardi, F., et al., "Over-The-Air Tests of 5G Devices with Direct Measurements at Reduced Distance and Parallax Compensation," *Proc. 43th Antenna Measurement Techniques Association (AMTA),* Florida, November 2021.

[23] Asghar, M. E., et al., "Comparative Analysis of Spherical Near-Field Automotive Antenna Measurement Facilities," *Proceedings of the 13th European Conference on Antennas and Propagation (EuCAP2019),* Krakow, Poland, 2019.

[24] Graham, A., and P. O. Iversen, "Rapid Spherical Near Field Antenna Test System for Vehicle Mounted Antennas," *AMTA 2004,* Stone Mountain Park, Georgia, GA, October 17–22, 2004.

[25] Dooghe, S., et al., "Spherical Near-Field Antenna Test Range for Automotive Testing from 70MHz to 6 GHz," *Proc. EuCAP 2006,* Nice, France, November 2006.

[26] Farina, D., and M. Pinkasy, "Full Scale Automotive Antenna Near Field and Far Field Range," *Antenna Measurements and Techniques Association Annual Conference,* November 1998.

[27] https://www.calearo.com/it/.

[28] Mercier, M., et al., "Evaluation of Automotive Antenna Over the Air Performance," *ISAP 2021,* Taipei, Taiwan, October 19–22, 2021.

[29] Mercier, M., et al., "Evaluation of Integral Quantities of Over the Air Automotive Antenna Measurements," *IEEE Antennas and Propagation Society International Symposium,* Marina Bay Sans, Singapore, December 4–10, 2021.

[30] Walkenhorst, B. T., et al., "A Survey of MIMO OTA Test Methodologies for Automotive Applications," *IEEE Antennas and Propagation Annual Conference,* July 2020.

[31] Saccardi, F., et al., "Experimental Comparison of Absorber and Conductive Floor Automotive Near Field Antenna Measurement Systems," *International Journal of Microwave and Wireless Technologies,* Cambridge University Press, September 2021.

[32] Asghar, M. E., Chr. Bornkessel, and M. Hein, "Experimental Determination of the Total Radiated Power of Automotive Antennas in the Installed State," *14th EuCAP,* Copenhagen, Denmark, March 15–20, 2020.

[33] 3GPP TR 37.977, Verification of radiated multi-antenna reception performance of User Equipment (UE), V14.5.0 (2018-06).

[34] Scannavini, A., et al., "OTA Throughput Measurements by Using Spatial Fading Emulation Technique," *Proceedings of the Fourth European Conference on Antennas and Propagation*, Eucap, 2010.

[35] Scannavini, A., et al., "OTA Measurement of Wireless Devices with Single and Multiple Antennas in Anechoic Chamber," *European Conference on Antennas and Propagation (EuCAP),* 2014.

[36] Scannavini, A., "2x2 MIMO downlink OTA measurement based on CTIA guidelines," *2015 IEEE International Symposium on Antennas and Propagation & USNC/URSI National Radio Science Meeting*, 2015, pp. 290–291.

[37] Szini, I., et al., "MIMO 2x2 Reference Antennas Concept," *6th European Conference on Antennas and Propagation, EUCAP 2012*, pp. 1540–1543.

[38] Saccardi, F., et al., "Status and New Challenges in Performance Testing of Wireless Automotive Communication Systems," *IEEE MTT-S International Conference on Microwaves for Intelligent Mobility (ICMIM)*, 2018.

[39] Ji, Y., et al., "Virtual Drive Testing Over-the-Air for Vehicular Communications," in *IEEE Transactions on Vehicular Technology,* Vol. 69, No. 2, Feb. 2020, pp. 1203–1213.

[40] Berlt, P., C. Bornkessel, and M. A. Hein, "Spatial Correlation of C-V2X Antennas Measured in the Virtual Road Simulation and Test Area," *Proc. 12th Eur. Conf. Antennas Propag.*, 2018, pp. 1–5.

[41] Saccardi, F., et al., "Virtual Drive Testing Based on Automotive Antenna Measurements for Evaluation of Vehicle-to-X Communication Performances" *AMTA 2019,* San Diego, CA, October 6–11, 2019.

CHAPTER 7

Advanced Post-Processing Techniques

Lucia Scialacqua, Francesco Saccardi, and Lars J. Foged

Recent years have brought a general modernization of automotive measurements and a widespread acceptance of NF measurement techniques including multiprobe systems. This development extends NF techniques to lower-frequency applications down and into FM bands. Consequently, antenna measurements systems have significantly improved in both measurement accuracy and speed. Embracing NF techniques that include a post-processing step to determine performance parameters such as FF and the use of advanced post-processing techniques as a means to gain further and better information have also achieved acceptance. During recent years, significant improvements have been brought to algorithms for the post-processing of the acquired data. New techniques based on modal or spatial filtering have been proposed to increase the quality of the antenna measurement results. NF/FF transformation algorithms have been improved to reduce acquisition time as discussed in Chapter 3. There has been a rapid evolution in advanced post-processing algorithms for measurement improvements, reducing the effect of echoes, noise, and leakage among other problems, and extrapolating the radiation pattern to angular ranges without acquired data. Finally, significant improvements have been made in the combination of numerical simulation and measurements that allow a more accurate and faster antenna design [1–3].

This chapter describes some of the advanced post-processing features pertinent to applications in the automotive community. The modal and equivalent current (EQC) expansions of the NF, as discussed in Chapter 3, are accurate mathematical expressions of the measured antenna and/or vehicle and as such can be manipulated in many useful ways in a post-processing step. An overview of general post-processing techniques can be found in [1]. The properties of spherical mode expansion can be used to mitigate errors in the measurement due to truncation of the measured field. It can also be used to retrieve the free-space radiation of an antenna and/or vehicle from measurements on a PEC ground plane. The information contained in the basis of the equivalent current expansion can be used for diagnostics and filtering to investigate specific contributions to the measurements. The technique can also be used in the extrapolation of truncated areas in some measurement scenarios and NF/NF transformations at any distance from the measured vehicle. Finally, the expansion base in terms of currents can be used to create very accurate Huygens box representations of the measured antennas suitable as sources in numerical simulations. This is an important feature to rapidly investigate antenna placement.

The use of the expansion sources in modes or currents to emulate different virtual drive scenarios are discussed in Chapter 8.

7.1 Post-Processing by the Equivalent Current Method

The EQC or equivalent source method is introduced in Chapter 3 as a diagnostic, filtering, and NF/FF expansion tool [4]. This technique is implemented in the software INSIGHT [5]. In the following paragraphs, the post-processing features of the equivalent source method in automotive applications is discussed. These features include the diagnostics of antennas for investigating malfunctions and their possible motivations; the extrapolation of truncation areas where it is not possible to measure samples of radiated field; the possibility to calculate the field in every position from the vehicle, even very close where it is not possible to measure using the traditional NF techniques; and finally the link between measurements and simulations.

To illustrate the post-processing features by the equivalent current method we will refer to the NF measurements on a realistic scaled model of an antenna mounted on a vehicle and measured in a NF system introduced in Chapter 4. The DUT is a 12:1 scaled vehicle (Morris Minor 1965) with a wide bandwidth, omnidirectional patch antenna in different positions on the vehicle body; see Section 5.1.3 for further details. The antenna is fed by a cable below the vehicle, and when scaling the measurement frequency, the dimensions of the antenna, vehicle, and multiprobe system become very close to realistic vehicles.

7.1.1 Diagnostics and Filtering

Designing antennas, a complex process consisting of several steps, generally involves the implementation and testing of prototypes before the final design can be reached. A significant insight into the antenna performance can be obtained from advanced simulation tools. However, only through prototyping and testing, including the antenna installation in the final scenario, vehicle, or platform, it is possible to have a faithful representation of the reality.

Often, the prototype measurements need interpretation or post-processing so we can investigate the pertinent features we are trying to enhance or avoid through the antenna design and to identify the components to be optimized at the next stage of the design process. Accurate diagnostics of the prototype measurements are therefore a valuable tool in the design process. Traditional methods based on plane-wave expansion can be adopted from measured radiation patterns to achieve a back propagation/holography on a plane in front of the DUT used, in particular, for diagnostics of planar antennas and/or platforms. However, for more complete diagnostics such as the case of nonplanar antennas or vehicle-mounted antennas, more advanced techniques are required. The equivalent currents method has been demonstrated over the years to be an extremely efficient tool for 3-D, rather than just planar diagnostics [7–9]. The method can be applied to all antennas without limitations of typology or shape.

The complete working procedure and results are shown in Figure 7.1. Colors represent the amplitude of the equivalent currents, and the arrows represent the correspondent instantaneous phase. The inputs for the post-processing by EQC are the measured NF pattern and the reconstruction geometry. Since in this case the diagnostics on both the antenna and the entire vehicle/car are desired, the reconstruction geometry encloses both the vehicle and the antenna. More detailed diagnostics can be obtained by adding more geometric details to the expansion surface.

As expected, the obtained electrical and magnetic currents in Figure 7.1 have higher intensity in the vicinity of the roof-mounted antenna. The coupling with the different parts of the vehicle has less intensity, also as expected. This indicates a strong local coupling between antenna and vehicle and a low level of distant scattering in the vehicle body. The position of this antenna is thus close to optimal for applications requiring omnidirectional coverage. The information that can be retrieved from diagnostics is thus very useful to the designer of the antenna, because it makes it possible to verify that the antenna is working well, and in what way the radiator interacts with the structure where it is installed. Knowing whether there are parts of the vehicle where radiation is stronger or weaker is crucial to understand how to position other antennas and to minimize their mutual coupling.

A more detailed diagnostic can be carried out, when currents are turned off (set to almost zero) on areas of the vehicle to isolate the radiation from the remaining parts. This functionality is called "filtering," and it has been demonstrated as an efficient tool for isolating, investigating, and suppressing undesired radiation on the antenna and on the mounting supports and spurious contributions coming from feeding cables or other parts in the testing scenario [10].

To illustrate the filtering principle, the currents representing the patch antenna and the radiation in the vicinity of the antenna on the roof have been selected and isolated. After selecting the currents to be spatially filtered, it is possible to

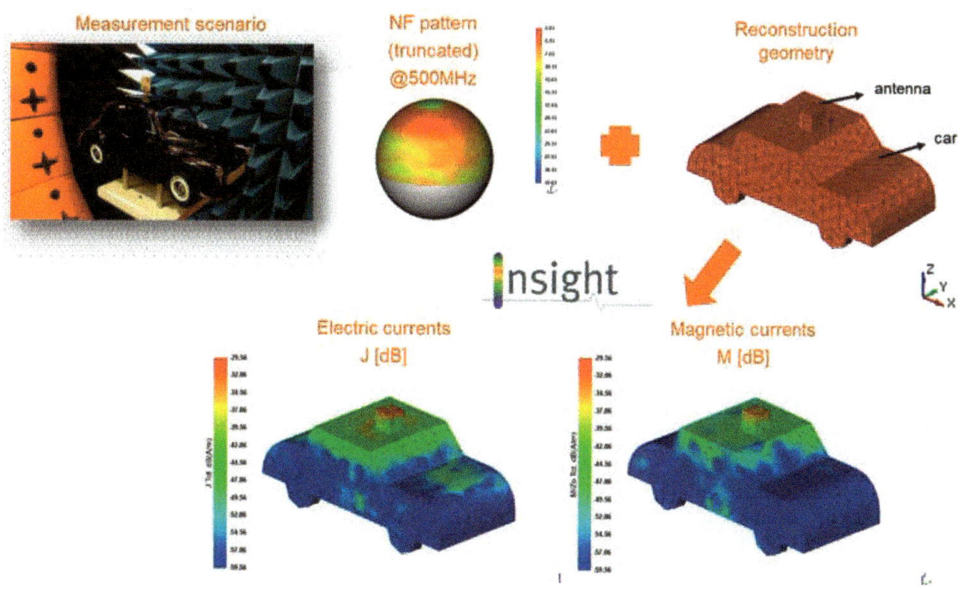

Figure 7.1 Example of diagnostics for an automotive application by the EQC method (top) and EQC on the antenna and the roof of the vehicle (bottom).

recalculate the radiation pattern from this new set of filtered currents, which can then be compared with the pattern before filtering. This comparison is very useful because it makes it possible to get insight into how much the filtered currents contribute to the overall radiation. In this case the filtered currents were the source of disturbance and the radiation pattern improves after filtering. The 1-D plot comparisons in Figure 7.2 show the effects of the spatial filtering on the radiation pattern for the measured example of the scaled vehicle. Patterns are shown before and after filtering for the two principal cut planes phi = 0° and phi = 90°, which correspond to the plane along the longest side of the vehicle and the shortest one, respectively. The original radiation pattern is illustrated by blue and green curves, while the pattern after filtering is shown in red and black. The area in the vehicle where the currents are filtered is denoted in grey.

The phi = 0° pattern, in the long plane of the vehicle, shows a ripple reduction in the filtered case when compared to the original measurement as the distance scattering is removed by the filtering. The ripple results from the reradiation from the vehicle, excluding the upper roof, which is stronger along the longer side of the car. Indeed, the ripple reduction after filtering is less evident for the phi = 90° cut plane that is along the short plane of the vehicle, because the scattering from the vehicle in this plane is limited. The selection of the areas where to apply the filtering of the equivalent currents is arbitrary depending on what the user wants to investigate.

7.1.2 Extrapolation of Truncated Areas

Modern vehicle antennas are often highly integrated. At lower frequencies, such as UHF and VHF, it is often the antenna and the entire vehicle contributing to the radiated field. A complete characterization of the full vehicle is thus typically required. For frequencies down to 70 MHz, a widely accepted and cost-effective measurement solution is a multiprobe SNF system, as described in Chapter 5. In

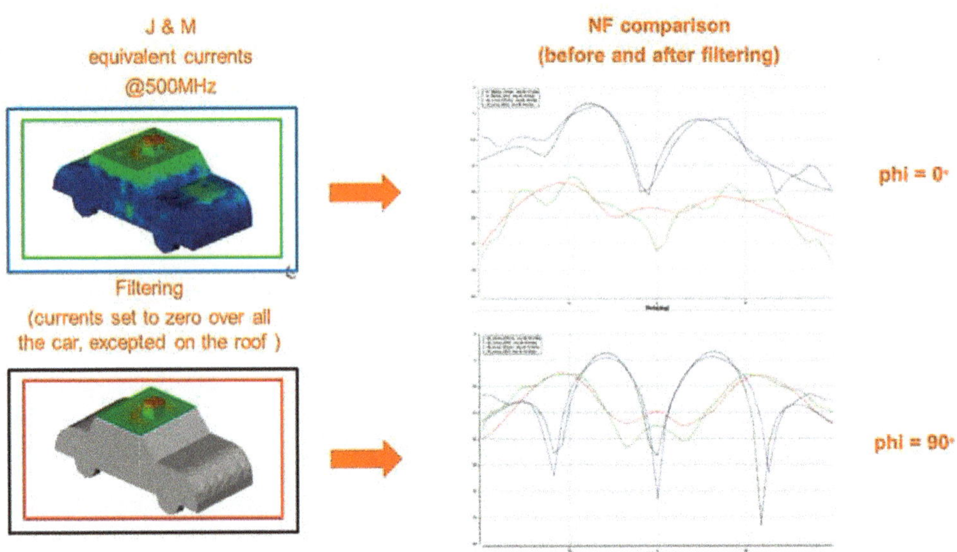

Figure 7.2 Example of spatial filtering of the equivalent currents to isolate only the radiation from the antenna and the roof of the vehicle.

such systems, the scanning area is truncated below the horizon to minimize system dimensions, as shown in Figure 7.3. As the NF acquisition or sampling cannot be performed on the full sphere enclosing the vehicle, due to the presence of the physical support where the vehicle is positioned and the mechanical limits of the measurement system itself, some truncation areas/zones occur. The truncation zone can be wide, typically in the range +/−70°/80°, as shown in the measurement scenario in Figure 7.3.

As described in Section 3.1.2.1, the partial NF acquisition could lead to truncation errors if the standard NF/FF transformation is applied. The EQC method can be used as a tool for reconstructing the field over the whole sphere allowing for mitigation of truncation error in general measurement scenarios [11]. An example procedure for extrapolation of the truncation area by the equivalent current method in the case of an automotive NF system is shown in Figure 7.3. Since the purpose of the EQC expansion in this case is not diagnostics, a detailed geometry of the vehicle and/or antenna is not needed. A simple box, fully enclosing the geometry, has been shown to be accurate and sufficient for truncation mitigation and general NF/FF transformation purposes as discussed in Chapter 3. It can be noted that the measured and thus truncated SNF dataset is directly used as input for the EQC expansion, omitting samples in the truncation region. No field padding is applied, which would result in a discontinuous and unphysical field. As the EQC computation is in fact a minimum energy operator on the truncated region, this allows for a direct computation of the currents, thereby avoiding errors caused by the processing of a discontinuous field. The FF can then be calculated directly from the expansion currents, or the NF can be determined on a full sphere. The NF/FF transformation can then be performed by standard spherical expansion techniques. Other truncation mitigation techniques are also available and are discussed in Chapter 3.

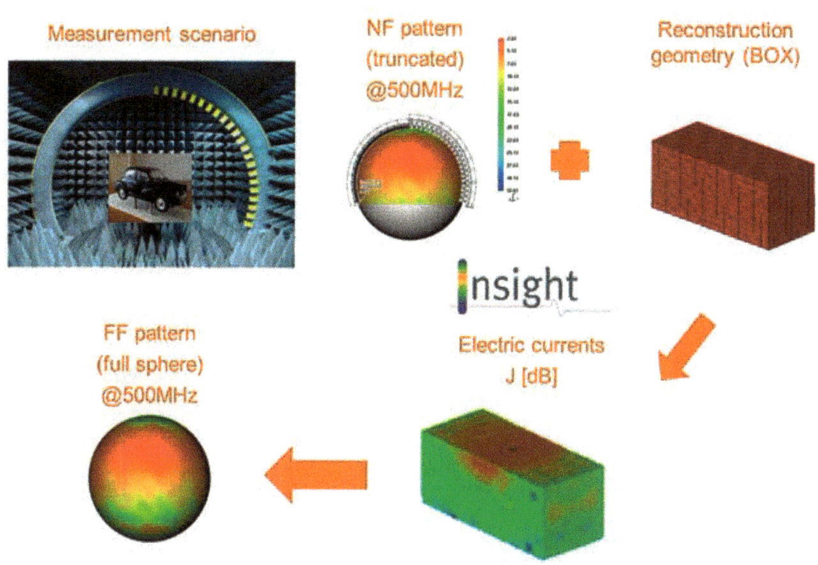

Figure 7.3 Example of extrapolation of truncated areas by equivalent current method for calculation of a full sphere FF radiation pattern.

7.1.3 NF Calculation from the Equivalent Current

During standard SNF acquisition, the radiated fields are measured at a certain distance from the antenna; then with the NF/FF transformation the FF radiation is calculated. In some applications there is a need to know the field values at distances closer to the antenna or different than the distance at which the NF is acquired. If such a need arises, the equivalent currents technique has been shown to be a very effective tool because of its ability to reradiate the currents to any position in space as long as it is outside the reconstruction surface where the currents have been defined. By this feature, it is possible to calculate the radiated field, including power densities at very close distances to the DUT.

A procedure for the calculation of the NF from the equivalent currents is shown in Figure 7.4. Here, a set of regular points enclosed into a sphere has been selected, but an irregular set of points in arbitrary locations can be also chosen. This kind of application is particularly useful for evaluation of the field for EMC testing purposes and for the evaluation of power densities close to the radiating antenna surface (measurement security/norms) [12, 13].

7.1.4 Link Between Measurements and Simulations

In automotive testing, measurements are generally performed in NF systems able to test the whole vehicle. However, such measurement systems are not always available, and sometimes only smaller or limited facilities capable of measuring the antenna in isolation can be used.

In this situation, a possible alternative solution is to measure the antenna only in isolation, using the EQC method to represent the radiator as a NF source or Huygens box and performing the placement of the source on the vehicle in a

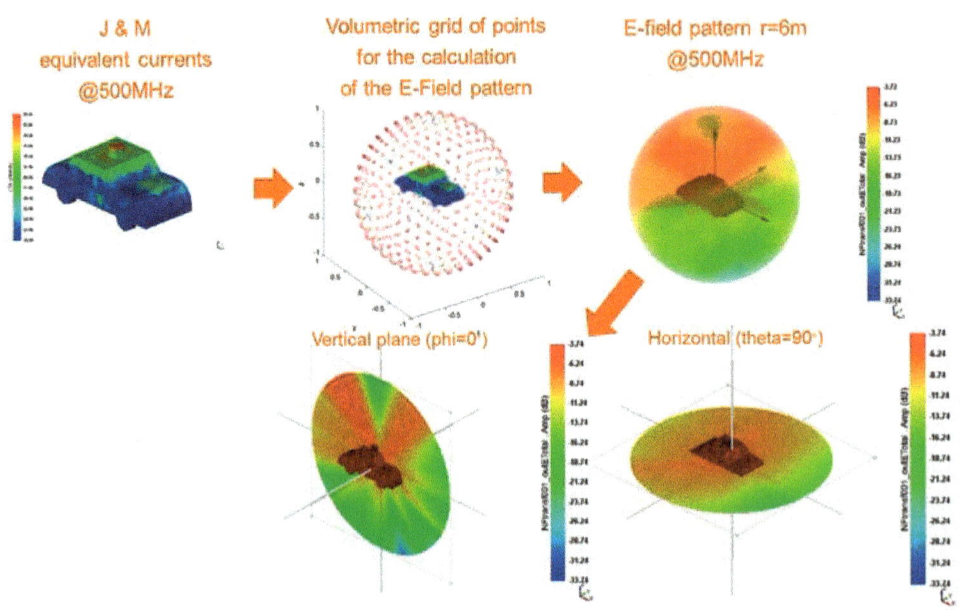

Figure 7.4 Example of NF calculation from the equivalent current method for an automotive scenario.

numerical simulation by a CEM tool. This approach is called the "link between measurements and simulations" [14–16].

The complete procedure implementing the link, shown in Figure 7.5, consists of three steps: (1) the radiating measurement of the AUT, (2) the preparation of the NF source/Huygens box in terms of equivalent currents, and (3) importing the NF source representing the antenna in a simulation tool for integration/placement in the vehicle and calculation of the predicted performance.

The AUT of the example depicted in Figure 7.5 is a shark fin antenna by the company Calearo; see Figure 7.6. The shark fin antenna has been measured in the GSM, LTE, and UMTS frequency bands from 698 to 2,690 MHz. The preparation of the measured Huygens box has been performed starting from the antenna measurement on a circular plate of 40-cm diameter in the MVG StarLab measurement system. A demonstration of the link between measurement and simulation for this antenna integrated on different vehicle models in different simulation tools from the various vendors is reported in Chapter 8.

The different steps of the link as a general approach for automotive processing are described in Sections 7.1.4.1–7.1.4.3.

7.1.4.1 Step 1: Antenna Measurement Equipment

A NF over a spherical, cylindrical, or planar scan surface or a FF measurement, is equally applicable in the preparation of the EQC representation to be used in a CEM simulation. Measurements of the isolated antenna in many cases do not necessarily require large measurement equipment, and compact NF measurement ranges can be used to perform accurate and efficient measurements of the source antenna. Compact and portable measurement systems are in fact quite advantageous in this case, as, like other instrumentation in the EM laboratory, they can be easily used in combination with and in proximity to the platforms where the simulations are carried out. In addition, if the system is a multiprobe system, measurement time can be proportionately reduced.

The measurements reported in this chapter and used for validation of "the link" have been performed in the compact and portable MVG SL measurement system [18] shown in Figure 7.6. The diameter of the SL probe array is 90 cm. In the SNF configuration, the DUT is positioned on a foam mast in the center of

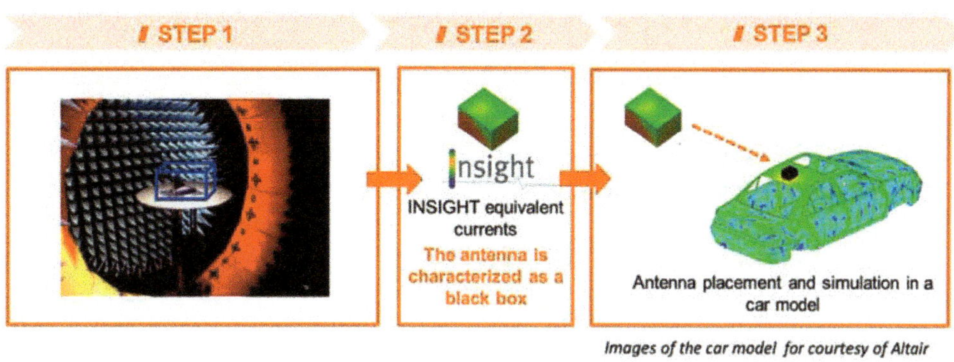

Figure 7.5 Example of link between measurements and simulations for an automotive scenario. Antenna by Calearo and image of the car model courtesy of Altair [22].

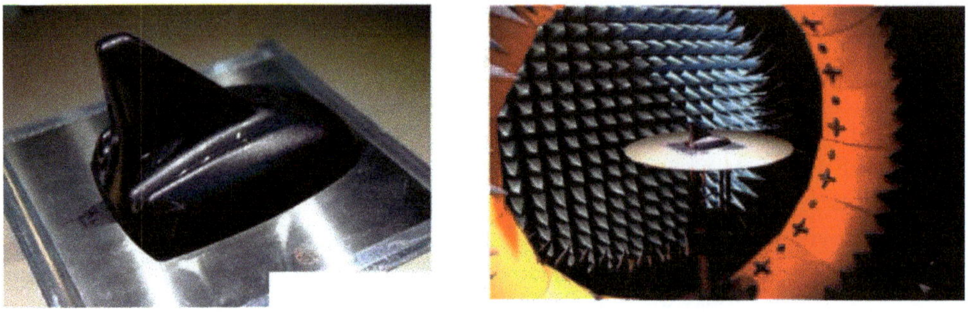

Figure 7.6 AUT: Shark fin antenna by Calearo measured in the MVG SL18-GHz spherical near field multiprobe system covering the frequency range from 690 MHz to 18 GHz [18].

the system made from very low dielectric, low loss material. The full sphere measurement is performed by electronically scanning the probe array in elevation and rotating the DUT 180° in azimuth. Consequently, a full 3-D measurement can be performed very rapidly.

7.1.4.2 Step 2: Equivalent Current Processing NF Source Modeling

An equivalent current representation of the isolated antenna, calculated by the inverse-source or equivalent current/source method (EQC) described in [4] and in Chapter 3 of this book, is proposed as an efficient measured NF model to represent the source antenna in antenna simulations with special attention to antenna placement problems. When equivalent current processing is applied to measured data for numerical modeling, an equivalent black box based on a Huygens' formulation, enclosing the antenna, provides enough accuracy for the procedure.

Since the EQC can produce a form-fitting representation of the antenna surface, the source antenna can be mounted more freely in any position on any large structure or final scenario. This method is recommended as opposed to SWE. In SWE, only the field radiated outside the minimal sphere of the source antenna can be computed [19]. This sphere of minimum radius fully enclosing the source antenna must not be intersected by the structure; therefore the source antenna cannot be installed too close to the structure. As a result, this approach can be applied only to a limited number of practical test cases.

7.1.4.3 Step 3: NF Source Modeling Exchange Data and Numerical Simulation in the CEM tools

The NF source model is provided to the CEM tools as a set of electric (J) and magnetic (M) currents values or electric (E) and magnetic (H) values defined on the vertex of the mesh, discretizing the Huygens' box. The E and H values are calculated starting from the J and M currents determined during the EQC computation. The field internal to the NF source is zero, in agreement with the Love's condition applied during the NF source calculation (EQC) [4].

The different numerical methods for simulation consider the source antenna (Huygens box) as a full representation of the antenna in any scenario required for automotive investigations. Applying the black box approach, the simulation is carried out with no more additional information. No extra information about the antenna is needed, and the NF source is a generic EQC representation that can be used in any scenario under test even the most complex. This makes it possible to overcome limitations in terms of intellectual property rights or confidentiality on antennas to enhance exchanging data. Since the radiating measurements can make it possible to determine the realized antenna gain and realized efficiency, NF sources could be also provided including the realized gain normalization so that the information of gain could be available and managed by the simulation tools. The equivalent current representation model can be imported by a number of commercial CEM [20–25].

7.2 Investigation and Mitigation of Truncation Errors in Free-Space Automotive Systems

After first discussing the applications of the method of the equivalent currents, this section provides some experimental results relative to an investigation regarding the truncation errors in free-space automotive measurements and their mitigation.

7.2.1 Case Study Description and Investigation of Truncation Errors Effect

To investigate the effect of the truncation errors in typical free-space, (absorber-based) automotive ranges similar to the one shown in Figure 5.4 have been considered.

As shown in Figure 7.1, the measurement system is the StarLab multiprobe system, and the considered car for the measurement is the 12:1 scaled model of the Morris Minor 1000. The dimensions of the scaled car are approximately: L × W × H = 31.3 × 12.9 × 12.7 cm (original dimensions: L × W × H = 3.76 × 1.55 × 1.52m). A wideband patch antenna with an omnidirectional pattern installed on the rear hood is considered as the AUT. The scaled vehicle has been measured in the 1–18-GHz band, approximately corresponding to the 84–1,500-MHz scaled frequency band.

Reference FF data have been obtained applying the standard NF/FF transformation to the measured spherical NF neglecting the errors caused by the relatively small truncation area of the StarLab systems (just ±22.5° in elevation).

The SWC obtained from the full StarLab measurement have been scaled in frequency with the previously mentioned 12-times scaling factor, and a typical free-space truncated acquisition has been emulated. Like in the automotive multiprobe system shown in Figure 7.3, the measurement distance is set to 4m. Low-frequency measurements (~90–400MHz) have been emulated by taking into account a ±72.5° scanning area truncation and an angular sampling step of $\Delta\theta = \Delta\phi = 5°$. High-frequency (HF) measurements (~400–1,500 MHz) have been emulated with a ±70.0° scanning area truncation and a sampling step of $\Delta\theta = \Delta\phi = 1°$.

The emulated truncated measurements have been processed with the standard NF/FF transformation performing a zero-padding in the truncated area. To evaluate the loss of accuracy due to the truncation, the FF obtained from the truncated measurement has been compared to the reference FF. To do that, the ENL, defined in (5.7), has been used. The ENL has been evaluated only on the measured portion of the sphere.

The ENL at each (scaled) frequency is shown in Figure 7.7. The truncation errors are higher at low frequencies (up to 400 MHz) where, as explained in Section 3.1.2.1, the number of SWCs used to represent the truncated scenario is smaller. As expected, the impact of the truncation gets lower at higher frequencies. In fact, the higher sampling density makes it possible to compute more SWCs and thus to better represent the introduced discontinuity. In particular, beyond 600 MHz the ENL is in this case below −35 dB, indicating that the truncation error can even be considered negligible at such frequencies.

The different effect of the truncation of the scanning area can be appreciated in Figures 7.8 and 7.9 where the elevation and azimuth gain pattern cuts, with and without the truncation, are shown at 91.7 MHz and 1,330 MHz, respectively. The shadowed areas in the plots with elevation cuts indicate the truncated regions. Again, the impact of the truncation is more evident at the lower frequency where the deviations with respect to the reference are observed all over the pattern. On

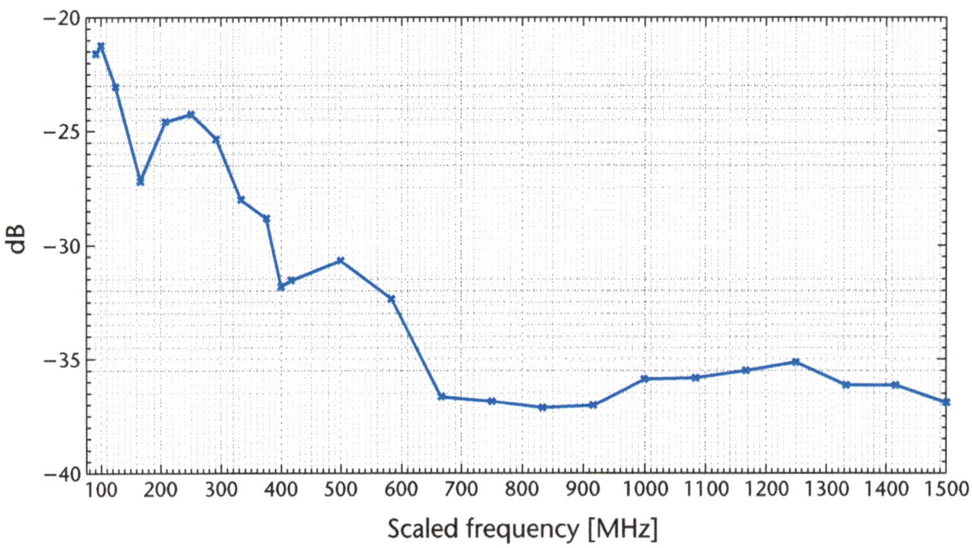

Figure 7.7 ENL with frequency in typical automotive measurements using zero-padding.

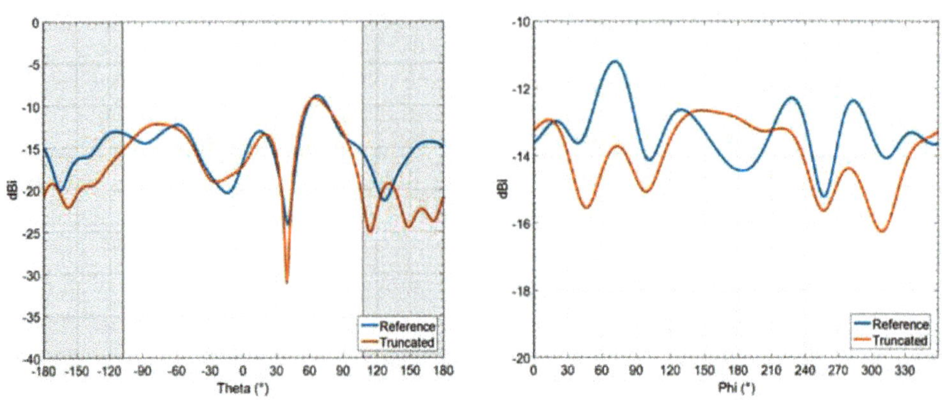

Figure 7.8 Gain pattern from the reference and truncated measurements at 91.7 MHz (scaled frequency): (a) Vertical cut at $\phi = 0°$ (left), and (b) horizontal cut at $\theta = 90°$.

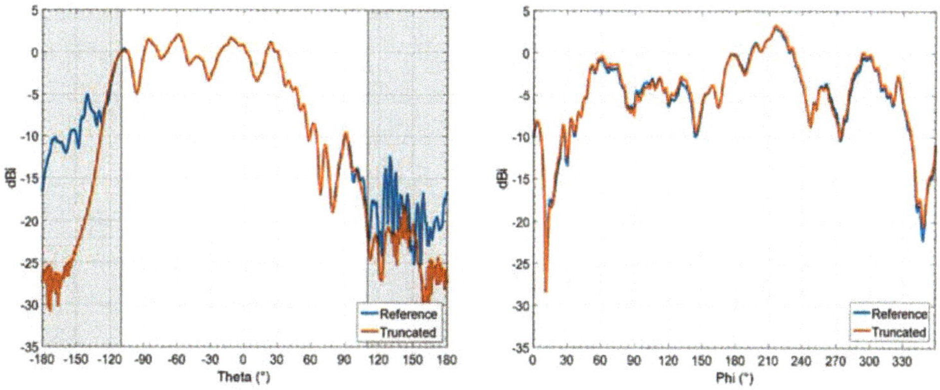

Figure 7.9 Gain pattern from the reference and truncated measurements at 1,330 MHz (scaled frequency): (a) Vertical cut at $\phi = 0°$ (left), and (b) horizontal cut at $\theta = 90°$.

the other hand, at the higher frequency a good agreement with the reference pattern cuts is obtained in the untruncated part of the pattern.

7.2.2 Mitigation of Truncation Errors

To reduce the truncation errors at lower frequencies, two post-processing techniques already introduced in this book have been applied to these scaled measurements at low frequencies (91.7–400.0 MHz), where the ENL is approximately bounded between –20 and –30 dB. The two processing techniques are the iterative modal filtering (IMF), introduced in Section 3.1.2.1 and the EQC technique, whose use as field extrapolator in the truncated areas is described in Section 7.1.2.

The IMF technique has been applied performing the spherical modal filtering at mode indexes corresponding to radius of the DUT minimum sphere (e.g., R_{min} = 20 cm corresponding to 2.4m of the equivalent full-size vehicle) and performing 603 iterations (ENL NF threshold level set to –60 dB).

The EQC technique has instead been applied considering an equivalent box for the current reconstruction slightly larger than the scaled vehicle. The equivalent box has been discretized with a mesh made of triangles with side length $\lambda/4$. It should be noted that the number of unknowns increases with the frequency leading to different computational efforts at the considered frequencies (see Table 7.1.)

A comparison in terms of ENL between the FFs obtained with the zero-padding (blue trace), IMF (orange trace), and EQC (Insight) techniques (green trace) is shown in Figure 7.10. As before, the ENL has been computed only at angles corresponding to the untruncated region. Both post-processing techniques improve the radiation pattern obtained with the simple zero-padding of the truncated area. It is clear that the achieved performance of the EQCs is better with respect to the IMF at almost each scaled frequency.

A gain radiation pattern comparison at 91.7 MHz is reported in Figure 7.11 for the vertical cut at $\phi = 0°$ (left) and for the horizontal cut at $\theta = 90°$ (right). The reported traces are the total field cuts of the reference pattern (black trace), the pattern obtained with zero-padding (blue trace), the IMF technique (orange trace), and the EQC technique (green trace). The shadowed areas drawn in the vertical

Table 7.1 Computational Time of the EQC and IMF Technique+

Frequency (MHz)	*EQC*	*IMF*
91.7	~4s	~90s
100.0	~5s	
125.0	~9s	
166.7	~19s	
208.3	~37s	
250.0	~63s	
291.7	~81s	
333.3	~112s	
375.0	~172s	
400.0	~188s	
Total	~11.5 min	~1.5 min

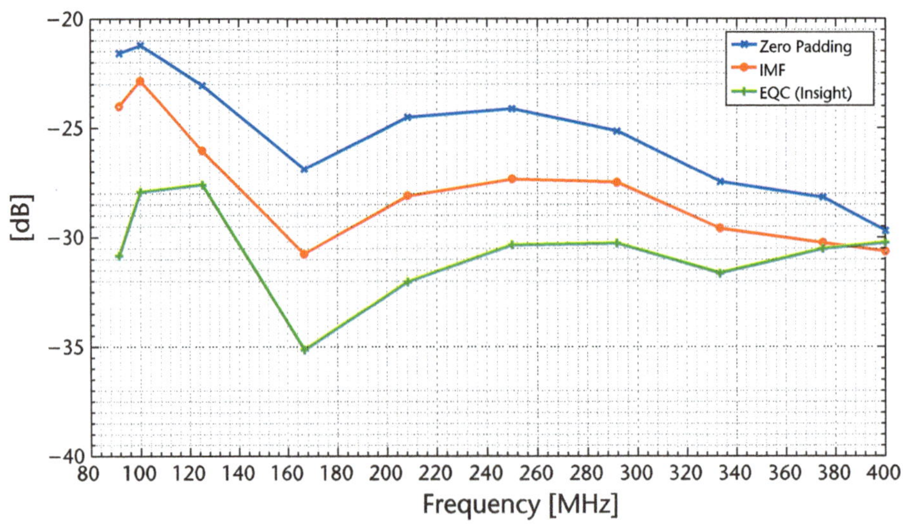

Figure 7.10 ENL with frequency obtained with the different processing techniques.

cut illustrate the truncated regions. Again, it should be noted that the truncation errors are not just present in the unreliable region, but all over the pattern. The improvements obtained with the EQC technique are clearly observable in the reliable region of the vertical cut and in the horizontal cut. Instead, as previously observed, the performance of the IMF technique is worse at this test frequency for both cuts.

Gain pattern comparisons like the previous ones are reported in Figure 7.12 for the test frequency of 250 MHz. Even in this case, the truncation effect is visible all over the zero-padded pattern cuts. The improvements obtained with the two techniques are remarkable and very similar in the $-70° < \theta < +80°$ angular range of the vertical cut. In the same cut, it is observed that the EQC technique can give better performance at θ-values close to the truncation. The high performance of

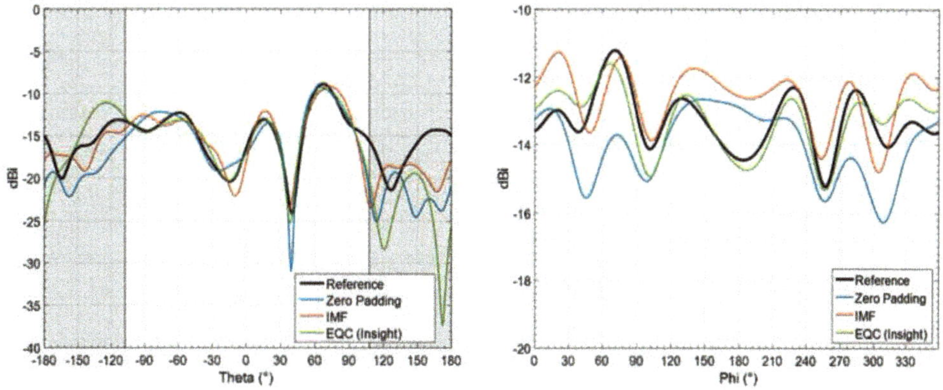

Figure 7.11 Gain pattern comparison among the different processing at 91.7 MHz (scaled frequency) (a) vertical cut at $\phi = 0°$, and (b) horizontal cut at $\theta = 90°$. ©2017 IEEE. (Reprinted with permission.) [11].

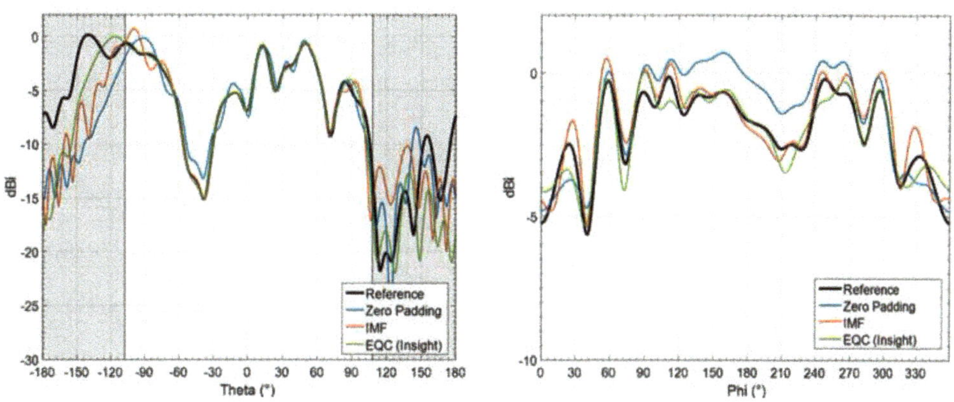

Figure 7.12 Gain pattern comparison among the different processing at 250 MHz (scaled frequency): (a) vertical cut at $\phi = 0°$, and (b) horizontal cut at $\theta = 90°$. ©2017 IEEE. (Reprinted with permission.) [11].

the two techniques is also noticed in the horizontal cut, where they both give good and similar results.

The computational times of the EQC and IMF technique, both obtained with the same standard laptop computer, are shown in Table 7.1. The complete IMF multifrequency processing (10 frequencies in total) runs in approximately 90 seconds. As can be seen, the EQC running time increases with frequency leading a much more demanding total process with respect to the IMF technique.

Summarizing, the application of both advanced post-processing techniques to the scaled automotive measurement has shown significant improvements with respect to the conventional zero-padding approach. The computational cost of the EQC increases with the frequency and is, in general higher, with respect to the IMF. The EQC has shown better performance with respect to the IMF at lower frequencies, where its computational cost is not very high. The achieved performance of the two techniques is instead very similar at higher frequencies, where is thus convenient to use the IMF due to its faster speed.

7.3 Free-Space Response Retrieval in PEC-Based Automotive Systems

PEC-based automotive systems were introduced in Chapters 3–5. The use of a conductive floor allows for two main advantages, namely, the ease of vehicle installation (the vehicle is in fact simply parked in the center of the system for testing) and the mitigation of truncation errors [26, 27]. In PEC-based systems the NF/FF transformation is in fact usually performed by mirroring the data acquired on the upper hemisphere to the backward hemisphere obtaining a quasi-full spherical "virtual" acquisition (application of the PEC boundary condition as described in Section 3.1.2.2).

On the other hand, the main disadvantage of PEC-based systems is that a conductive ground is only in a few cases an accurate representation of a real driving ground. As discussed in Chapter 8, emulations of realistic driving grounds, such as different types of asphalts and soils, can be easily performed if the free-space response of the vehicle is available [28, 29]. For this reason, it is sometimes necessary to retrieve the free-space (or quasi-free-space) response from PEC-based measurement (or in other words to remove/attenuate the effect of the conducting ground) [30]. Moreover, the end user may simply require that the vehicle be measured under free-space conditions because of, for example, comparison purposes or so that it is compliant with a specific test procedure.

Because of the strong coupling between the conductive ground and the vehicle, the free-space response retrieval is usually a challenging task. However, different spatial filtering techniques combined with a proper handling of the car can be used for this purpose.

7.3.1 Spatial Filtering Techniques for PEC Removal

Spatial filtering techniques to remove the metallic ground include the SWE-based modal filtering and the EQC technique.

SWE-based modal filtering [31] takes advantage of the spatial filtering properties of the SWE [19] of the measured field. The scattered field components, being highly oscillating, are attributed to higher-order modes that can be eliminated by filtering the modal spectrum. As described in Chapter 3, the minimum number of spherical modes required to fully represent the antenna depends on the physical size of the minimum sphere or the cylinder fully enclosing the antenna. This type of filtering is implemented in the commercially available MV-Echo tool [31, 32], which makes it possible to minimize the AUT minimum sphere/cylinder with proper translation/rotation operations, resulting in a more effective low-pass spatial filtering.

The specific workflow required to retrieve the free-space response from measurements over the PEC is schematized in Figure 7.13. The input scenario is the FF of the vehicle above an infinite PEC ground. Due to the image theory [33, 34], the input scenario is equivalent to the original vehicle plus its image (mirrored replica of the car with respect to the horizontal plane). The image can be seen as an echo source. By moving the reference system to the center of the car, the image is translated and therefore represented by higher-order modes of the modal spectrum, as shown in Figure 7.14. These modes can be removed with low-pass filtering, applied in the form of a cylinder or a sphere. It should be noted that due to the coupling

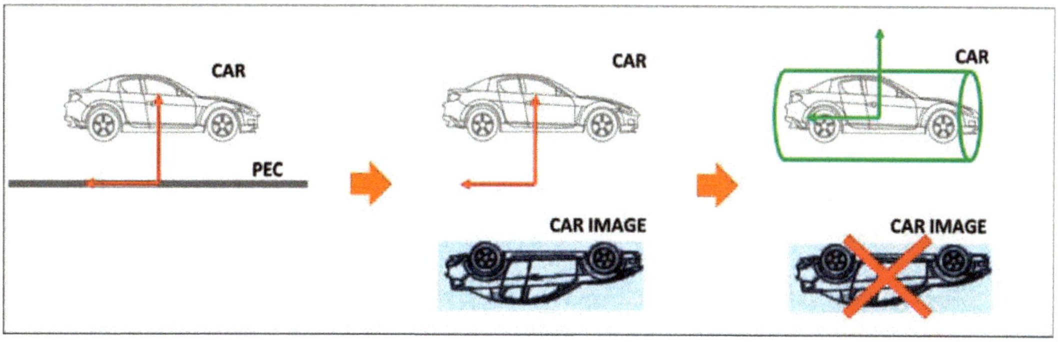

Figure 7.13 Schematization of the free-space response retrieval from PEC-based measurement using the SWE-based spatial filtering. ©2019 IEEE. (Reprinted with permission.) [37].

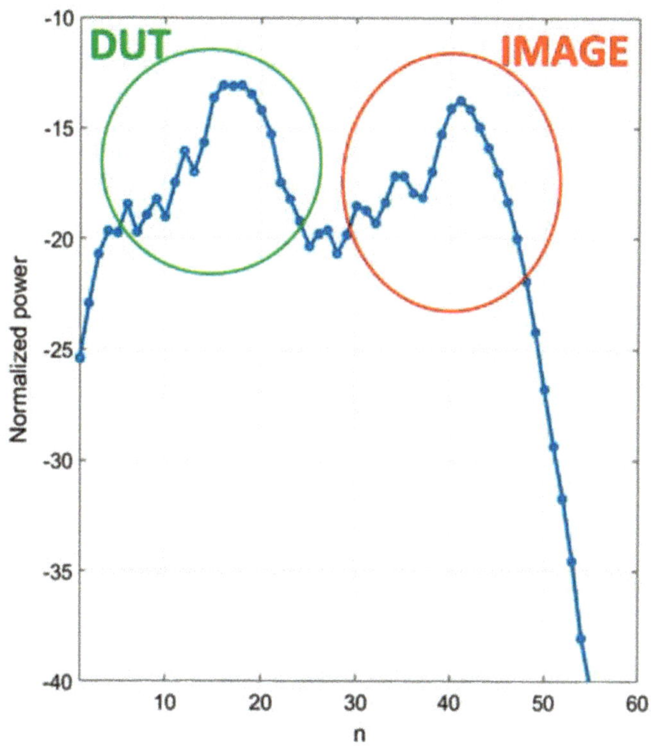

Figure 7.14 Example of the spherical wave spectrum of PEC-based measurement with a raised car after the translation of the reference system to the geometrical center of the car. The image of the car due to the PEC is identified, and it can be filtered out.

with the PEC, the image cannot be fully removed, resulting in residual error. Such error can be reduced by increasing the distance between the PEC floor and the vehicle. This procedure is thus expected to work better at higher frequencies because of the larger electrical separations that can be achieved.

The second spatial filtering technique is the EQC technique, implemented in the commercially available software INSIGHT [5] and widely described in this chapter. In this case the workflow to remove the effect of the conductive floor is very similar to the one shown in Figure 7.13. First, the measured FF over PEC is represented as the superposition of the vehicle and its image. The EQCs are then reconstructed on an equivalent box including the vehicle and its image as shown on the left side of Figure 7.15. The currents associated with the image are then "switched-off" as shown in Figure 7.15 (right), and the FF is recomputed from the filtered currents.

The advantage of the use of the EQC technique is that a finer reconstruction geometry can be defined, allowing for a more effective spatial filtering when compared to the SWE-based modal filtering technique (where the spatial filter can be defined only with spheres and cylinders). This could allow for a better rejection of the image, even when the vehicle is located directly on the floor.

7.3.2 Example with Scaled Measurements

A scaled automotive measurement setup has been implemented to compare the effectiveness of the two spatial-filtering techniques in the removal of the effect of the conductive floor. The advantage of the scaled-model method in this situation is the possibility to have access to full-spherical free-space data that can be used as a baseline/reference.

The same 12:1 scaled-car model (Morris Minor 1000 of 1965) fed by a patch antenna is considered also in this case. Such a scaled-car has been measured in the StarLab-18GHz (SL18GHz) multiprobe system in different configurations as shown in Figure 7.16. Three similar wideband patch antennas have been installed in three different positions: close to the windshield, on the rear part of the roof, and on the hood of the scaled vehicle, respectively. In each measurement, only one patch is fed while the other two are terminated on a matched load. Measurements have been performed in the frequency band 1.008–8 GHz. With the considered scaling factor, the performed measurements are equivalent to the ones of a full-size vehicle (with the real dimensions being L × W × H = 3.76 × 1.55 × 1.52 m) measured in the 84–1,500-MHz band in a system with a 5.4-m radius. (The real radius of the SL18GHz is 45 cm.)

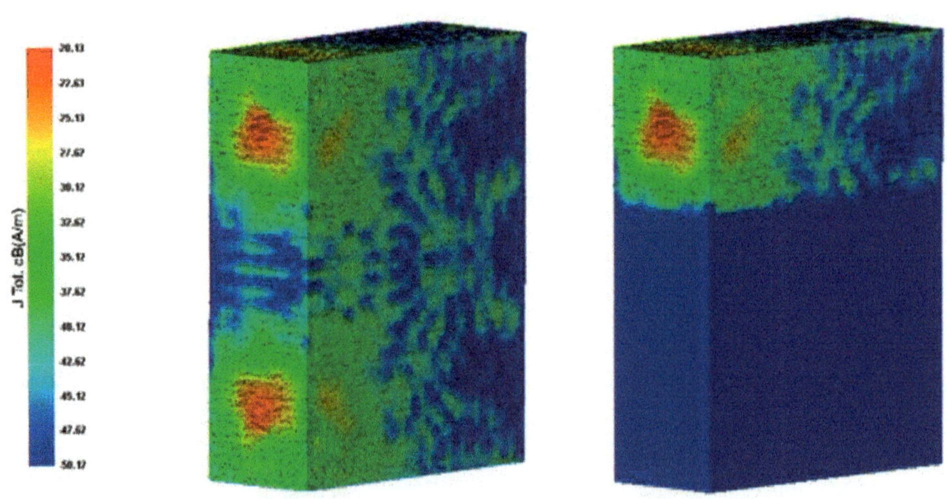

Figure 7.15 EQC of the vehicle and its image (left); and filtering of the vehicle image (right). ©2019 IEEE. (Reprinted with permission.) [37].

In this analysis, only the measurements relevant to the patch installed in the hood position have been considered. It should be noted that the chosen antenna location represents a worst-case scenario, since it is closer and thus more coupled with the conductive floor.

Beforehand, a free-space, (quasi) full-spherical acquisition has been performed, as shown on the left side of Figure 7.16, and considered as a reference.

To emulate a PEC-based automotive system scenario, a conductive floor has been introduced inside the system. As depicted in Figure 7.16 (center and right), such a floor is composed of a 75-cm-diameter metallic turntable that rotates with the vehicle, and by a fixed metallic part that extends outside the system. Conductive contacts have been included in the junction between the two metallic parts to ensure the electrical continuity. In the setup, the floor is placed 11 cm below the center of the scanner (corresponding to a position of 1.32 m in real dimensions) to emulate real automotive systems where the top of the car is located close to the center of the spherical scan. Such displacement allows for measurements to approximately 10° below the horizon (approximately 100° of elevation scanning). The scaled vehicle has been measured in two different configurations: 7.5 cm (corresponding to 0.9m) raised from the floor ("raised" configuration), and at the ground level ("floor" configuration), as shown on the center and right side of Figure 7.16, respectively. In the raised configuration the electrical separation between the vehicle and the ground is increased, and thus the coupling is reduced, allowing for a better rejection of the conductive floor with both techniques. Both filtering techniques have been analyzed in both configurations, "raised" and "floor."

The scaled free-space spherical acquisitions have been processed with the conventional NF/FF transformations while the PEC-based measurements have been processed with the TSWE NF/FF transformation (detailed in Section 3.1.3.5) to properly account for the position of the metallic ground, which was not at Z = 0. The obtained FF has been normalized in gain, taking advantage of the system gain calibration performed with the substitution method [35], as described in detail in [36].

The gain-normalized FF pattern coming from the PEC-based measurements has been used as input to MV-Echo and INSIGHT to remove the effect of the conductive floor. MV-Echo has been applied considering the smallest cylinder, including the vehicle, as a spatial filter, as shown in Figure 7.13. Insight has instead been applied considering a box enclosing both the vehicle and its image as reconstruction

Figure 7.16 Different scenarios considered in the scaled experiment: free-space scenario (left) and PEC-based measurement scenarios with the car raised from the floor (center) and with the car on the floor (right). ©2019 IEEE. (Reprinted with permission.) [37].

geometry. An example of the reconstructed EQC at 500 MHz in the case of the raised configuration is shown in Figure 7.15 (left). As shown in Figure 7.15 (right), the currents associated to the image have then "switched off."

Considering the vehicle raised from the metallic floor, comparisons of the gain pattern cuts at 84, 240, and 500 MHz (scaled frequencies) are reported in Figure 7.17 from top to bottom, respectively. The plots in the first column show the pattern cuts in the $\phi = 0°$ elevation cut (the one along the car's longest dimension), while the plots in the second column show the comparison along the $\theta = 85°$ azimuth cut (the one being 5° above the horizon). The black-dashed traces are the reference free-space pattern while the orange-dashed traces are the "raw" pattern measured over the PEC. (Values in the lower hemisphere are just a mirrored replica of the upper hemisphere resulting from the application of the PEC boundary condition.) As expected, the comparison of these two patterns highlight large differences between the two different measurement scenarios. The blue traces are the retrieved free-space (FS) patterns obtained with MV-Echo. The agreement with the reference is good at 240 and 500 MHz approximately up to $|\theta| = 90°$ (the forward hemisphere). At 84 MHz, the electric separation between the vehicle and the metallic floor is too small (approximately 0.25 wavelengths), and it does not allow for

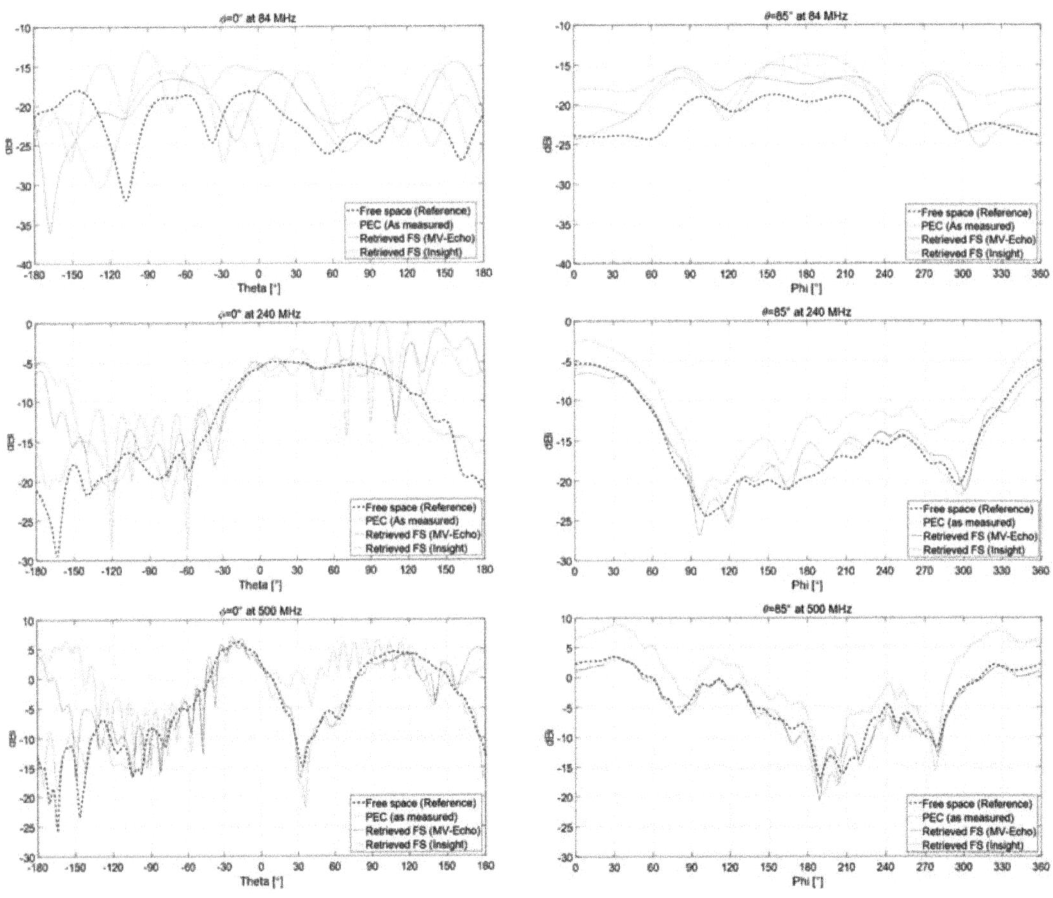

Figure 7.17 Elevations (first column) and azimuth (second column) gain pattern comparison with the vehicle raised from the floor. ©2019 IEEE. (Reprinted with permission.) [37].

a good rejection of the vehicle image. The green traces are the retrieved FS pattern obtained with Insight. Even in this case, the rejection of the image at 84 MHz is poor, because of the reduced separation distance. At 240 and 500 MHz, the agreement with the reference pattern on the upper hemisphere is good and similar to the one obtained with MV-Echo. However, on the lower hemisphere, much better results are obtained with Insight.

A similar comparison is repeated in Figure 7.18, considering this time the measurements of the scaled vehicle placed at floor level. Also, in this case, the deviations between the free-space and the "raw" PEC-based measurements are very large. In this case, MV-Echo (blue traces) in this case is not capable of effectively removing the impact of the metallic floor, and only in reduced angular regions does the retrieved free-space pattern agree well with the reference. The car and its image are now electrically very close, and their contributions are not well separated in the spherical mode domain; hence poor performance is obtained with the modal filtering. In comparison, the performance obtained with Insight is much better at all the analyzed frequencies. Surprisingly, at 84 MHz the performance of the EQC technique is even better than in the previous case when the car was raised from the floor. As previously noted, the better results obtained with Insight are due to the

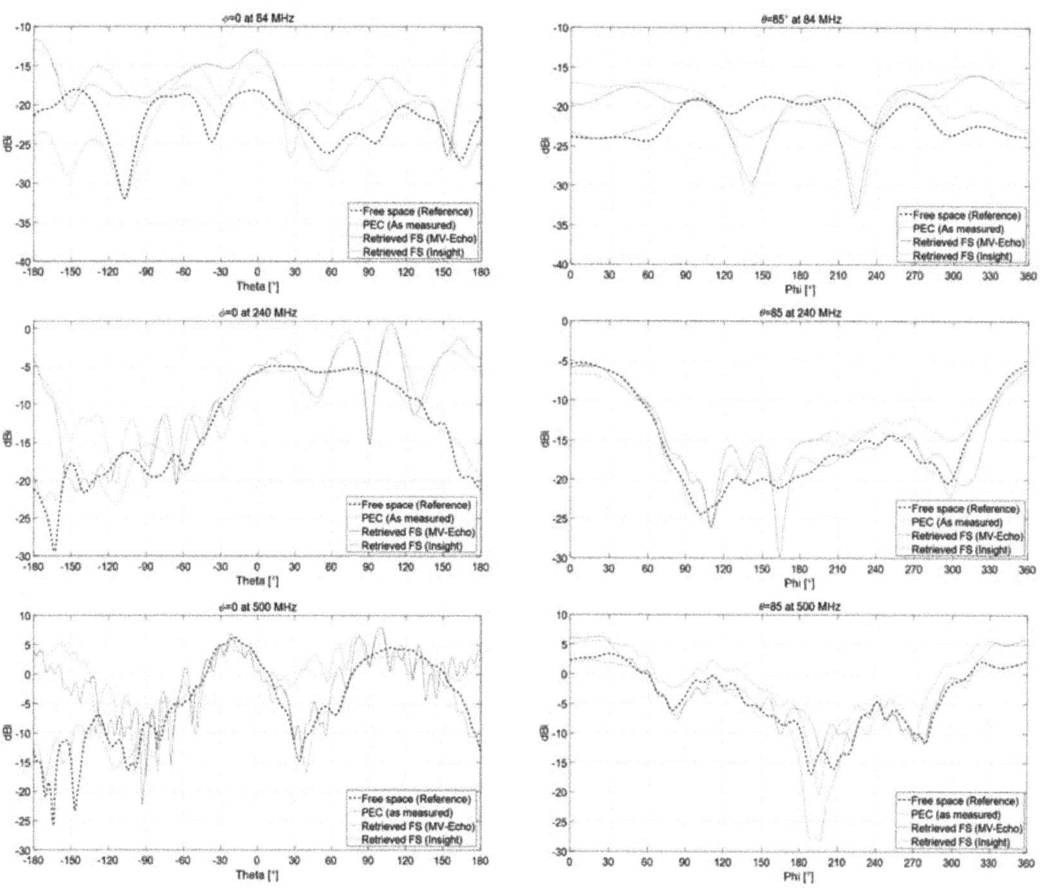

Figure 7.18 Elevations (first column) and azimuth (second column) gain pattern comparison with the vehicle on the floor. ©2019 IEEE. (Reprinted with permission.) [37].

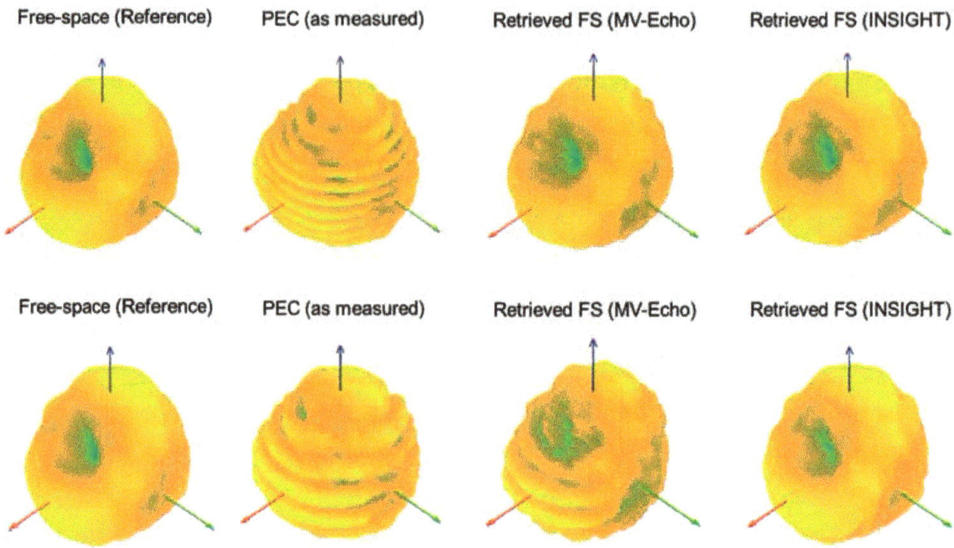

Figure 7.19 3-D pattern comparison at 500 MHz, with the vehicle raised from the floor (top) and the vehicle on the floor (bottom).

fact that a reconstruction geometry closer to the vehicle can be defined, allowing for more effective filtering and thus better rejection of the image.

To summarize, and globally appreciate the effectiveness of the two spatial filtering techniques, Figure 7.19 compares the 3-D patterns at 500 MHz (scaled frequency) in the "raised" and "floor" configuration.

In general, it is observed that the EQC technique allows for better rejection of the conductive floor, either with the vehicle in the raised or floor configuration. The price to be paid for such better performance is a higher computational time. The EQC technique implemented in Insight is based on an inverse MoM [7], whose computational cost increases according to the dimension of the problem with respect to the wavelength.

Using a standard laptop, the Insight processing time for the considered cases was around half minute at 84 MHz, and approximately seven minutes at 500 MHz. Conversely, the SWE implemented in MV-Echo takes advantage of the FFT, strongly reducing the computational cost and processing time.

References

[1] Sierra-Castaner, M., Foged L. J. (eds.), *Post-Processing Techniques in Antenna Measurement,* London: SciTech Publishing, The Institution of Enigneering and Technology, 2019.

[2] Huang, Y., *Antennas: From Theory to Practice*, 2nd Edition, Hoboken, NJ: Wiley, 2021.

[3] Hong Loh, T. (ed.), *Metrology for 5G and Emerging Wireless Technologies,* London: The Institution of Engineering and Technology, 2021.

[4] Quijano, J. L. A., and G. Vecchi, "Near- and Very Near-Field Accuracy in 3-D Source Reconstruction," *IEEE Antennas and Wireless Propagation Letters,* Vol. 9, 2010, pp. 634–637.

[5] Insight software website: https://www.mvg-world.com/en/products/field_product_family/antenna-measurement-2/insight.

[6] Scialacqua, L., et al., "Practical Application of the Equivalent Source Method as an Antenna Diagnostics Tool," AMTA Symposium, Englewood, Colorado, October 2011.

[7] Quijano, J. L. A., "3D Spatial Filtering Applications in Spherical Near Field Antenna Measurements," *AMTA 2010 Symposium,* Atlanta, Georgia, October 2010.

[8] Salghetti Drioli, L., et al., "Analysis of Spacecraft Antenna Farm Interaction with Equivalent Current Technique," *9th European Conference on Antennas and Propagation (EuCAP),* Lisbon, Portugal, April 12–17, 2015.

[9] Scialacqua, L., et al., "Diagnostics on Electrically Large Structures by a Nested Skeletonization Scheme Enhancement of the Equivalent Current Technique," *The 14th European Conference on Antennas and Propagation* (EuCAP 2020 – vitual format), 2020.

[10] Araque Quijano, J. L., et al., "Suppression of Undesired Radiated Fields Based on Equivalent Currents Reconstruction from Measured Data," *IEEE Antenna and Wireless Propagation Letters,* Vol. 10, 2011, pp. 314–317.

[11] Saccardi, F., et al., "Truncation Error Mitigation in Free-Space Automotive Partial Spherical Near Field Measurements," *39th Annual Symposium of the Antenna Measurement Techniques Association,* AMTA, Atlanta, Georgia, October 15–20, 2017.

[12] Foged, L. J. , L. Scialacqua, and G. Bit-Babik, "Experimental Near-Field Method for Validating Simulation Antenna Models," *IEEE Antennas and Propagation Society International Symposium,* Vancouver, BC, Canada, July 19–24, 2015.

[13] Mioc, F., et al., "Power Density Evaluation of Simulated and Measured Data Based on Equivalent Currents Method," *The 15th European Conference on Antennas and Propagation* (EuCAP 2021 – vitual format), 2021.

[14] Foged, L. J., et al., "Bringing Numerical Simulation and Antenna Measurements Together," *IEEE Antennas and Propagation Society International Symposium,* July 6–11, 2014.

[15] Foged, L. J., et al., "Antenna Placement Based on Accurate Measured Source Representation and Numerical Tools," *IEEE Antennas and Propagation Society International Symposium,* July 19–24, 2015.

[16] White paper, https://www.mvg-world.com/en/news/newsroom/ white-paper-finding- missing-link.

[17] Calearo's website: https://www.calearo.com/it/.

[18] Foged, L. J., and A. Scannavini, "Efficient Testing of Wireless Devices from 800MHz to 18GHz," *Radio Engineering Magazine,* Vol 18, No 4, December 2009.

[19] Hansen, J. E. (ed.), *Spherical Near-Field Antenna Measurements*, Peter Peregrinus Ltd., on behalf of IEE, London, 1988.

[20] www.3ds.com/products-services/simulia/products/cst-studio-suite/ (CST Studio Suite).

[21] www.ansys.com (Ansys HFSS and Savant).

[22] https://www.altair.com (Altair FEKO).

[23] www.idscorporation.com/space (ADF, IDS).

[24] www.wipl-d.com, (WIPL-D d.o.o.).

[25] https://www.mvg-world.com/en/products/antenna-measurement/software/satsim.

[26] Mauermayer, R. A. M. , and T. F. Eibert, "Spherical Field Transformation for Hemispherical Antenna Measurements above Perfectly Conducting Ground Planes," *AMTA,* October 2016, Austin, Texas.

[27] Mauermayer, R. A. M., and T. F. Eibert "Spherical Field Transformation Above Perfectly Electrically Conducting Ground Planes," *IEEE Transactions on Antennas and Propagation,* Vol. 66, No. 3, March 2018.

[28] Saccardi, F., et al., "Estimation of the Realistic Ground Effect in Free-Space Automotive Measurements," *40th AMTA Symposium,* November 2018, Williamsburg, Virginia.

[29] Saccardi, F., et al., "Estimation of Real Ground Effect in Absorber-Based Automotive Measurements: Experimental Validation," *APS 2020,* July 5–10, 2020, Montréal, Canada.

[30] Saccardi, F., et al., "Comparative Investigation of Spatial Filtering Techniques for Ground Plane Removal in PEC-Based Automotive Measurements," *AMTA 2019,* October 6–11, 2019, San Diego, California.

[31] Foged, L. J., et al., Echo Suppression by Spatial Filtering Techniques in Advanced Planar and Spherical NF Antenna Measurements," *AMTA Symposium,* October 2012, Seattle, Washington.

[32] https://www.mvg-world.com/en/products/field_product_family/antenna-measurement-2/mv-echo.

[33] Foged, L. J., et al., "Infinite Ground Plane Antenna Characterization from Limited Ground-plane Measurements,“ *APS 2010,* July 11–17, 2010, Toronto, ON, Canada.

[34] Balanis, C. A., *Advanced Engineering Electromagnetics*, Second Edition, Hoboken, NJ: John Wiley & Sons, 2012.

[35] ANSI/IEEE Std 149-1979, IEEE Standard Test Procedures for Antennas.

[36] Saccardi, F., et al., "Accurate Calibration of Truncated Spherical Near Field Systems with Different Ground Floors using the Substitution Technique," to be published at AMTA 2019.

[37] Saccardi, F., et al., "Comparative Investigation of Spatial Filtering Techniques for Ground Plane Removal in PEC-Based Automotive Measurements" *AMTA 2019,* October 6–11, 2019, San Diego, CA.

CHAPTER 8

Virtual Drive Testing

Lars J. Foged, Alessandro Scannavini, Lucia Scialacqua, and Francesco Saccardi

Vehicle performance requirements are increasingly stringent, while time available for automotive antenna and antenna system development is diminishing. The complexity of these systems requires added testing for placement and performance verification on the full vehicle. However, exhaustive testing consumes significant amounts of the available development time. It is therefore quite natural to look to alternative methods to reduce costs and find time for new developments. An efficient development process is a crucial necessity to meet performance targets, especially in a world in which the decrease of the time between idea and serial production of products is gaining more importance. Changing parts of the vehicle development process to occur on a virtual level is an important step to keep up with modern-day product development demands in a progressive customer orientated market. In this chapter we introduce virtual drive testing (VDT). This concept was introduced and briefly discussed in Chapter 6 as part of the two-stage testing method. Various methods with different levels of numerical simulation and/or emulation by measurement or hardware simulation are presented.

8.1 VDT

Numerical modeling is currently being used in various automotive development disciplines to reduce the amount of actual physical testing by simulating these tests, rather than performing them. Chapter 7 shows how antenna measurements on early prototypes can be used with modern numerical modeling tools in the accurate prediction and thus placement investigation of antenna systems on vehicles, long before the actual vehicle has been fully defined. Virtual development tools, such as numerical modeling of physical phenomena, are used to improve the development process. Such tools find increasing application in the development of modern vehicles. They allow simulations focusing on different aspects of the vehicle as well as the buildup of virtual prototypes. The term virtual prototype stands for a mathematical description of both the design state and the vehicle properties. It can be beneficial to use simulation technologies before the construction of physical models, as the building of a virtual model concept needs fewer resources than the preparation of the physical prototype. Some production aspects and testing characteristics are easily investigated already on a virtual level. The virtual models allow for the testing of many different design iterations over a shorter time frame, which leads to lower costs in the development process. Another advantage of virtual prototypes is the design optimization at an early stage of the vehicle's development process. As a result of the above-mentioned benefits, the amount of different, expensive prototypes for testing can be reduced. Only after the simulations are concluded is it necessary to validate the obtained results on a physical model.

Virtual testing is also applicable to antenna and antenna system testing levels. Although virtual testing encompasses situations in which the full system is emulated/simulated before any manufacturing has taken place, we will limit this discussion to scenarios in which the antenna and/or antenna system exists and in which the vehicle is available at least in conceptual form. This is illustrated in Figure 8.1, where different real-world communication scenarios are simulated or emulated starting from measurements of the physical device.

The ultimate scope of VDT is to arrive at emulation by hardware or full hardware-in-the-loop testing of the final industrialized vehicle, preferably without having to modify the device for testing purposes. Such testing is discussed in Chapter 6 in the case of direct MIMO OTA testing. The testing is generally a time-limited replica of a standard/reproducible test scenario in which the vehicle antenna system response is determined. Such testing enables a traceable evaluation of the antenna devices on the vehicle that can be repeated in facilities with similar equipment using standardized or custom scenarios.

Alternative VDT scopes are to perform the vehicle testing on various levels of readiness during, and as part of, the vehicle development cycle. Excluding simulations only, the most basic level of VDT has a strong simulation component and/or only with the antenna(s) from measurements. This technique is described in Chapter 7 as the link between antenna measurements and simulations. Using the link technique, the antennas measured in isolation or on a representative mock-up are post-processed in some form and imported in numerical simulation environments in which different antenna placement scenarios can be investigated. Once the

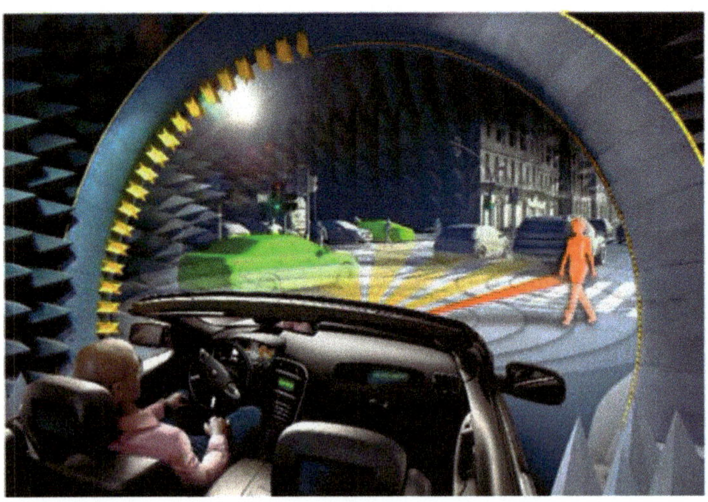

Figure 8.1 Conceptual illustration of VDT using automotive measurements. Through measurements of the physical devices, different scenarios can be simulated/emulated.

optimum placement has been determined, considering performance, mechanical, and manufacturing constraints, the resulting numerical model of the antenna(s) and antenna system(s) can be used to emulate different performance scenarios.

A more complex implementation of VDT is an extension of the two-stage method introduced in Chapter 6. As before, the antennas, measured in isolation or on a representative mock-up are post-processed and used in different representations of realistic environments. Various implementations of such VDT scenarios are illustrated in Figure 8.2.

In Figure 8.2, the various levels are illustrated, in order from left to right, with decreasing use of numerical simulation and increasing emphasis on emulation by measurement or hardware simulation. This is also summarized in Table 8.1. In this content, a post-processing step is generally computationally less intensive and time-consuming than numerical computation/simulation. Section 8.2 discusses some of the different VDT implementations in further detail.

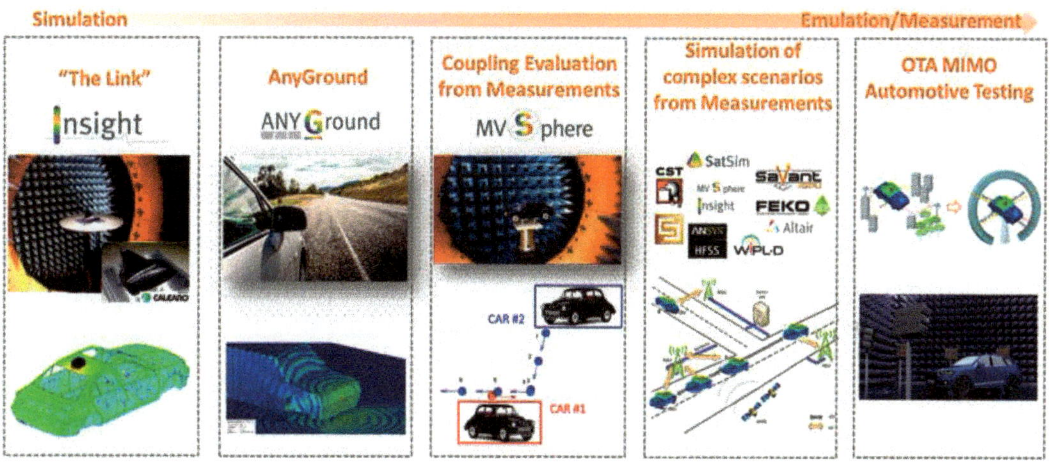

Figure 8.2 Different levels of VDT. The levels apply at different levels of the vehicle and/or antenna system development. The ultimate scope is to arrive at full hardware-in-the-loop testing of the final industrialized vehicle.

Table 8.1 Different Levels of VDT Including Dependence on Numerical Simulation or Hardware Emulation

Technique	*Numerical Simulation*	*Hardware Simulation*
Antenna(s) placement using measured antennas and antenna systems.	High	None
Measured antennas on vehicle with different ground emulation.	Post-processing	Low
Evaluation of antenna coupling such as V2V or V2X scenarios including different ground parameters.	Post-processing	Low
Simulation of different V2V or V2X scenarios using the measured vehicle-mounted antennas or antenna systems.	Post-processing	Low
Hardware-in-the-loop testing of the final industrialized vehicle. Examples are MIMO testing of the receivers, complex transmit/receive V2X scenarios, and automotive radar testing.	Low	High

8.2 The Link Between Measurement and Simulation

The link between measurements and simulations [1–3] of typical workflow for antenna placement in automotive is illustrated in Figure 8.3 and described in Chapter 7. This section discusses a practical application of the link where the radiating device is a shark fin antenna by the company Calearo [4] that has been simulated on various models of vehicles by different commercial electromagnetic tools.

The shark fin antenna has been measured in the GSM, LTE, and UMTS frequency bands from 698 to 2,690 MHz. The preparation of the measured Huygens box has been performed starting from the antenna measurement on a circular plate of 40-cm-diameter in the MVG StarLab 18GHz SNF multiprobe system [5] by the equivalent current (EQC) method [6] implemented in [7].

The determination of the electromagnetic model of the source antenna for flush-mounted applications, as this kind of scenario under test, is more difficult than a situation where the antenna is detached from the scattering structure. The proximity of scattering modifies the current distribution on the antenna itself. In a flush-mounted antenna situation, an infinite ground plane boundary condition is a good approximation for the placement, but such a condition cannot be directly obtained on a realistic measurement scenario.

However, this condition can be emulated from measurements of the source antenna on a finite ground plane. Then the desired infinite boundary can be obtained by an automatic post-processing that eliminates the diffractive contributions from the edge of a finite ground plane called the infinite plane boundary condition (IPBC) [8]. A circular ground plane with minimum 2-λ diameter is considered adequate for most measurements of these source antennas.

After post-processing, to eliminate edge diffraction, the 3-D electromagnetic model in the form of equivalent electric and magnetic currents associated with the source can be evaluated by the inverse source method on a box representing the antenna.

In the last step, the NF source is placed in the final scenario for simulation. The complete workflow of the link as described above is shown in Figure 8.4.

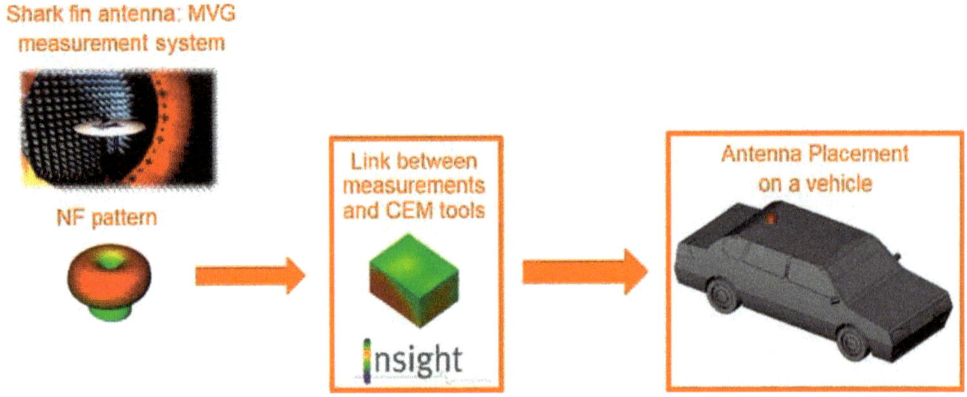

Figure 8.3 The workflow of the link between measurements and simulations; example of antenna placement in automotive.

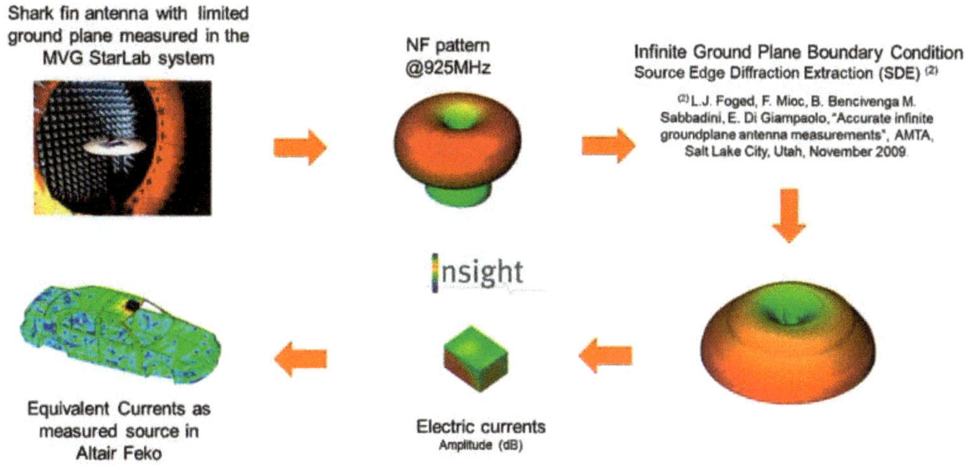

Figure 8.4 Complete workflow of the link and example of antenna placement [9].

The same NF source has been used in antenna placement problems with different vehicle models and simulation tools, including CST Studio Suite [9], Altair FEKO [9], and Ansys HFSS [11].

In the example by CST Studio Suite a crude and simplified model have been used. That is why this example is only for illustration purposes. With CST Studio Suite, users can perform accurate simulation of realistic antenna placement studies with the application of a more realistic model of vehicles.

The 3-D radiation patterns of the shark fin antenna on the vehicle models at 925 MHz simulated by [9–11] are illustrated in Figure 8.5, while the induced surface currents and the E-field generated by the antenna on the vehicle body are shown in Figure 8.6.

Finally, the variation in directivity pattern because of the antenna's location on the car at all the tested frequencies is shown in Figure 8.7.

8.3 Simulation of Complex Scenarios from Measurements

Simulations of complex scenarios in automotive applications from measurements can be performed starting from the measurement of the antenna only (in isolation)

Figure 8.5 3-D radiation pattern amplitude [dBi] at 925MHz [9–11].

Figure 8.6 Induced surface currents at 925 MHz [9–11].

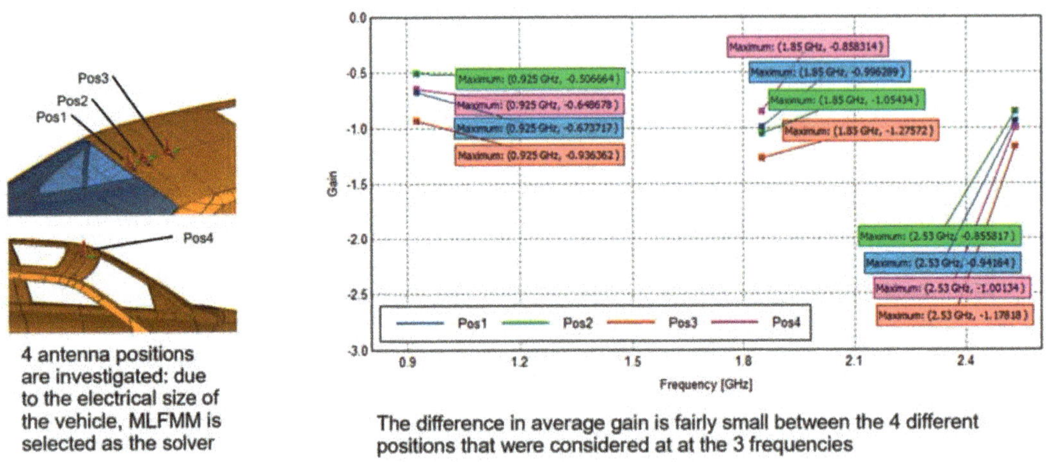

Figure 8.7 Shark fin antenna on vehicle; effect of antenna position on radiation pattern at the different tested frequencies [9].

or directly from the antenna(s) integrated on the full vehicle. The simulation includes performance evaluation in a more elaborate and complete scenario such as outdoors, on realistic roads, with the presence of more vehicles, in time-varying conditions, and generally complex environments. The input of the simulation of the complex scenario is a FF pattern that can come from the link between measurement and simulation (described in Section 8.2), or it could be directly the measured FF pattern of the full vehicle. An overview of VDT capabilities in terms of simulations of a complex scenario from measurements using MVG measurements and the Altair tools Feko and Winprop [9] is shown in Figure 8.8.

For simulation of complex scenarios in automotive applications, ray-tracing numerical techniques are more appropriate for accurate computation of large environments. The software used for the demonstration reported in this paragraph and implementing a ray-tracing technique is Altair Winprop [9]. The workflow consists of the following:

1. Definition of scenario (e.g., highway, urban, other traffic, and real test route);
2. Selection of air interface (definition of the network parameters);

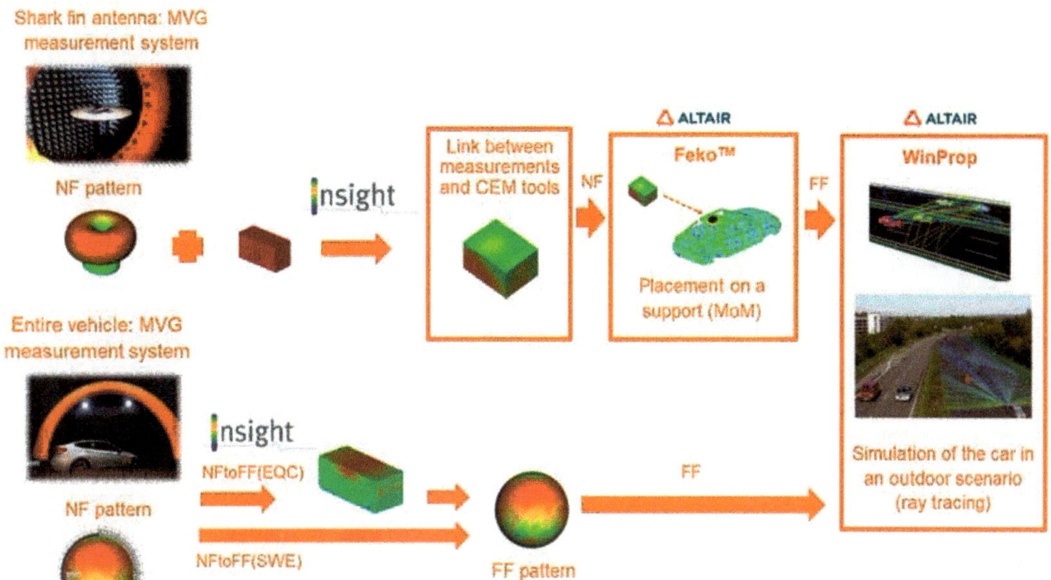

Figure 8.8 Example of post processing combining measurements and simulations of a complex scenario, Illustrative example with the Altair tools [9].

3. Computation of radio waves impinging on car antenna along trajectory;
4. Superposition of 3-D car antenna patterns (measured with MVG and, computed with Altair Feko or by combining MVG measurements with Feko simulations);
5. Evaluation of possible antenna configurations based on performance indicators.

For the specific case of the shark fin antenna on a vehicle model (antenna placement in Section 8.2), an evaluation of received power by the antenna along a trajectory for a suburban scenario is shown in Figure 8.9. The vehicle covers 5 km, and the colors in the map represent the amplitude of the received power along the path.

Finally, thanks to the possibility of defining the air infaces in the simulations, the range and connectivity for V2V scenarios together with data rate and throughput for different air interfaces and antenna configurations can be analyzed even for very complex scenarios.

Such an analysis is shown in Figure 8.10 for two scenarios (line-of-sight and non-line-of-sight). The data rates reported in the 1-D plot of Figure 8.10 refer to the line-of-sight scenario.

8.4 Different Ground Emulation of Measured Vehicle Antennas

As extensively described in this book, SNF ranges able to host real vehicles are usually considered in automotive antenna measurements [12–14]. Accounting for the entire vehicle under test (or at least a significant portion of it) is very important to

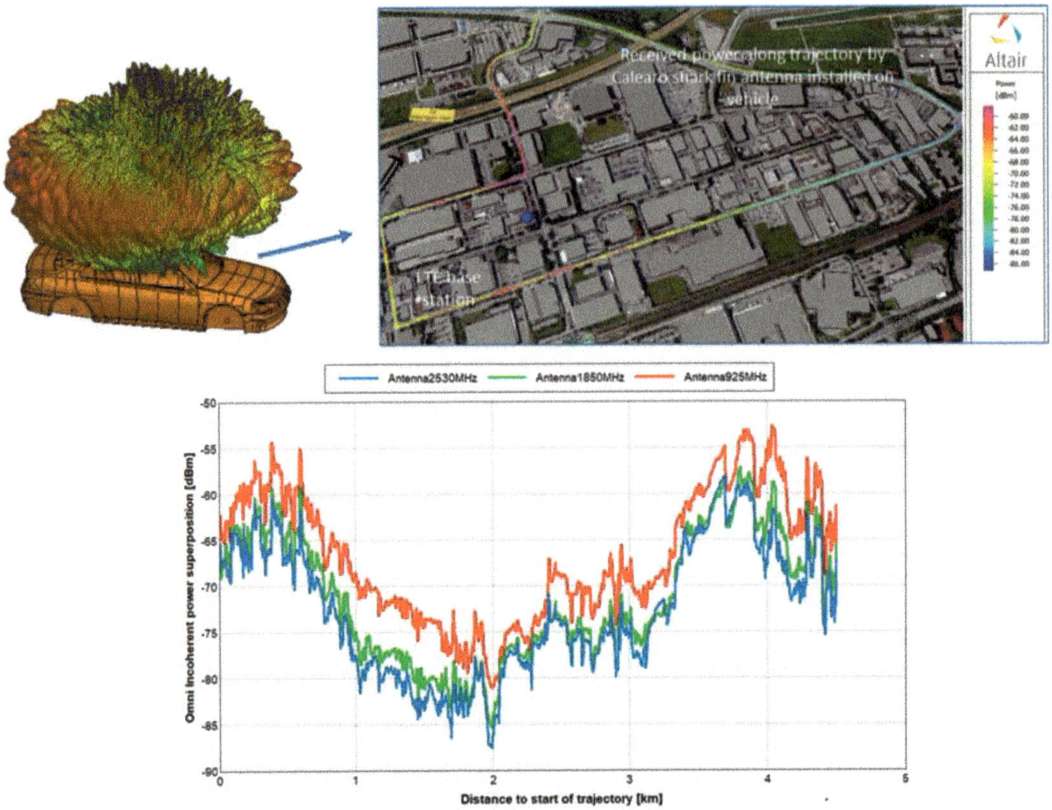

Figure 8.9 VDT for V2I LTE shark fin antenna evaluation: (upper) test drive in suburban scenario with about a 5-km length around the LTE base station and (lower) received power by antenna installed on vehicle against distance for different frequencies [9].

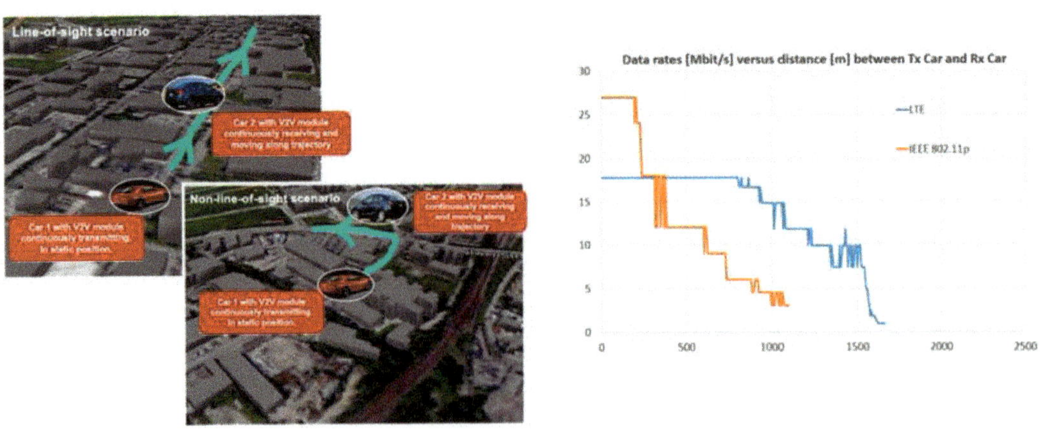

Figure 8.10 V2V communication analysis: (a) possible scenarios, and (b) data rate and throughput for different air interfaces and antenna configurations (line-of-sight scenario) [9].

include the effect of antenna coupling with the structure [14–16]. Other than the radiation performance of the antenna mounted on the vehicle, the impact of a real arbitrary ground such as asphalt or soil is often additional information that is sought.

As discussed in Chapter 5, common implementations of automotive measurement systems are based on hemispherical or truncated SNF measurements in which the vehicle is placed on a conducting or absorbing surface. To obtain the correct response of the vehicle with realistic ground parameters, measurement should be performed on different types of grounds. However, it is often unfeasible to physically change the ground material in a test range due to economical and practical constraints. Moreover, the processing of NF data acquired on a generic dielectric ground significantly complicates the NF/FF transformation process [17].

The conducting and absorbing ground conditions represent two extremes in ground material parameters. The conducting ground is far from being representative of a realistic automotive environment such as asphalt that is strongly dielectric. The fully absorbing ground is often the preferred solution as the free-space condition is closer to realistic grounds but nevertheless an approximation. Examples of possible ground parameters of interest are illustrated in Table 8.2 [18].

In terms of conductivity, but also in the effect on the vehicle-radiated pattern, the perfectly conducting ground plane and free space are effectively the two extreme conditions (see measurements examples in Chapter 5).

The response of an antenna to different operational environments can be determined in post-processing by combining the measurement of the stand-alone antenna with a simulation of the ground or surrounding structure. In the case of a flat surface, such as in most automotive applications, the procedure can be reduced to a direct application of image theory (IT) [19–20] that can be implemented directly on the radiation pattern or on the spherical wave coefficients [21–22]. In particular, from the measured radiation pattern ($E_{\theta,\phi}(\theta,\phi)$) the response over an arbitrary ground ($E_{\theta,\phi}^{Ground}(\theta,\phi)$) can be emulated considering the mirrored replica of the same pattern weighted with reflection coefficients ($\Gamma_{h,s}(\theta)$) that depends on the electrical properties of the ground, as shown in (8.1) and sketched in Figure 8.11.

$$E_{\theta}^{Ground}(\theta,\phi) = E_{\theta}(\theta,\phi) + \Gamma_{h}(\theta)E_{\theta}(180^\circ - \theta,\phi)$$
$$E_{\phi}^{Ground}(\theta,\phi) = E_{\phi}(\theta,\phi) + \Gamma_{s}(\theta)E_{\phi}(180^\circ - \theta,\phi) \tag{8.1}$$

To define the reflection coefficients, we consider the case of a horizontal flat plane, with normal direction $\hat{n} = \hat{z}$ (located at $z = 0$) as the interface between two materials, as shown on the left side of Figure 8.11. The orthogonal (hard, Γ_h) and

Table 8.2 Different Ground Parameters in Terms of Permittivity and Conductivity with PEC and Free Space Representing the Two Extreme Conditions

	Relative Permittivity (ε_r)	*Conductivity* (σ)
Wet Soil	10	0.01
Asphalt	4.61	0.0004
Soil	2.25	0.16
Free space	1	0
PEC	—	∞

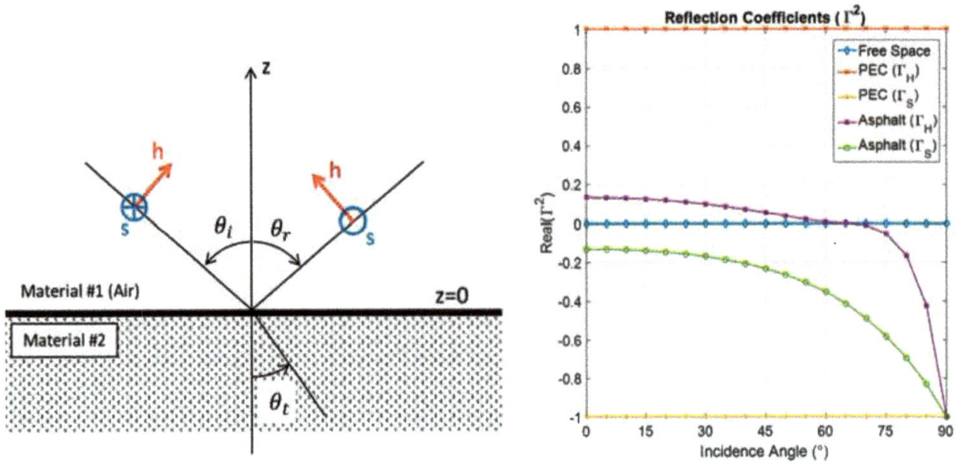

Figure 8.11 Image principle for orthogonal (hard) and parallel (soft) reflection (left). Comparison of reflection coefficients for three materials: Free space, PEC and asphalt, assuming material parameters $E_r = 4.61$, $\sigma = 0$ (right).

parallel (soft, Γ_s) reflection coefficients are the ratio between the reflected ($E^i_{\theta,\phi}$) and incident field ($E^r_{\theta,\phi}$) [19] and are defined in (8.2).

$$\Gamma_h = \frac{Z_o cos\theta_i - Z_{eq}\cos\theta_t}{Z_o cos\theta_i + Z_{eq}\cos\theta_t}$$

$$\Gamma_s = \frac{Z_{eq}\cos\theta_i - Z_o cos\theta_t}{Z_{eq}\cos\theta_i + Z_o cos\theta_t} \tag{8.2}$$

where θ_i is the incidence angle formed by the unit normal vector $\hat{n}$ of the reflecting surface and the incidence direction; θ_t is the transmission angle; Z_{eq} is the equivalent impedance of the reflecting surface; and $Z_0 = 120\pi$ is the free-space impedance. The equivalent impedance is defined by

$$Z_{eq} = Z_o\sqrt{\frac{\mu_r}{\varepsilon_r - \frac{j\sigma}{\omega\varepsilon_o}}} \tag{8.3}$$

where ε_r is the relative permittivity, μ_r the relative permeability, and σ the conductivity. Finally, the transmission angle is defined by

$$cos\theta_t = \sqrt{1 - \left(\frac{k_o}{k_{eq}}\right)^2 sin\theta_i^2} \tag{8.4}$$

where k_{eq} is the equivalent material wave number

$$k_{eq} = k_o \sqrt{\left(\varepsilon_r - \frac{j\sigma}{\omega \varepsilon_o} \right) \mu_r} \tag{8.5}$$

If the material is a perfect electric conductor (PEC), $\Gamma_h = 1$ and $\Gamma_s = -1$.

It should be noted that after the superposition of the image to the original data, the resulting pattern is valid only on the forward hemisphere [22, 23], as shown in Figure 8.12.

Examples of computed reflection coefficients as a function of the incident angle are shown in the chart in Figure 8.11 (right). Beside the two extreme scenarios (the free space and the PEC) an asphalt example ($\varepsilon_r = 4.61, \sigma = 0$) is reported. It should be noted that for smaller incident angles (e.g., $\theta_i < 60°$ the reflection coefficients of the asphalt are quite close to that of the free space, suggesting that the latter can in some cases be considered a good approximation of a realistic ground. At larger angles instead, more deviations from the free space are observed; accordingly, the described IT-based technique should be used to better estimated the response over realistic grounds.

From the above discussion it is evident that realistic grounds can be emulated with the IT-based technique if the free-space response of the vehicle is available. The absorber-based automotive systems described in Chapter 5 are ideal for this purpose because the absorbers emulate the free-space condition. On the other hand, when PEC-based systems are used, the free-space response of the car must be retrieved before applying this technique. This can be done with the spatial-filtering techniques described in Section 7.3 but at the cost of additional processing time and loss of accuracy due to the residual errors caused by the nonperfect cancelation of the influence of the conductive floor.

When free-space systems are considered, the truncation of the scanning area plays an important role in the reliability of the estimated car behavior over the ground [21]. A full SNF acquisition would make it possible to achieve a reliable radiation pattern all over the 3-D space, and, consequently, the estimated car behavior over the ground will be reliable on the whole upper hemisphere. As seen in Chapter 5, most SNF automotive systems are truncated at a maximum scanning angle $\theta = \theta_{max} > 90°$, which limits the reliability of the measured radiation pattern to approximately the same angle. Consequently, the output pattern, being obtained

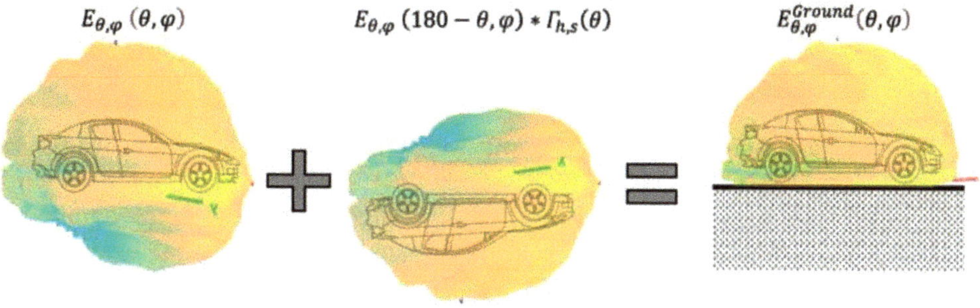

Figure 8.12 Illustration of the IT-based technique applied to vehicle antenna measurements using different ground plane parameters.

from the superposition of the image pattern (affected by the mirrored truncation) will also be reliable only on a limited portion of the upper hemisphere, namely $(180° - \theta_{max}) < \theta < 90°$ [21]. From these considerations we clearly understand the importance of performing SNF acquisitions on more than just a hemisphere. In fact, the field samples below the horizon are fundamental to estimate the car behavior at angles close to the ground. Of course, the truncation error mitigation techniques described in Chapter 7 are very useful tools to improve the accuracy of the data close to the truncation and hence to increase the reliability of the final processed data [21].

Typical multiprobe systems are characterized by a θ_{max} ranging from 100° to 110°, giving rise to reliable estimations of the car behavior over arbitrary grounds up to 10°, 20° above the horizon. It should be noted that at higher elevation angles (i.e., from $\theta = 70°$ to $\theta = 0°$) the estimation of the ground effect is usually less important because, as seen on the chart on the right side of Figure 8.11, in many cases, the free space response can already be considered a good approximation of realistic grounds.

To illustrate the potential of the IT technique in the evaluation of the effect of arbitrary grounds, the full-wave simulation of a common passenger car with an IFA at 600 MHz mounted on the bumper is considered, as show in Figure 8.13. Three scenarios have been considered: car in free space, infinite PEC, and infinite asphalt ($\varepsilon_r = 4.61, \sigma = 0$). The picture on the top of Figure 8.13 shows the simulated electric currents when the asphalt is considered, while the one at the bottom is the comparison of the gain azimuth pattern (averaged in the range). The differences between the asphalt scenario (blue trace) with respect to the free space (orange trace) and the PEC (dashed black trace) are evident and confirm how the latter two represent extreme situations.

The free-space simulation has been used to emulate a typical truncated SNF acquisition with a measurement radius of 4m and a maximum scanning angle of $\theta_{max} = 110°$. From the simulated SNF measurement, the NF/FF transformation has been performed, and the IT technique has been applied to include the effect of the asphalt. The obtained radiation pattern, shown in green in Figure 8.13, agrees very well with the full-wave simulation over the asphalt [21].

To validate the IT-based technique with real measurements, the same 10:1 scaled measurement scenario considered in various chapters of this book (e.g., Section 4.3.4) is also taken into account here, as shown in Figure 8.14. The objective of this validation is to emulate the effect of a specific ground with the IT technique starting from a measurement over absorbers and to compare the achieved result with a real measurement performed over that actual ground. To simplify the measurement setup, a conductive ground is considered for the validation. Such a ground could be considered representative of some types of wet asphalt.

The antenna positioned in the hood of the scaled car has been measured over 4-inch absorbers in the 1.008–18-GHz frequency range as show on the top-left side of Figure 8.14. Such a scenario is equivalent to the measurement of the full-size car over 48-inch absorbers at 84–1,500 MHz. The considered maximum scanning angle is $\theta_{max} = 100°$. The same scaled car has then been measured in the reference scenario with the conductive floor as shown on the top-right side of Figure 8.14.

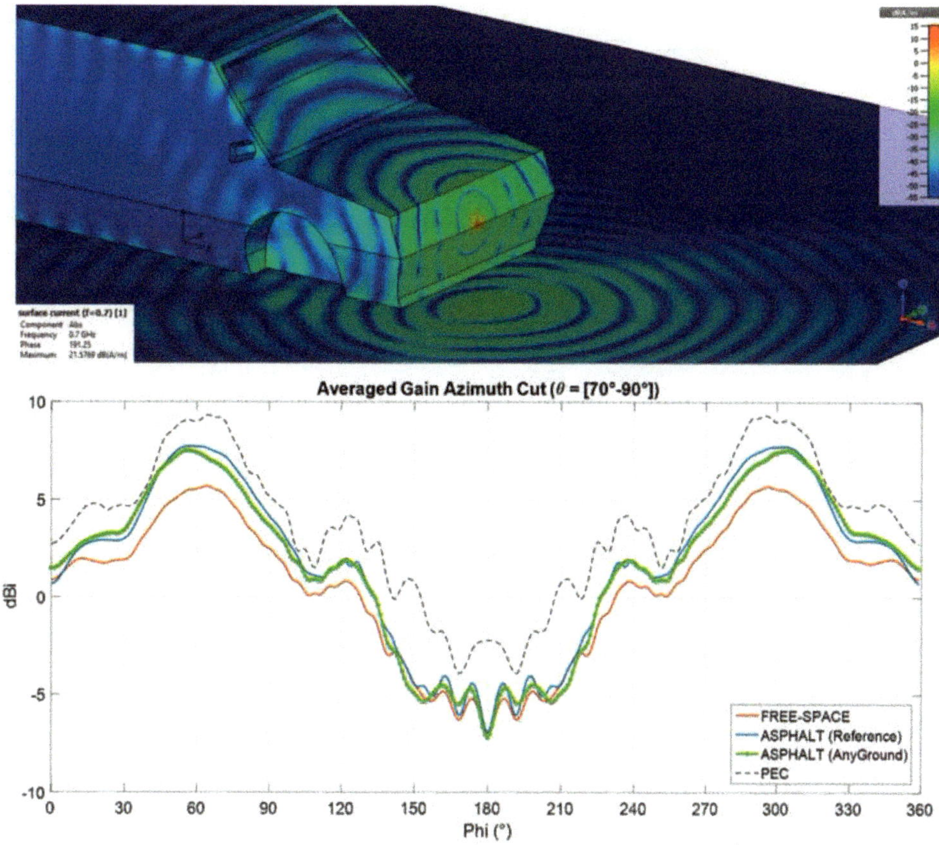

Figure 8.13 Simulation of a car with an IFA at 600 MHz on the bumper and different types of grounds. The response over asphalt is emulated also from the free space with the IT-based technique (green trace) and compared to the full-wave simulation (blue trace).

Azimuth gain pattern cuts at $\theta = 85°$ (5° above the horizon) at 470 and 1,500 MHz (equivalent) frequencies are shown on the bottom-left side of Figure 8.14. The black (reference) and orange-dashed traces are the patterns derived from the measurements over the conductive and absorber floor, respectively. The green traces are instead the patterns obtained applying the IT technique to the absorber-based measurements. A good correlation between the actual measurement over the conductive floor and the emulation of the PEC is obtained. Differences are mainly due to the mutual coupling between the antenna and the considered ground, which is not considered in the processing. The average gain pattern comparison at $\theta = 85°$ is shown on the bottom-right side of Figure 8.14 for all the tested frequencies. For (equivalent) frequencies beyond 186 MHz, the maximum gain deviation of the processed data from the reference is 1 dB, and for most of the points, the deviations are within 0.5 dB. Because of the mutual coupling, which is not included in the IT-based technique, at lower frequencies larger gain deviations are obtained. Given that the conductive floor is a worst case with regard to the mutual interaction, the achieved results suggest that sufficiently accurate evaluations of the effect of realistic grounds can be obtained with the IT-based technique.

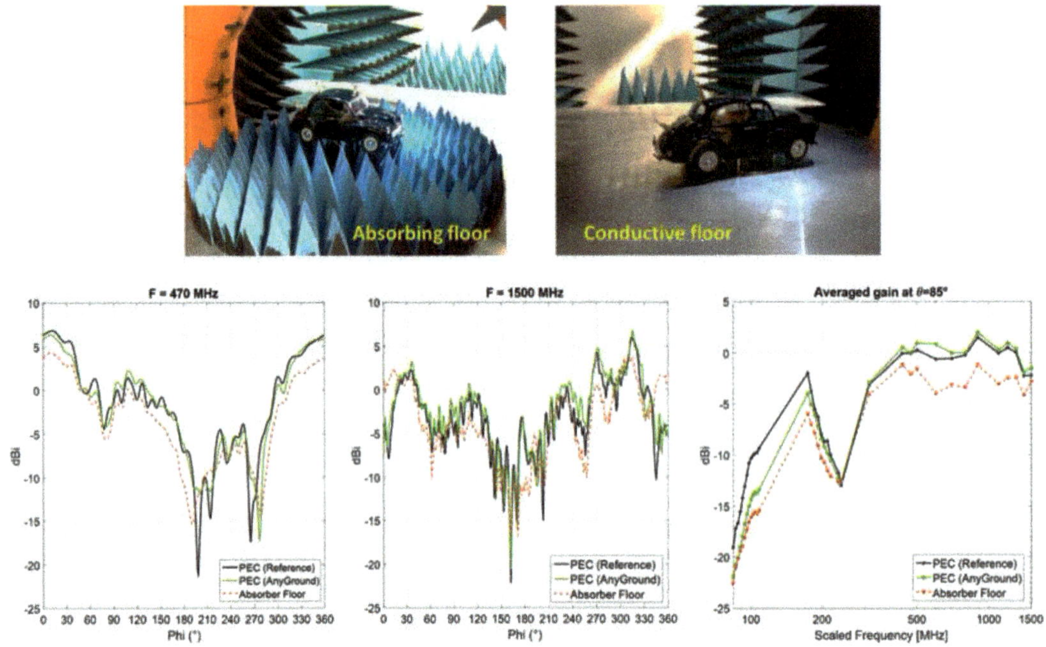

Figure 8.14 Validation of the IT-based technique with an automotive 10:1 scaled measurement scenario. The car model is measured with an absorbing floor, and a PEC ground is emulated. Scaled measurements over a conductive floor are used as a reference. ©2020 MVG. (Reprinted with permission.) [28].

8.5 V2V and V2X Coupling Evaluation

V2V and more generic V2X communications are more and more employed in modern cars. The performance of such types of communications can be evaluated already in the preliminary stages of the development cycle computing the coupling between two antennas installed on different cars (in the case of V2V) or between a vehicle antenna and a second arbitrarily placed antenna (V2X). A relatively easy and convenient way to accurately evaluate the V2V/V2X coupling is to exploit the SWCs, known from the SNF measurements of the antennas, and the transmission formula. The procedure is illustrated in Figure 8.15.

As seen in Chapter 3, the transmission is widely used for probe correction purposes in SNF measurements [12–13, 24–26]. However, the formulation can also be employed to emulate and thus accurately determine the coupling between two generic transmitting/receiving devices such as antennas on vehicles/cars as shown in Figure 8.15. The SWC of the first car are denoted with $Q_{smn}^{(\text{car}\#1)}$ while those of the second car, or of an arbitrarily placed antenna on a vehicle, with $R_{\sigma\mu\nu}^{(car\#2)}$. In this case the translation and rotation operators [13] of the transmission formula are needed to properly place and orient the second car with respect to the first one. The first car (car #1) is placed in the origin of the (x, y, z) coordinate system, where normally the DUT is considered in the conventional use of the transmission formula. The second car, car #2, is instead located in the primed (x', y', z') coordinate system, where normally the probe is described, and different positions along a trajectory are considered. As can be seen the z-axis always points toward the origin of (x, y, z); hence, to maintain the correct direction of the second car, different rotations

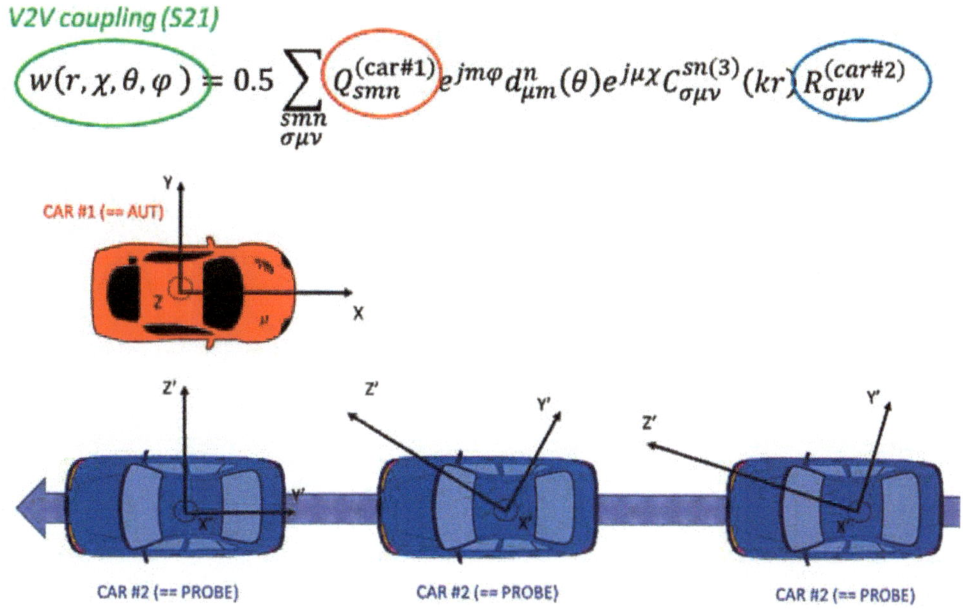

Figure 8.15 Use of the spherical wave-based transmission formula for the evaluation of the V2V coupling and coordinate systems in the xy-plane. ©2019 IEEE. (Reprinted with permission.) [29].

must be considered. Such rotations can be implemented efficiently directly in the spherical wave domain by using the Euler's angles as described in [13].

As an example, the V2V coupling has been evaluated for the 12-times scaled measurement of the Morris Minor fed by the patch antenna installed on the hood shown in Figure 8.16 [27]. The measurement frequency is 18 GHz, which scales to 1.5 GHz in the 12:1 scenario. (See more details in Section 4.3.4.) The measured averaged azimuth pattern is shown on the right-hand side of Figure 8.16 (blue trace). From the measured pattern, three types of grounds have been emulated with the IT-based technique described in Section 8.4, namely PEC, asphalt, and soil. (See the considered properties of the material in Table 8.2.)

The emulated V2V communication scenario is illustrated in Figure 8.17 (left). For simplicity, the same car/antenna configuration has been considered for the first

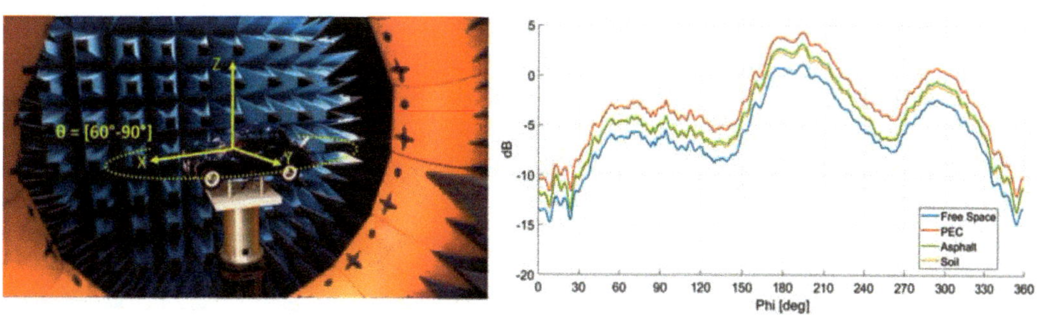

Figure 8.16 Measurement of a 12-times scaled car in the StarLab multiprobe system (left). Averaged azimuth pattern in the $\theta = [60°–90°]$ range: directly from the free-space measurement (blue trace) and including the effect of different grounds (right). ©2019 IEEE. (Reprinted with permission.) [29].

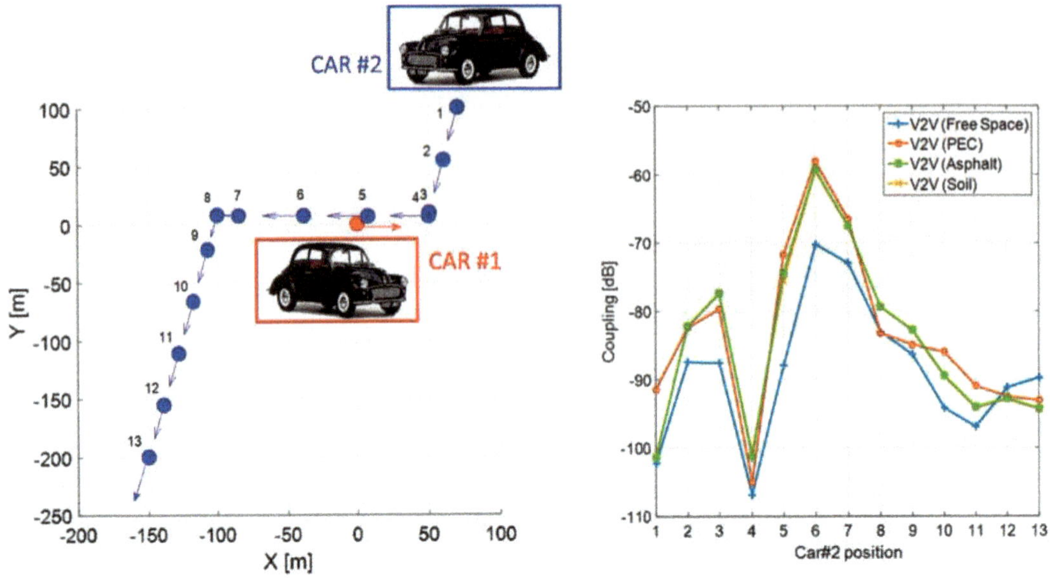

Figure 8.17 Evaluation of the V2V coupling between two cars: scaled car model fed by a patch antenna in the hood at 18 GHz (1.5 GHz). (a) Emulated scenario and (b) coupling results for different types of grounds. ©2019 IEEE. (Reprinted with permission.) [29].

and second car. The path traveled by the second car is illustrated by the blue points and arrows. The coupling evaluated at 13 different points along the path is shown on the right side of Figure 8.17. The emulated free-space V2V coupling is shown in blue. The maximum of the coupling is observed at the sixth position. At such a position, in fact, the two cars "look at each other" at an azimuth angle close to $\phi = 180°$ where, as observable in Figure 8.16, there is a peak of the azimuthal radiation pattern. Conversely, the weakest coupling is observed at the fourth position where the two cars face approximately at $\phi = 0°$, where the azimuthal radiation pattern has the lowest level.

The evaluation of the V2V coupling of the measured scaled vehicle has been repeated including the effect of the three grounds mentioned above. The effect of the different ground materials on the azimuth radiation pattern is shown on the right side of Figure 8.16. The V2V coupling for the different ground materials is shown in Figure 8.17(b). As can be seen, differences up to 10 dB with respect to the free-space case are observed.

This simple example shows that V2V/V2X performance with different ground materials can be evaluated directly from the measurement of the vehicle-installed antennas. This makes it possible to identify possible criticalities in the early stages of the development and hence accelerate the deployment of the system.

References

[1] Foged, L. J., et al., "Bringing Numerical Simulation and Antenna Measurements Together," *IEEE Antennas and Propagation Society International Symposium*, July 6–11, 2014.

[2] Foged, L. J., et al., "Measurements as Enhancement of Numerical Simulation For Challenging Antennas," *EUCAP 2015*, April 12–17, 2015.

[3] White paper: https://www.mvg-world.com/en/news/newsroom/white-paper-finding-missing-link.

[4] Calearo's website: https://www.calearo.com/it/.

[5] Foged, L. J., and A. Scannavini, "Efficient Testing of Wireless Devices from 800 MHz to 18 GHz," *Radio Engineering Magazine,* Vol. 18, No. 4, December 2009.

[6] Quijano, J. L. A., G. Vecchi, "Improved-Accuracy Source Reconstruction on Arbitrary 3-D Surfaces," *IEEE Antennas and Wireless Propagation Letters,* Vol. 8, 2009, pp. 1046–1049.

[7] Insight software website: https://www.mvg-world.com/en/products/field_product_family/antenna-measurement-2/insight.

[8] Foged, L. J., et al., "Infinite ground Plane Antenna Characterization from Limited Ground-plane Measurements," *Antennas and Propagation Society International Symposium APSURSI,* July 11–17, 2010.

[9] www.altair.com/feko/.

[10] www.3ds.com/products-services/simulia/products/cst-studio-suite/ (CST Studio Suite).

[11] www.ansys.com (Ansys HFSS and Savant).

[12] IEEE Std 1720-2012, Recommended Practice for Near-Field Antenna Measurements.

[13] Hansen, J. E. (ed.), *Spherical Near-Field Antenna Measurements*, London: Peter Peregrinus Ltd., on behalf of IEE, 1988.

[14] Noren, P., Ph. Garreau, and L. J. Foged, "State of the Art Spherical Near-Field Antenna Test Systems for Full Vehicle Testing," *6th EuCAP,* Prague, Czech Republic, March 2012.

[15] Graham, A., and P. O. Iversen, "Rapid Spherical Near Field Antenna Test System for Vehicle Mounted Antennas," *AMTA 2004,* Stone Mountain Park, Georgia, October 17–22, 2004.

[16] Dooghe, S., et al., "Spherical Near-Field Antenna Test Range for Automotive Testing from 70MHz to 6 GHz," *Proc. EuCAP 2006,* Nice, France, November 2006.

[17] Eibert, T. F., and R. A. M. Mauermayer, "Equivalent Sources Based Near-Field Far-Field Transformation Above Dielectric Half Space," *2018 AMTA Proceedings,* 2018, pp. 1–6.

[18] Gaoyuan-Ci, G. C., X. Kuo-Jun, and B. Jin-Fu, "Waveguide Method for Measuring Dielectric Constant of Asphalt Concrete at 2.45GHz," *Proceeding of the 7th WSEAS International Conference,* Athens, Greece, 2007.

[19] Balanis, C. A., *Advanced Engineering* , Second Edition, Hoboken, NJ: John Wiley & Sons, 2012.

[20] Foged, L. J., et al., "Accurate Infinite Groundplane Antenna Measurements," *AMTA,* Salt Lake City, Utah, November, 2009.

[21] Saccardi, F., et al., "Estimation of the Realistic Ground Effect in Free-Space Automotive Measurements," *2018 AMTA Proceedings,* Williamsburg, VA, November 4–9, 2018, pp. 1–5.

[22] Saccardi, F., et al., "Estimation of Real Ground Effect in Absorber-Based Automotive Measurements: Experimental Validation," *2020 IEEE International Symposium on Antennas and Propagation and North American Radio Science Meeting*, Montreal, Canada, 2020, pp. 1437–1438.

[23] Iversen, P. O., et al., "Experimental Comparison of Vehicular Antenna Measurements Performed over Different Floors," *2020 14th European Conference on Antennas and Propagation (EuCAP)*, Copenhagen, Denmark, 2020, pp. 1–5.

[24] Castaner, M. S., and L. J. Foged, *Post-Processing Techniques in Antenna Measurement*, London: SciTech Publishing, Institution of Engineering and Technology, 2019.

[25] Huang, Y., *Antennas: From Theory to Practice,* Second Edition, Hoboken, NJ: Wiley, 2021.

[26] Hong Loh, T., Metrology for 5G and Emerging Wireless Technologies, London: SciTech Publishing, The Institution of Engineering and Technology 2021.

[27] Saccardi, F., et al., "Virtual Drive Testing based on Automotive Antenna Measurements for Evaluation of Vehicle-to-X Communication Performances,"*2019 IEEE International*

Symposium on Antennas and Propagation and USNC-URSI Radio Science Meeting, Atlanta, GA, 2019, pp. 927–928.

[28] Saccardi, F., et al., "Experimental Comparison of Vehicular Antenna Measurements Performed over Different Floors," *EuCAP 2020*, Copenhagen, Denmark, March 15–20, 2020.

[29] Saccardi, F., et al., "Virtual Drive Testing Based on Automotive Antenna Measurements for Evaluation of Vehicle-to-X Communication Performances," *AMTA 2019*, San Diego, CA, October 6–11, 2019.

CHAPTER 9

In-Situ Acquisition Systems for Automotive Measurements: Drone and Handheld Approaches

Jaime Laviada Martínez, Yuri Álvarez López, María García Fernández, Guillermo Álvarez Narciandi, and Fernando Las-Heras

9.1 Introduction

In recent years, the advent of new wireless technologies has fostered the development of novel antenna measurement techniques. Thanks to advances in georeferring and positioning techniques based on GNSS, optical tracking, and LIDAR sensors, in-situ acquisition systems for antenna measurement have been developed.

Automotive antenna measurements have been traditionally conducted using anechoic chambers. For example, [1] presents a study where the AUT was placed at six different locations on the vehicle. Similarly, in [2] a global positioning system (GPS) receiving antenna was tested at four on-vehicle mounting locations. Mobile communication technologies such as 4G and 5G have introduced several improvements, such as massive MIMO, which requires new antenna measurement strategies, as discussed in [3].

Most of the existing systems for automotive antenna measurement are based on hemispherical acquisition systems [4–6]. In some implementations, the vehicle is mounted onto a rotary platform, whereas the probe antenna, usually an ultrawideband (UWB) log-periodic antenna, is placed several meters away from the platform (e.g., [1]). Modern antenna measurement facilities make use of a robotic arm to perform the scan, thus avoiding the need for a rotary platform for the vehicle. In the examples presented in [1, 4–6] measurements were conducted from a few megahertz up to 5 GHz, as this is the frequency range where most communication systems involved in automotive applications work.

Thus, conventional approaches are performed under ideal conditions, which emulate free-space or half-plane setups, enabling standard and easily comparable measurements. However, these kinds of measurements also require a complex and expensive infrastructure. In addition, testing the antennas under more realistic and challenging conditions is also advised to gauge the performance in complex situations. For this purpose, the use of in-situ measurement setups becomes a sensible choice offering several advantages over anechoic chamber facilities, described as follows:

- The impact of the surrounding environment on the antenna performance can be assessed (e.g., the effect of multipath contributions, or interference from other radiofrequency devices).
- The cost and complexity associated with an anechoic chamber facility is greatly reduced.
- In some cases, antenna measurement in difficult-to-access scenarios can be conducted.

Some desirable features for in-situ setups are the potential for easy deployment even in outdoor scenarios and a lower overall cost and level of skill required by the operator. These kinds of setups must be tailored based on the type of antennas under consideration. On one hand, millimeter-wave systems, such as the one required by the new generation of mobile communications as well as radar sensors, can afford a relatively close measurement, since the reactive coupling and reflections are limited due to the high propagation losses. For this reason, handheld approaches based on an operator moving a handheld system around the setup are a reasonable choice; these methods avoid heavy and bulky positioning systems. Moreover, the size of the millimeter-wave equipment makes it possible to implement all the hardware in a compact and lightweight device.

On the other hand, other antenna measurement techniques that have gained attention in the last years are those ones based on the use of unmanned aerial vehicles (UAVs). Similar to handheld antenna measurements, improvements in positioning and georeferring subsystems together with the huge development of UAV technology have made possible the implementation of airborne-based antenna measurement systems capable of providing enough accuracy to perform in-situ antenna characterization and diagnostics for lower frequencies—limited only by the accuracy of the positioning system.

However, it is important to know that several issues, listed as follows, degrade the performance of in-situ antenna measurement systems:

- The upper working frequency is limited by the accuracy of the positioning and georeferring system.
- In general, only the amplitude of the radiated field is acquired. Thus, in case of NF measurements, amplitude-only antenna measurement techniques are required, introducing additional uncertainty in the measurement.

This chapter focuses on the possibilities that in-situ antenna measurement techniques (mainly handheld and UAV-based techniques) bring to automotive antenna testing.

9.2 Airborne-Based Antenna Measurement Techniques

9.2.1 An Overview of Airborne-Based Antenna Measurement Techniques

In the last few years, UAVs have been successfully introduced in the field of radiofrequency applications, ranging from synthetic aperture radar (SAR) mapping and ground penetrating radar (GPR)-based inspection, to antenna measurement and radioelectric coverage analysis [7, 8].

Several architectures for UAV-based antenna measurement have been developed, proposing different solutions and approaches. For example, in [9, 10] the transmitter was placed on the UAV while the grounded AUT worked as the receiving antenna. Concerning positioning and georeferring methods, a GPS receiver was used in [11] whereas a more accurate positioning system relying on a laser tracker was proposed in [9]. The prototypes presented in [9–11] operated in the FF region of the AUT, which, given the working frequency bands (from 150 to 650 MHz) and the AUT size, resulted in flights conducted tens of meters away from the AUT.

The possibility of performing UAV-based NF measurements was proposed in [12–14], enabling flights in the vicinity of the AUT. Thanks to this, two-dimensional (2-D) acquisition domains can be defined (overcoming the limitation of measuring just the main cuts of the radiation pattern, as done in [9, 11]). Furthermore, NF measurements allow conducting antenna diagnostics, as shown in [14] to detect faults in base transceiver station (BTS) antennas. NF measurements were also carried out in [15] for antenna calibration in the high-frequency (HF) frequency band.

Most of these UAV-based antenna measurement systems are designed to operate in the radioelectric spectrum ranging from the HF band up to the super-HF (SHF) band (more precisely, up to 5 GHz), which encompasses the majority of broadcasting and mobile telecommunication networks. In [16], first steps toward the use of UAV-based antenna measurement systems in Q-band (33–50 GHz) are presented.

Other applications where UAV-based antenna measurement systems have been introduced include beam calibration of radio telescopes [17], assessment of radioelectric coverage and path loss characterization [18], and over-the-horizon (OTH) radar antenna testing [19].

In general, the architecture of a UAV-based antenna measurement system is composed of several subsystems [13, 20], described as follows:

- The UAV itself, whose main elements are the propellers, the batteries, and the UAV controller.
- The positioning and georeferring subsystem. UAVs usually include a standard GNSS receiver, an inertial measurement unit (IMU), a compass, and a barometer. To achieve centimeter-level accuracy positioning, and taking into account that these systems operate outdoors, UAVs can be fitted with a GNSS-real-time kinematics (GNSS-RTK) receiver, together with a laser rangefinder or LIDAR.
- The radiofrequency subsystem onboard the UAV. For example, in [13] the radiofrequency subsystem consisted of a monopole antenna connected to a power detector. In [9], the UAV was used as a radiofrequency source, so the

grounded AUT was connected to a spectrum analyzer that registered the measurements.

- A ground station (e.g., a laptop) that records the georeferred measurements and post-processes the measurements (e.g., for NF-FF transformation and antenna diagnostics).

A picture of a prototype for antenna measurement is shown in Figures 9.1 and 9.2. As indicated in [21], the positioning and georeferring subsystem of this prototype is mainly based on a dual-band GNSS-RTK receiver. To measure the NF in two acquisition surfaces at the same time, the UAV is fitted with two probe antennas connected to a dual-channel power detector. Given the size of the UAV, it was possible to place the two probe antennas 80 cm away, which is the distance between the two NF acquisition surfaces. Each probe antenna consisted of a UWB monopole antenna [22], whose characterization was presented in [13].

In the field of automotive measurements, UAVs can be of interest especially in the case of large vehicles (e.g., trucks) whose characterization in anechoic chambers could be challenging. In general, VHF and UHF antennas placed in vehicles usually exhibit low-directive radiation patterns that can be affected by the metallic structure and elements of the vehicle. For this reason, hemispherical acquisition (upper plot of Figure 9.3) seems to be the most suitable acquisition scheme [4–6] as it avoids the truncation errors intrinsic to planar and cylindrical acquisitions, which are more noticeable in the case of low-directive antennas. It is worth mentioning

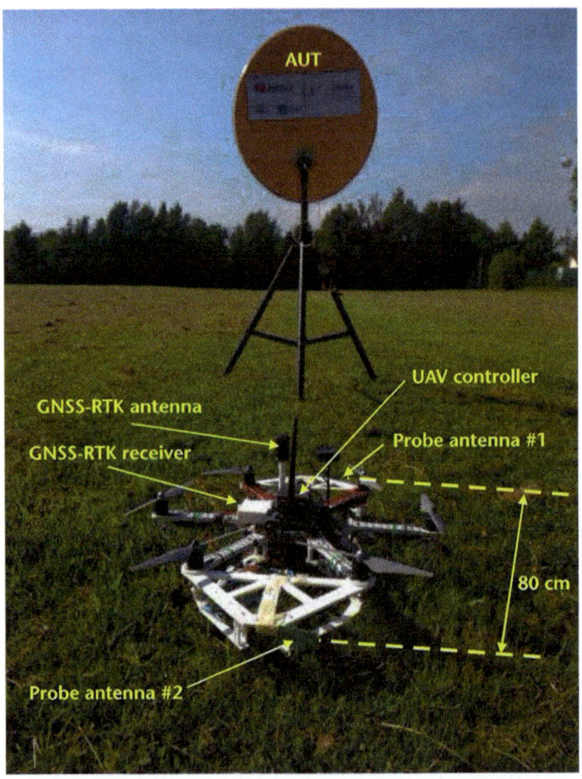

Figure 9.1 Picture of the UAV indicating some of the devices and subsystems onboard. The AUT, a reflector antenna, is shown in the background [21].

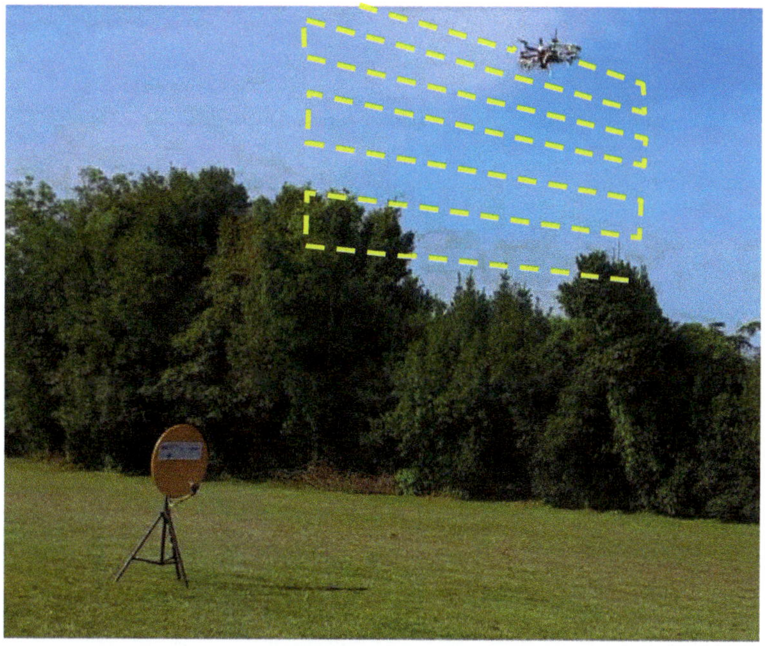

Figure 9.2 Picture of the UAV while performing the measurement of an offset reflector antenna. The dashed yellow line represents the flight path followed by the UAV (a planar domain in this case). (Adapted from: [21]. ©2019 University of Oviedo. Reprinted with permission.)

that the spiral path introduced for hemispherical acquisition systems [23, 24] enables faster, nonredundant field measurements.

Practical implementation of a hemispherical acquisition using UAVs would require additional hardware to tilt the probe antenna toward the center of the sphere to acquire the two tangential components (E_θ, E_ϕ). A trade-off solution between hardware complexity and antenna measurement accuracy could be the use of a planar acquisition domain parallel to the ground (Figure 9.3, lower plot), as it would require little or no modification of current state-of-the-art UAV architectures for antenna measurement (at the expense of larger truncation error). Another solution for hemispherical measurements could be the use of a probe made of three monopoles oriented along three orthogonal directions (e.g., to measure E_x,E_y,E_z), thus avoiding the need for probe antenna tilting.

9.2.2 Antenna Characterization by Means of Equivalent Currents

The main challenge of UAV-based antenna measurement systems is related to the use of canonical acquisition grids (planar, cylindrical, spherical). Due to inherent positioning errors caused by UAV vibration and wind gusts, acquisition surfaces do not perfectly fit a canonical, uniform grid. Thus, antenna measurement techniques for arbitrary-geometry acquisition domains, mostly based on antenna characterization using equivalent currents, are required.

Equivalent current techniques for antenna measurement provide an equivalent model of the AUT that radiates the same electromagnetic fields as the AUT outside the equivalent currents domain. These equivalent currents are reconstructed from the NF radiated by the AUT. Several equivalent currents techniques are described

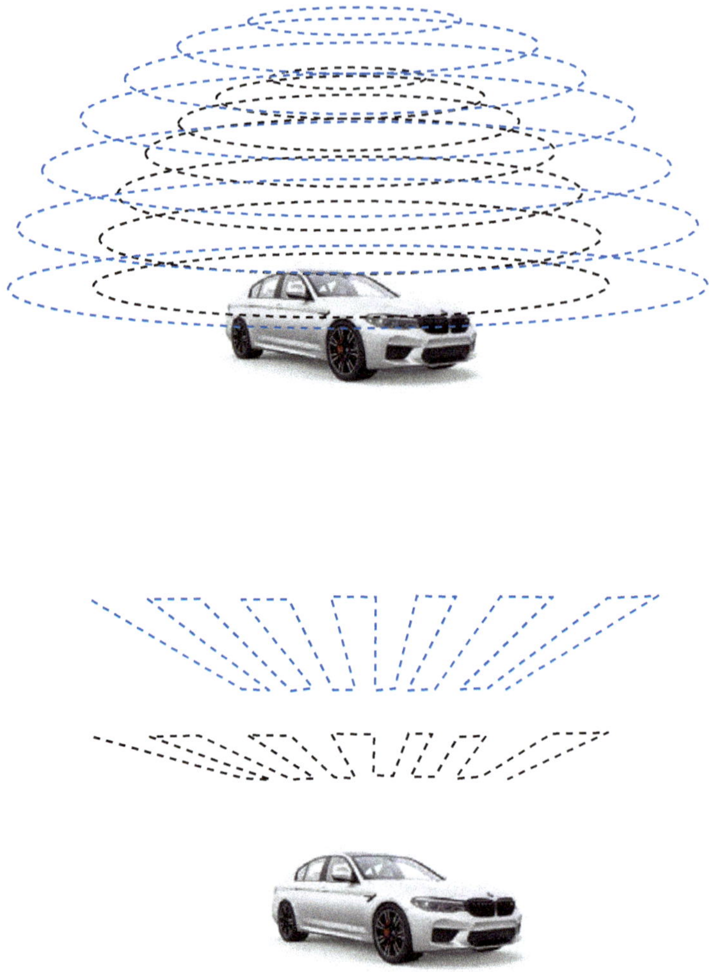

Figure 9.3 Possible acquisition schemes for airborne-based automotive antenna measurements using amplitude-only information of the radiated field (upper plot: hemispherical acquisition; lower plot: planar acquisition).

in the literature. For example, [25, 26] present a formulation based on the Love's equivalent principle, forcing the equivalent currents to be coincident to the physical currents (provided the domain fits the geometry of the AUT). This implementation, along with the ones presented in [27–29], are able to work with arbitrary-geometry domains (both for the NF acquisition and the equivalent currents domains).

Another challenge faced by airborne-based antenna measurement systems is the fact that the probe onboard the UAV and the AUT on the ground cannot share the same radiofrequency source or reference (unless using a tethered UAV). This is especially critical in the case of phase measurements, as a phase reference is needed. Thus, the proposed solution to enable UAV-based NF measurements consists of using phaseless antenna measurement techniques, which overcomes the need for a common reference for the AUT and the probe.

Among the different strategies to conduct phaseless antenna measurements, iterative phase retrieval techniques provide a trade-off solution between accuracy and hardware complexity. An overview of the setup and hardware required is shown in Figure 9.4. Basically, the amplitude of the field radiated by the AUT is measured on two acquisition surfaces using a probe antenna connected to a power meter. The simplicity of the hardware makes these methods suitable not only for UAV-based antenna measurement, but also for handheld antenna measurement systems [30]. These iterative phase retrieval techniques also avoid the need for an additional antenna acting as a reference, as proposed in [31] for UAV-based antenna measurement using amplitude and phase information.

Iterative methods for phaseless antenna measurement recover the phase of the field radiated by the AUT from the NF amplitude acquired on two measurement surfaces [32, 33]. Formulation for different kinds of acquisition domains has been presented in [34, 35] (planar domains), [36, 37] (cylindrical domains), [38] (bipolar domains), [39] (spherical surfaces), and [28, 40–42] (arbitrary surfaces).

The flowchart of the phaseless sources reconstruction method (pSRM) [28, 42] for antenna characterization using amplitude-only information of the measured NF is shown in Figure 9.5. Here, the pSRM algorithm is particularized to the case of a hemispherical acquisition domain (Figure 9.3, top), with E_θ,E_ϕ being the tangential components whose amplitude is acquired. To recover the phase of the NF, an auxiliary equivalent currents distribution is reconstructed on another domain enclosing the AUT. In the case of hemispherical acquisition, the electromagnetic second equivalence principle can be applied, simplifying the inverse problem to the reconstruction of an equivalent magnetic currents distribution (M_x,M_y) on the aperture plane of the AUT [35, 42]. Once the phase of the measured NF is recovered, the sources reconstruction method (SRM) for amplitude and phase NF measurements [27, 28] is applied. The output of the SRM is an equivalent currents distribution that characterizes the AUT. Besides providing antenna diagnostics information, the retrieved equivalent currents distribution can be used to compute the radiation pattern of the AUT.

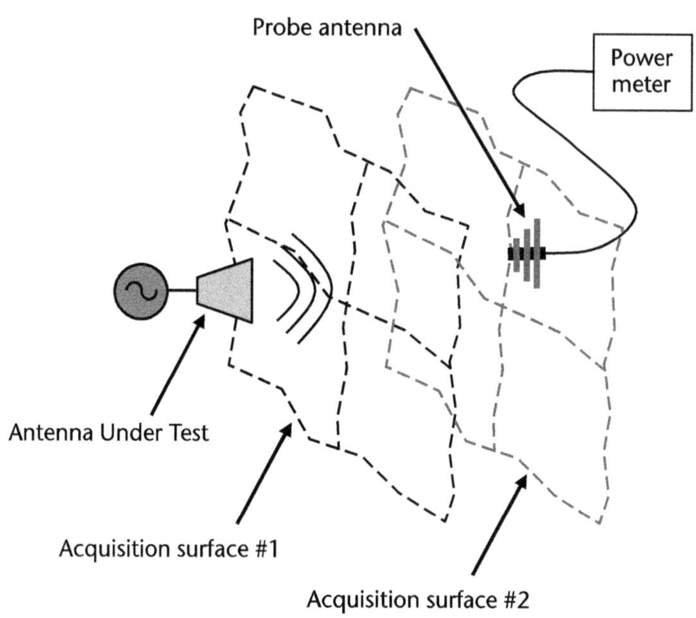

Figure 9.4 Typical scheme for antenna measurement using the amplitude of the NF acquired on two surfaces.

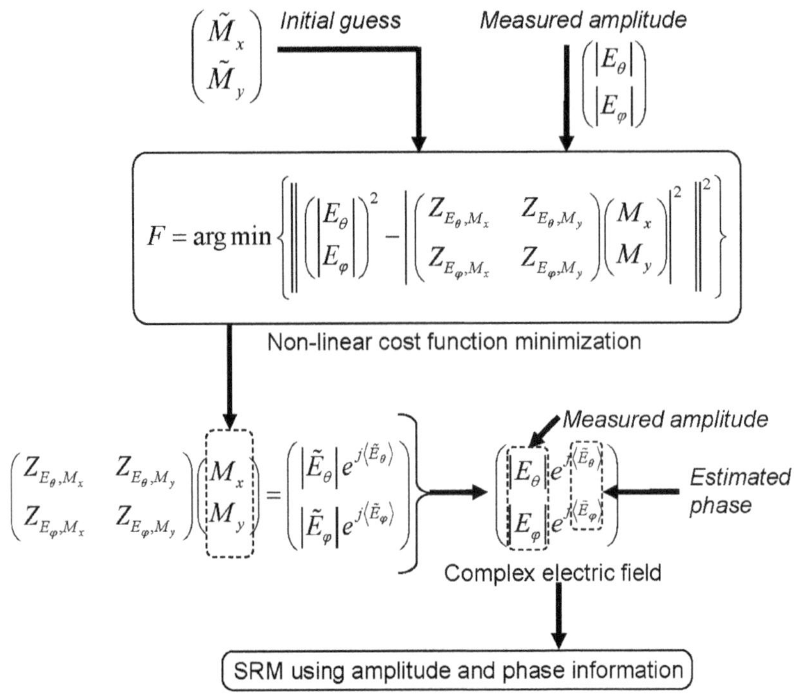

Figure 9.5 Flowchart of the SRM for amplitude-only NF measurements, particularized for hemispherical NF acquisition domains.

The pSRM can be modified to work with the amplitude of the E_x,E_y, and E_z components acquired on the two hemispherical surfaces that, as indicated in Section 9.2.1, could be of interest to avoid the need for probe antenna tilting in UAV-based antenna measurement systems. However, preliminary assessment based on this three-component formulation provided worse convergence of the iterative phase retrieval algorithm than the one based on acquiring the two tangential components.

9.2.3 Application Example

To illustrate the possibility of conducting in-situ automotive measurements using an airborne-based antenna measurement system, a simulation-based example is presented. The AUT, intended for digital audio broadcasting (DAB) reception and tuned to 200 MHz (the wavelength, λ, is 1.5m), is integrated within the windshield of the car as shown in Figure 9.6.

In UAV-based antenna measurement systems that make use of iterative phase retrieval techniques, the proper selection of the distance between the AUT and the acquisition surfaces depends on several factors:

- A safe distance between the UAV and the placement of the AUT must be kept. For example, this distance was around 3–4m in [13, 14, 21]. In hemispherical domains, this also limits the angular margin along the θ-coordinate to avoid the UAV flying too close to the ground.

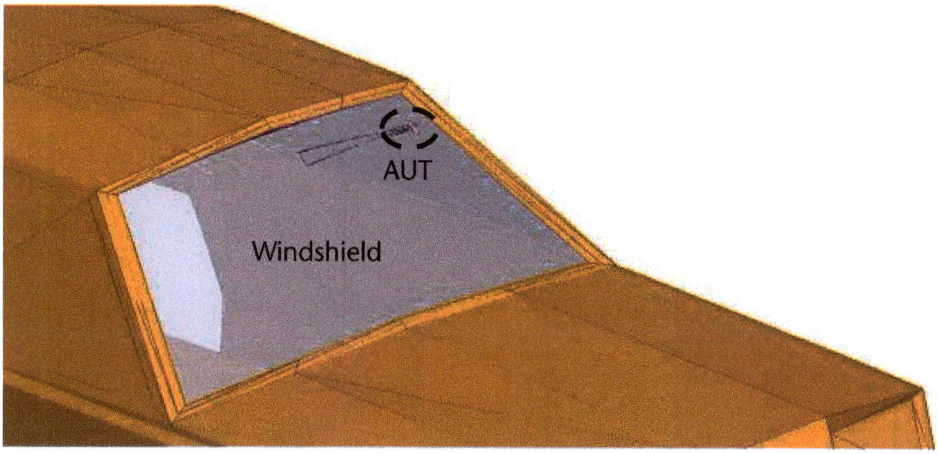

Figure 9.6 Electromagnetic modeling of the vehicle with a windshield-integrated antenna for DAB reception.

- The selection of the two acquisition surfaces influences the convergence of the iterative phase retrieval method. Parametric studies for different kinds of acquisition domain geometries have been conducted in [21] (pSRM, planar domains) and in [43] [phaseless NF-FF using spherical wave functions (i.e., spherical domains)]. In the case of spherical acquisition domains, it has been reported in [43] that best results were achieved when the radii of these domains was 10%–15% and 20%–30% of the FF or Rayleigh distance,[1] respectively.

In the example presented in this section, the amplitude of the electric field radiated by the AUT is acquired on two hemispherical acquisition domains whose radii are R_1 = 4 m and R_2 = 6 m, as shown in Figure 9.7. For this example, the FF distance is approximately 15m, so R_1 and R_2 are 27% and 40% of the FF distance, respectively, which is on the same order of magnitude as the results presented in [43].

From a practical point-of-view, hemispherical acquisitions carried out with UAV-based systems would require circular sweeps (in ϕ-coordinate) for different values of the θ-coordinate. Consequently, sampling in θ would be sparser than in ϕ. In this example, given the working frequency of the AUT and the acquisition distances, a $\theta = 10^\circ$ sampling step is chosen. As depicted in Figures 9.7 and 9.8, θ ranges from 5° to 65°, so the circular sweep corresponding to $\theta = 65°$ and $R_1 = 4$m is 1.7m above ground.

The amplitude of the acquired NF electric field is depicted in Figure 9.8(a,b) for the $R_1 = 4$m acquisition domain. It can be observed that the maximum of the NF is not centered, as the AUT is placed on a tilted surface (the windshield of the vehicle).

The amplitude of the acquired NF is introduced into the pSRM algorithm to recover the phase of the NF. Once the phase of the NF is recovered, antenna

1. Rayleigh or FF distance must fulfill the following criteria: 1) it must be greater than $2D^2/\lambda$ (with D the size of the AUT) and 2) it must be much larger (at least 10 times) than the wavelength.

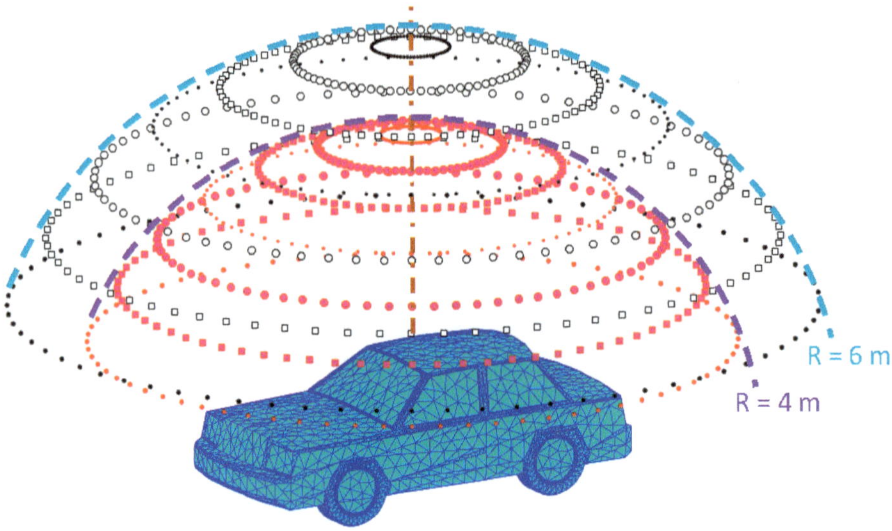

Figure 9.7 Hemispherical acquisition domain of the NF radiated by the DAB antenna.

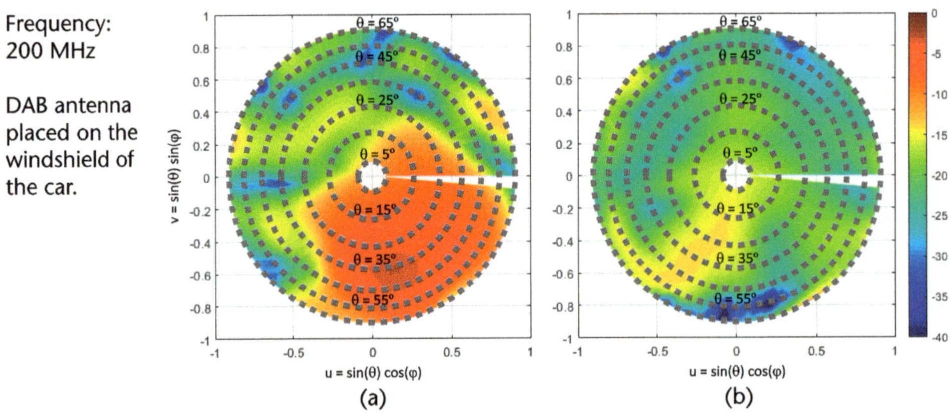

Figure 9.8 Simulated NF on the hemispherical acquisition domain: (a) co-polar component evaluated at $R_1 = 4$ m, and (b) cross-polar component evaluated at $R_1 = 4$m. The dashed gray lines denote the acquisition sweeps along the ϕ-coordinate for different θ angles between 5° and 65°, in 10°-steps.

diagnostics and NF-FF transformation are conducted (see Figure 9.5). Similar to the examples described in [13, 21], the SRM is used to recover an equivalent currents distribution on a surface that, in this example, encloses both the AUT and the chassis of the vehicle, as the metallic structure of the vehicle could interact with the AUT. The reconstructed equivalent electric currents are shown in Figure 9.9, where it can be observed that the maximum is located approximately at the true location of the AUT on the windshield.

Taking advantage of the fact that, in this simulation-based example, the phase of the NF is also available at the acquisition points, the SRM has been applied using the complex NF information on the two acquisition planes. Results are depicted in Figure 9.10, where it can be noticed that the level of the equivalent currents is maximum at the position of the AUT, as in the case of the reconstructed currents from amplitude-only NF information.

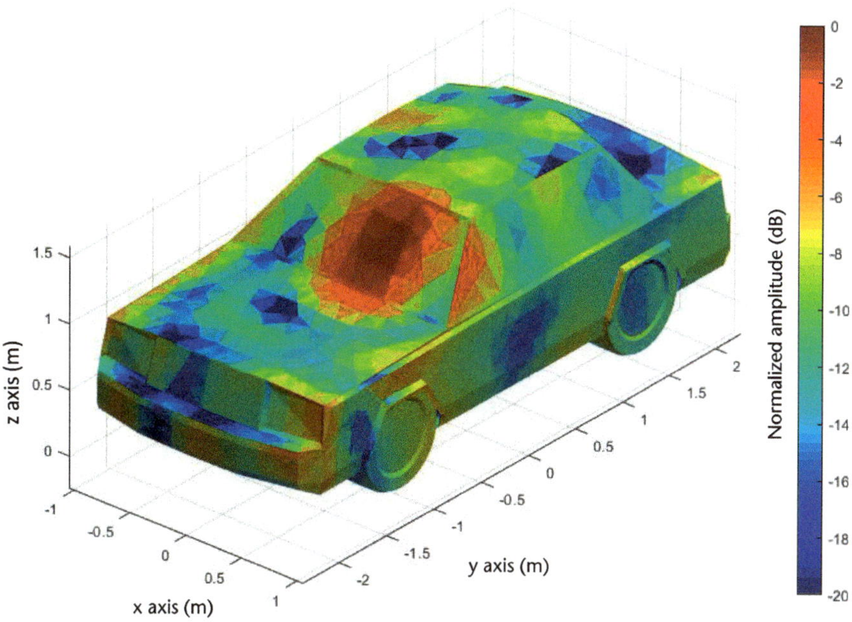

Figure 9.9 Reconstructed equivalent electric currents on a surface fitting the AUT and the vehicle. Results from amplitude-only information of the NF.

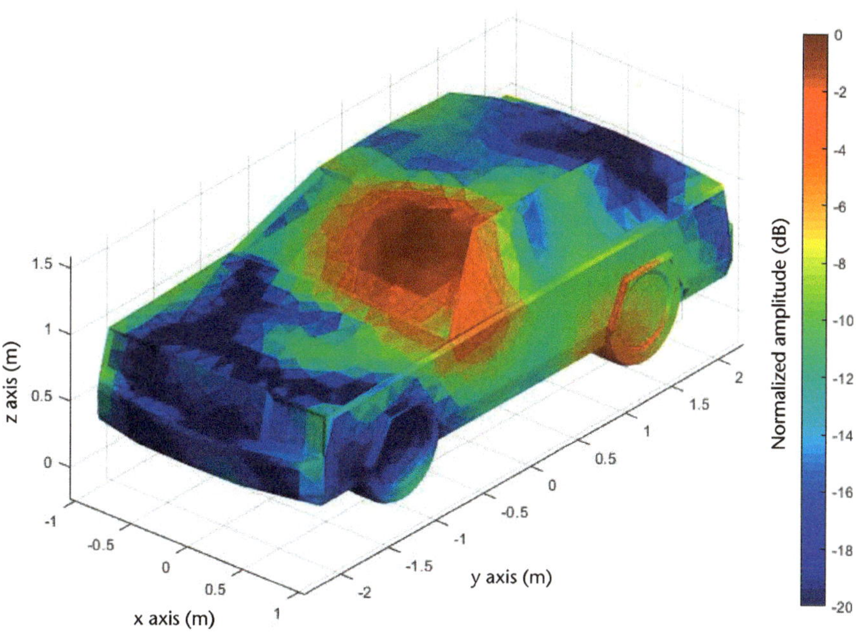

Figure 9.10 Reconstructed equivalent electric currents on a surface fitting the AUT and the vehicle. Results from amplitude and phase of the NF.

From the reconstructed equivalent currents on the AUT and the chassis of the vehicle, the radiation pattern is calculated. Results are depicted in Figure 9.11, where the 2-D radiation patterns calculated directly from simulations (the reference ones), from amplitude and phase NF samples, and from amplitude only NF

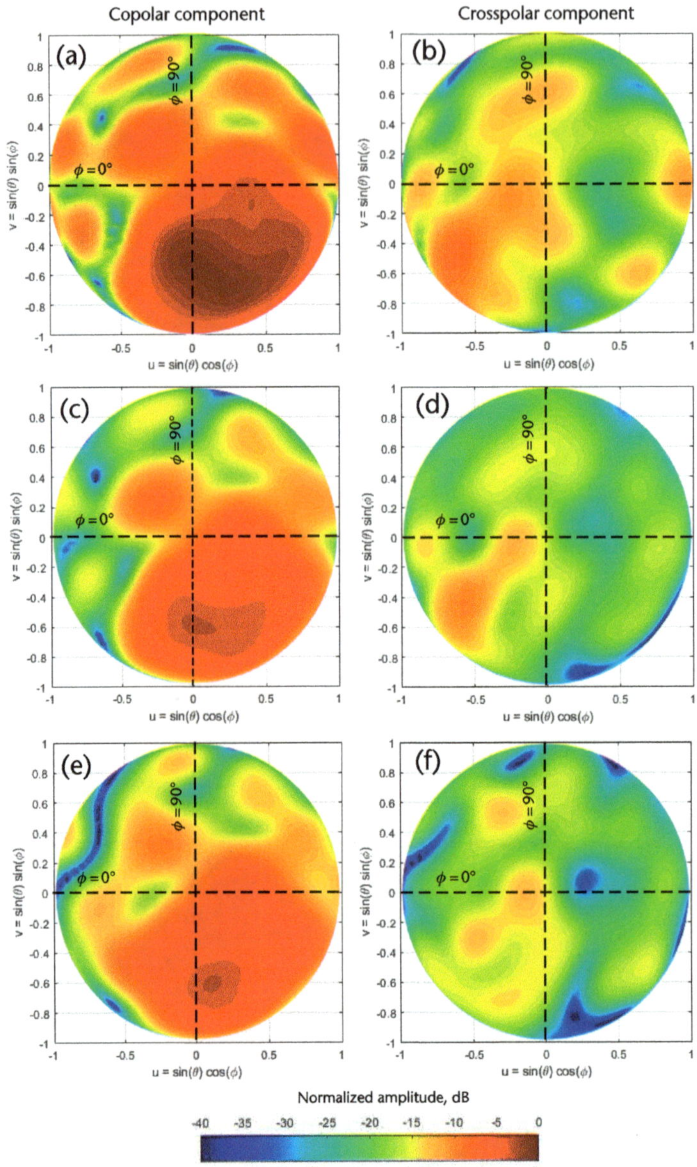

Figure 9.11 Radiation pattern of the DAB antenna, u-v representation: (a) Co-polar and (b) cross-polar components of the radiation pattern calculated from the electromagnetic model, (c) co-polar and (d) cross-polar components of the radiation pattern calculated from the reconstructed equivalent currents using amplitude and phase information, and (e) co-polar and (f) cross-polar components of the radiation pattern calculated from the reconstructed equivalent currents using amplitude-only information.

information, are compared. For a better analysis, the main cuts of the radiation pattern are depicted in Figures 9.12 and 9.13 (for $\phi = 0°$ and $\phi = 90°$, respectively).

In the $\phi = 0°$ cut of the FF pattern (Figure 9.12), agreement between the reference FF pattern (calculated from electromagnetic simulation) as well as the calculated ones from NF measurements (considering amplitude and phase, and amplitude-only information) is better than 3–4 dB from $\theta = -25°$ to $+65°$. For the $\phi = 90°$ cut, the angular margin where agreement between FF patterns is better than 3–4

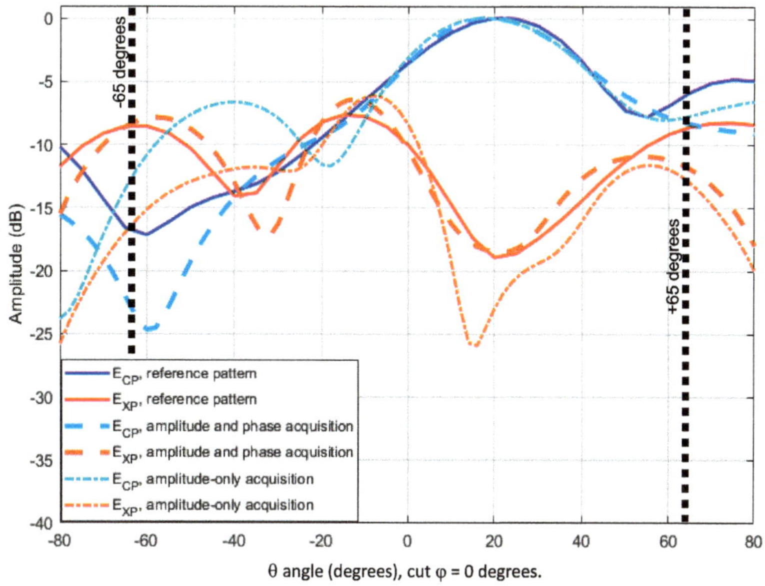

Figure 9.12 Radiation pattern, cut $\phi = 0^{\circ}$. Dotted black lines indicate the angular margin along the θ-coordinate where the NF was acquired.

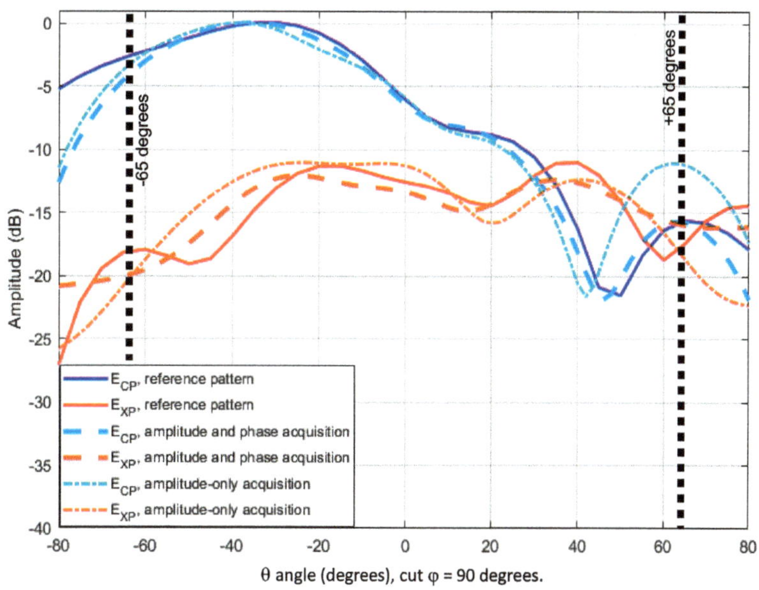

Figure 9.13 Radiation pattern, cut $\phi = 90^{\circ}$. Dotted black lines indicate the angular margin along the θ-coordinate where the NF was acquired.

dB ranges from $\theta = -65^{\circ}$ to $+65^{\circ}$ ($+40^{\circ}$ for amplitude-only acquisition). Theoretically, the FF pattern margin of validity[2] should be coincident with the size of the hemispherical NF acquisition domain, which, for this example, should range from

2. The FF pattern margin of validity is defined as the angular range where the FF pattern can be calculated from NF measurements on truncated acquisition domains.

$\theta = -65^\circ$ to $+65^\circ$, for all ϕ angles. However, that would be the case if the AUT were centered at the origin of the coordinate system. As this is not the case, the FF pattern margin of validity changes for different cuts of the radiation pattern.

9.3 Handheld Systems

9.3.1 An Overview of the Handheld System

Millimeter-wave communication systems are available in current cars for radar sensing, enabling a wide range of aids in the context of advanced driver-assistance systems (ADAS) [44–46]. Moreover, due to the advent of high-bandwidth communications exploiting the millimeter-wave band [47], more systems at those frequencies are expected to be integrated in incoming car models.

In contrast to conventional communication systems working at lower frequencies, millimeter-wave devices enable a high level of integration thanks to their compact size [48]. This fact makes it possible to develop small devices with the capacity to acquire electromagnetic fields. Moreover, if correctly designed, these systems can be easily moved by hand, providing the possibility of implementing millimeter-wave measurement systems by means of handheld systems exploiting freehand sweeps [49–51].

Thus, the possibility of in-situ antenna measurements without the need for mechanical positioning systems becomes feasible in a similar fashion to the previous UAV approach [30, 52, 53]. However, two relevant differences must be taken into account. First, in contrast to the UAV approach, the measurement cannot rely on GNSS systems as they cannot provide a submillimeter accuracy as required by these millimeter-wave systems. Consequently, another kind of positioning system, such as those used for motion capture tracking, which have become very affordable in the last decade, must be used. The second major difference is that a handheld system cannot cover a number of waypoints, as in the case of UAV measurement, because no human operator would be able to achieve such an accuracy. This entails a number of challenges that must be overcome.

Figure 9.14 illustrates an overview of the handheld approach. An operator holds a probe antenna that moves in front of the area in which the transmission system is located acquiring field samples. In addition, the movement of the probe is monitored by a tracking system (e.g., a motion capture system). After processing the data, the 3-D radiation pattern along with the reconstructed currents for diagnosis purposes are computed.

9.3.2 Handheld System Description

The architecture of the handheld system is depicted in Figure 9.15; it comprises the following elements:

- *Acquisition subsystem*: Includes the RF components to acquire the power of the electromagnetic field. In this case, an open-ended rectangular waveguide is used as a probe, and a spectrum analyzer is used as a power meter. Other choices such as using a portable vector network analyzer are also possible. In addition, a flexible cable connecting both devices is used. Depending on

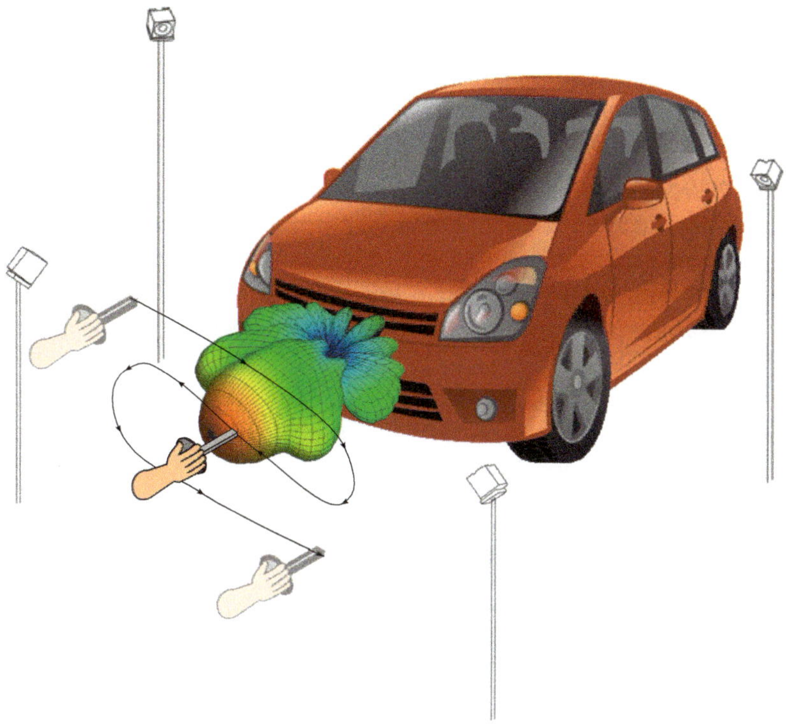

Figure 9.14 Freehand measurement for automotive applications.

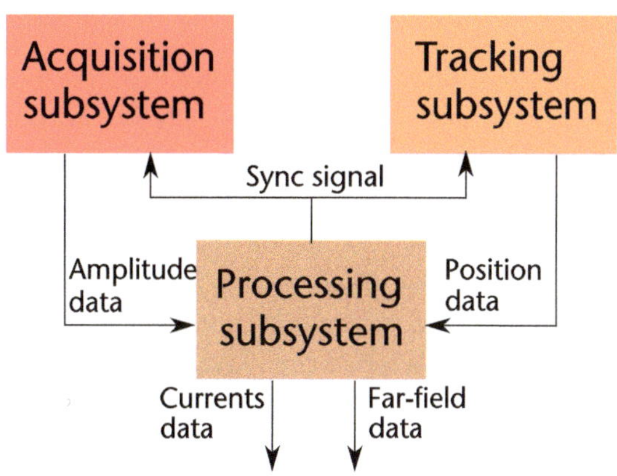

Figure 9.15 Overall scheme of the subsystems comprising the handheld system.

the setup, an external mixer can be used between the probe antenna and the power meter. This device is used if the losses are too high or if the upper frequency of the power meter is too low.

- *Tracking subsystem*: Several highly accurate options are possible for the tracking system. Laser tracking is one of the most robust and accurate ones, but it is also significantly expensive. As an alternative, motion capture (MoCap) systems can provide affordable tracking capabilities with an accuracy under 1 mm, which is usually enough for the millimeter-wave band [54]. These systems are based on a number of infrared cameras (three or more), which are deployed at the scene before the measurements, and reflective markers, which are attached to the probe antenna, so that the position is found by triangulation. It is relevant to note that in order to track not only the position, but also the attitude of the probe, it is necessary to use at least three markers so a rigid body can be defined.

- *Processing subsystem*: This system can be implemented by means of a regular laptop, which must accomplish three main tasks. First, it must perform a synchronized acquisition of the EM field and tracking data. As discussed in Section 9.3.3, this system must also discard some samples to keep a correct sampling balance as well as to minimize errors due to an incorrect positioning calibration. The third task is the computation of the FF pattern from the amplitude data. For this purpose, the same NF to FF algorithms as in the case of the UAV measurements are used, and, therefore, the field must be acquired at two (or more) different planes.

9.3.3 Particular Characteristics of the Handheld System

In the case of the handheld system, there are some peculiarities that must be taken into account. First, one cannot expect hand movements to follow a specific path so the scanning trajectory, which ideally should cover at least two complete parallel planes, cannot be so well defined as in the UAV case. For example, the operator will probably repeat movements around the same area, resulting in redundant samples. Moreover, the movements will not be done at a constant hand speed so, if regular time sampling is considered, the points will have different separations between them.

An approach to mitigate the previous negative points is to split both planes into 3-D cells. Thus, points outside the cells are automatically discarded in order to guarantee that the considered samples are close enough to the planes. In general, the height of the cells is set to $\lambda/2$. In addition, a maximum number of acquisitions per cell is set. The purpose of this constraint is to avoid the oversampling of some areas. If the horizontal size of the cell is $\lambda/2 \times \lambda/2$, then one acquisition per cell is usually enough. However, it could be convenient to increase the size of the cell in some cases. One of them is when the wavelength is too small so moving into a number of specific cells becomes too complex due to the accuracy of the movements. In this case, the number of acquisitions per cell is increased proportionately to their size, so the overall sampling density is approximately one per $\lambda/2$.

An example of how to choose the samples from the ones available in a freehand trajectory is shown in Figure 9.16. In this example, a maximum of one sample per cell is used. Thus, several samples are discarded in those cells corresponding to slow movement areas. Consequently, oversampling is avoided, improving the overall balance. It is also relevant to note that some cells could have no sample due to the irregular trajectory. These cells should be marked on a map to help the operator know which areas have not been completely sampled.

Another relevant difference with respect to the UAV measurement is the need to correct the offset between the center of coordinates of the rigid body defined by the markers and the aperture of the probe antenna (see Figure 9.17). This can be easily estimated by different techniques such as (temporarily) placing a marker in

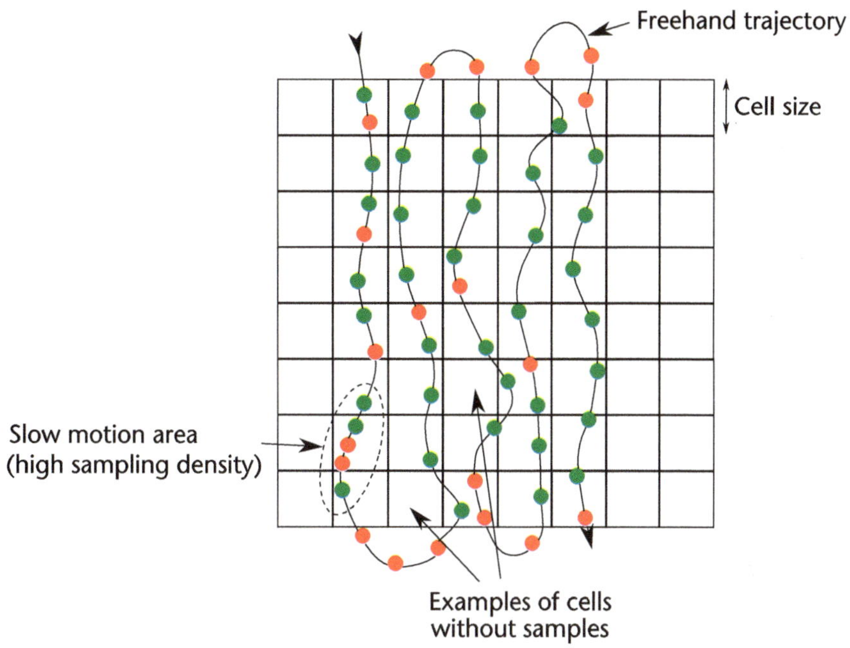

Figure 9.16 Choice of samples from some freehand sweeps.

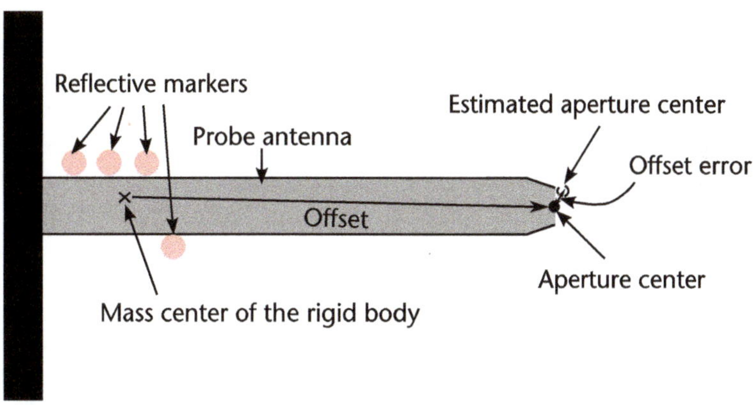

Figure 9.17 Offset between the rigid body and the antenna aperture, including the error that can happen during the estimation of this offset.

the aperture of the antenna. However, one must be aware that this offset is never perfectly computed. If one would not rotate the probe along the scan, this offset error would not have a relevant impact as all the measurements would suffer from the same spatial offset without major consequences in the FF estimation. However, hand movements are imperfect, and some rotation is always expected. If the spatial offset is not perfectly corrected, the larger the rotation the larger the positioning error. In order to mitigate the impact of this error, rotations are usually limited. In our research, we have found that 5° for each Euler's angle is a reasonable limit that a human can keep when scanning. Analogously to the previously described method to ensure a proper distribution of the acquisitions, if the operator rotates the probe beyond the predefined threshold, a warning is displayed so the attitude of the probe can be corrected.

9.3.4 Application Example

In order to illustrate the handheld in-situ measurements, the system shown in Figure 9.18 is used. It comprises a WR-10 probe antenna with four attached markers. A 3-D printed part is used to hold the markers. The probe antenna is connected to an external subharmonic mixer, which remains steady during the measurement. The intermediate frequency output of the mixer is used to feed a spectrum analyzer, which is controlled by a laptop.

In this example, the measurement of a radar with electromagnetic characteristics equivalent to those used in automotive radars is considered. In particular, an IWR1443 by Texas Instruments [55] is employed, which is functionally equivalent to the automotive radar AWR1443 [56], except for some automotive qualifications. The radar is mounted on an evaluation board (IWR1443 Boost) in order to ease the control.

The measured radar has three Tx channels and four Rx channels (see Figure 9.19), though only the transmissions capabilities (i.e., the Tx channels) are tested with this technique. The radar works under the frequency-modulated continuous-wave (FMCW) principle working from 76 to 81 GHz. However, the working frequency is set to 76 GHz during the measurements.

The acquired electric field amplitude for the three transmitters is shown in Figure 9.20. In addition, the average acquisition distance is shown for both acquisition planes. The equivalent magnetic currents reconstructed on the radar board plane are shown in Figure 9.21, revealing a good agreement with the real position of each transmitter and, therefore, enabling diagnosis to detect potential failures in any of them. Finally, the FF patterns are shown in Figure 9.22 revealing clear differences in the directivities of the E- and H-planes.

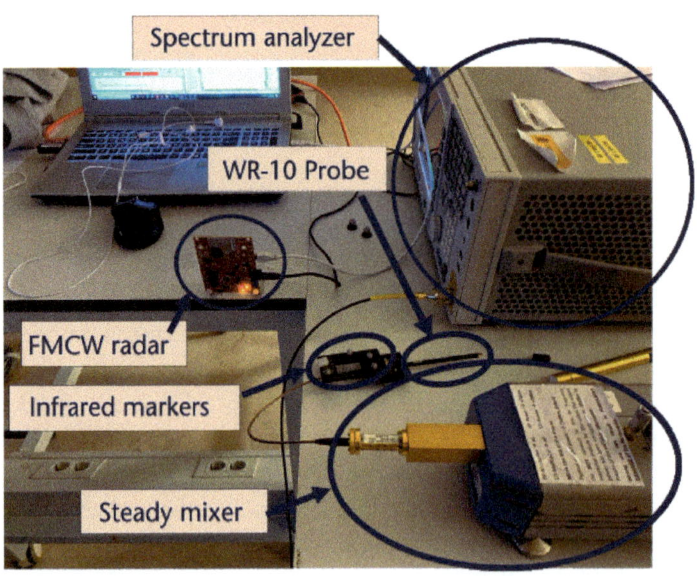

Figure 9.18 Handheld system.

Figure 9.19 Tx channels of the considered millimeter-wave radar (IWR1443).

9.4 Final Remarks

This chapter has presented the potential applications for in-situ antenna measurement systems in the field of automotive antenna measurements.

In the case of UAV-based antenna measurement systems, there are some limitations and challenges that must be faced in further developments and architectures; they are described as follows:

- Hemispherical acquisition domains cannot be extended up to $\theta = 90°$ due to the need to keep a safe distance between the UAV and the ground. Thus, the FF is partially affected by truncation errors, which can have some impact in the case of low-directive antennas.
- The implementation of hemispherical NF acquisition with current state-of-the-art UAV antenna measurement systems requires a modification of the payload. For example, the probe antenna can be mounted on a gimbal to ensure proper acquisition of the two tangential components of the NF on the hemispherical surface, E_θ and E_ϕ.
- Concerning phase retrieval techniques, although holographic or interferometric techniques are, in general, more accurate than iterative phase retrieval methods, they require additional hardware that would increase the complexity of the UAV-based antenna measurement systems. Hence, iterative phase retrieval techniques have been adopted for UAV-based antenna measurement systems, but at the expense of less accurate phase recovery. The possibility of

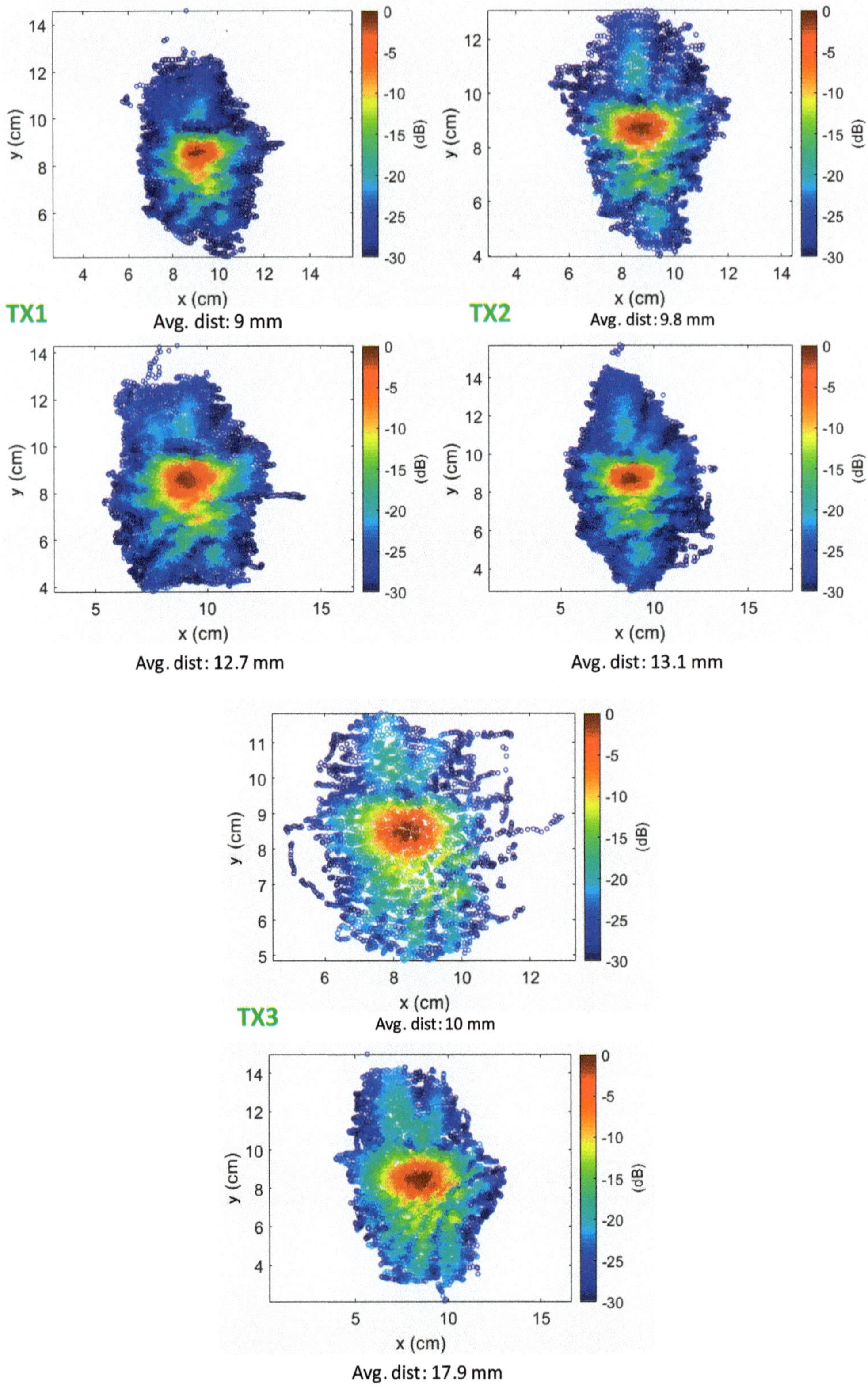

Figure 9.20 Acquired amplitude at two different planes for each Tx channel.

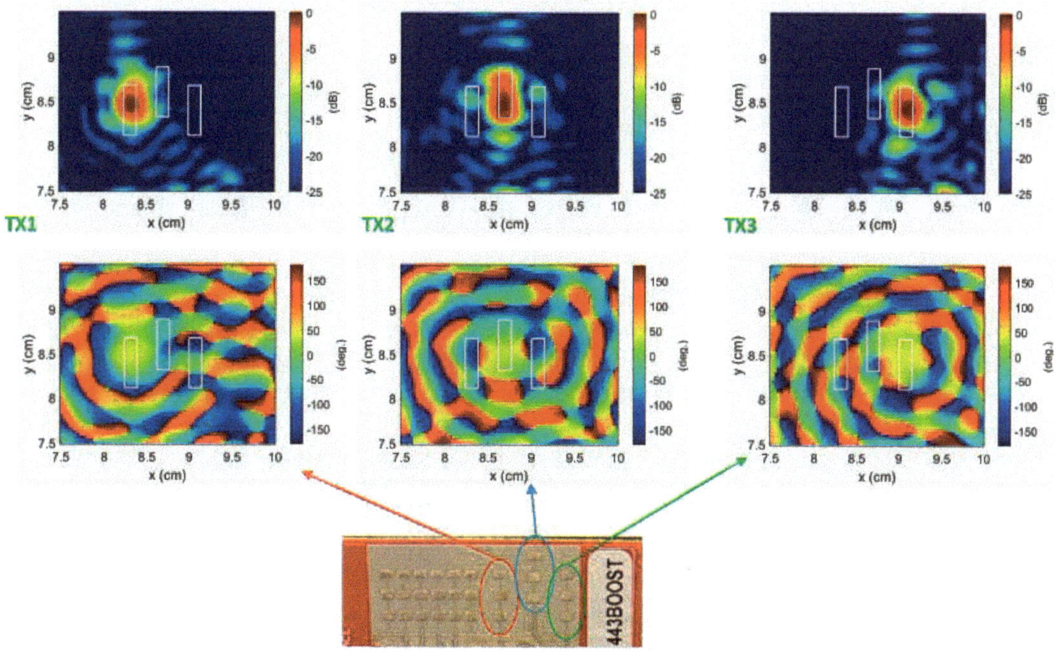

Figure 9.21 Reconstructed equivalent magnetic currents for each channel of the millimeter-wave radar.

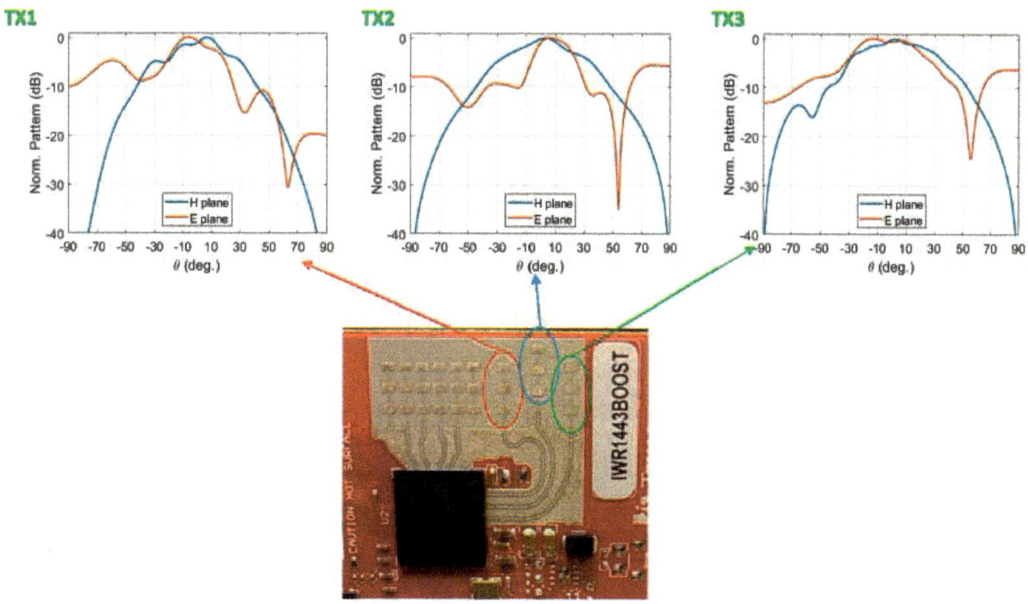

Figure 9.22 FF patterns from the magnetic currents for each channel of the millimeter-wave radar.

using a common reference source for the probe and the AUT would enable amplitude and phase measurements.

- Minimization of the georeferring uncertainty, which is tied to the accuracy of the positioning and georeferring subsystem onboard the UAV [13]. Devices capable of providing subcentimeter positioning accuracy would enable the use of UAV-based architectures for in-situ antenna measurement at millimeter-wave frequency bands.

On the other hand, the handheld approach is devoted to characterizing new transmission technologies available in novel cars, and it is suitable for *in-situ* measurements. In particular, it works well at the millimeter-wave band because 1) the effect of the reflections is minor so the operator does not have a significant impact on the measurements, and 2) the positioning accuracy can be achieved by means of affordable and portable systems.

The handheld system relies on pieces of hardware that are widely available with standard communication ports, enabling an easy integration. In a similar fashion to the UAV system, the amplitude of the electromagnetic field must be acquired in at least two different planes. However, the system is moved by means of free-hand sweeps, so techniques to overcome oversampled areas and large positioning errors must be implemented. As in the UAV case, the reconstructed currents in the aperture of the antenna as well as the FF radiation pattern can be found, enabling a complete evaluation of the radiation characteristics.

References

[1] Jesch, R. L., "Measured Vehicular Antenna Performance," *IEEE Transactions on Vehicular Technology,*1985, Vol. 34, No. 2, pp. 97–107.

[2] Yegin, K., "On-Vehicle GPS Antenna Measurements," *IEEE Antennas and Wireless Propagation Letters,* Vol. 6, 2007, pp. 488–491.

[3] Berbeci, M., P. Pelland, and T. Leifert, "Challenges for the Automotive Industry on MIMO OTA Testing," *2020 Antenna Measurement Techniques Association Symposium (AMTA),* 2020, pp. 1–5.

[4] Iversen, P. O., et al., "Experimental Comparison of Vehicular Antenna Measurements Performed Over Different Floors," *2020 14th European Conference on Antennas and Propagation (EuCAP),* 2020, pp. 1–5.

[5] Pelland, P., et al., "Automotive OTA Measurement Techniques and Challenges," *2020 Antenna Measurement Techniques Association Symposium (AMTA),* 2020, pp. 1–6.

[6] Ferrara, F., et al., "A Non-Redundant Hemi-Spherical Scanning for Automotive Antenna Near-Field Measurements," *2020 XXXIII General Assembly and Scientific Symposium of the International Union of Radio Science,* 2020, pp. 1–4.

[7] Hugler, C., et al., "Radar Taking Off," *IEEE Microwave Magazine,* 2018, Vol. 19, pp. 43–53.

[8] Salari, A., and D. Erricolo, "Unmanned Aerial Vehicles for High-Frequency Measurements: An Accurate, Fast, and Cost-Effective Technology," *IEEE Antennas and Propagation Magazine,* 2021, pp. 2–12.

[9] Virone, G., et al., "Antenna Pattern Verification System Based on a Micro Unmanned Aerial Vehicle (UAV)," *IEEE Antennas and Wireless Propagation Letters,* 2014, Vol. 13, pp. 169–172.

[10] Paonessa, F., et al., "UAV-Based Pattern Measurement of the SKALA," *Proc. of the IEEE International Symposium on Antennas and Propagation (APSURSI),* 2015, pp. 1372–1373.

[11] Schreiber, J., "Antenna Pattern Reconstitution Using Unmanned Aerial Vehicles (UAVs)," in *IEEE Conference on Antenna Measurements and Applications (CAMA),* 2016, pp. 1–3.

[12] Alvarez, Y., et al., "In Situ Antenna Diagnostics and Characterization System Based on RFID and Remotely Piloted Aircrafts," *Sensors and Actuators A,* Vol. 269, 2018, pp. 29–40.

[13] Garcia-Fernandez, M., "Antenna Diagnostics and Characterization using Unmanned Aerial Vehicles," *IEEE Access,* Vol. 5, 2017, pp. 23563–23575.

[14] Fernandez, M. G., Y. A. Lopez, and F. L. Heras, "On the Use of Unmanned Aerial Vehicles for Antennas and Coverage Diagnostics in Mobile Networks," *IEEE Communications Magazine,* Vol. 5, 2018, pp. 72–78.

[15] Bolli, P., et al., "Near-Field Experimental Verification of the EM Models for the LOFAR Radio Telescope," *IEEE Antennas and Wireless Propagation Letters,* Vol. 17, 2018, pp. 613–616.

[16] Paonessa, F., et al., "Design and Verification of a Q-Band Test Source for UAV Based Radiation Pattern Measurements," *IEEE Transactions on Instrumentation and Measurement,* 2020, Vol. 69, No. 12, pp. 9366–9370.

[17] Chang, C., et al., "Beam Calibration of Radio Telescopes with Drones," *Publications of the Astronomical Society of the Pacific,* Vol. 127, 2015, pp. 1131–1143.

[18] Teng, E., et al., "Aerial Sensing and Characterization of Three-Dimensional RF Fields," *2nd International Workshop on Robotic Sensor Networks,* 2015, pp. 1–6.

[19] Henault, S., "Antenna Measurement Drone for Over-the-Horizon Radar," *2021 15th European Conference on Antennas and Propagation (EuCAP),* 2021, pp. 1–5.

[20] Alvarez-López, Y., et al., "Airborne System and Method for´ the Characterization and Measurement of Antennas or Radiating Systems," *Espacenet,* 2017, Patent WO2018158472 (A1), https://patents.google.com/patent/WO2018158472A1/es.

[22] Garcia-Fernandez, M., Y. Alvarez-López, and F. Las-Heras, "Dual-Probe Near-Field Phaseless Antenna Measurement System on Board a UAV," *Sensors*, Vol. 19, No. 21, 2019.

[23] TimeDomain. BroadSpec UWB Antenna Datasheet of Time Domain; 2017, http://www.timedomain.com/datasheets/TD.

[24] Fordham, J. A., and F. D'Agostino, "Sphiral Near-Field Scanning for Automotive Antenna Measurements," *2016 10th European Conference on Antennas and Propagation (EuCAP),* 2016, pp. 1–5.

[25] Sanandiya, V. H., and S. T. McBride, "Definition, Implementation, and Evaluation of a Novel Spiral Sampling Technique," *2020 Antenna Measurement Techniques Association Symposium (AMTA),* 2020, pp. 1–5.

[26] Araque, J. L., and G. Vecchi, "Improved-Accuracy Source Reconstruction on Arbitrary 3-D Surfaces," *IEEE Antennas and Wireless Propagation Letters,* Vol. 8, 2009, pp. 1046–1049.

[27] Foged, L., et al., "Application of the Dual-Equation Equivalent-Current Reconstruction to Electrically Large Structures by Fast Multipole Method Enhancement," *IEEE Antennas and Propagation Magazine,* Vol. 62, No. 4, 2014, pp. 2041–2051.

[28] Alvarez, Y., F. Las-Heras, and M. Pino, "Reconstruction of Equivalent Currents Distribution Over Arbitrary Three-Dimensional Surfaces Based on Integral Equation Algorithms," *IEEE Transactions on Antennas and Propagation,* Vol. 55, No. 12, 2007, pp. 3460–3468.

[29] Alvarez, Y., F. Las-Heras , and C. Garcia-Gonzalez, "The Sources Reconstruction Method for Antenna Diagnostics And Imaging Applications," in *Solutions and Applications of Scattering, Propagation, Radiation and Emission of Electromagnetic Waves,* A. Kishk (ed.), Rijeka (Croatia): InTech, 2012.

[30] Persson, K., et al., "Radome Diagnostics—Source Reconstruction of Phase Objects with and Equivalent Currents Approach," *IEEE Transactions on Antennas and Propagation,* Vol. 56, No. 5, 2014, pp. 264–273.

[31] Alvarez Narciandi, G., et al., "Freehand System for Antenna Diagnosis Based on Amplitude-Only Data," *IEEE Transactions on Antennas and Propagation,* Vol. 69, No. 8, 2021, pp. 4988–4998.

[32] Ciorba, L., et al., "Near-Field Phase Reconstruction for UAV-Based Antenna Measurements," *2019 13th European Conference on Antennas and Propagation (EuCAP),* 2019, pp. 1–4.

[33] Razavi, S. F., and Y. Rahmat-Samii, "A New Look at Phaseless Planar Near-Field Measurements: Limitations, Simulations, Measurements, and a Hybrid Solution," *IEEE Antennas and Propagation Magazine,* Vol. 49, No. 2, 2007, pp. 170–178.

[34] Brown, T., I. Jeffrey, and P. Mojabi, "Multiplicatively Regularized Source Reconstruction Method for Phaseless Planar Near-Field Antenna Measurements," *IEEE Transactions on Antennas and Propagation,* Vol. 65, No. 4, 2017, pp. 2020–2031.

[35] Bucci,O. M., et al., "Far-Field Pattern Determination from the Near-Field Amplitude on Two Surfaces," *IEEE Transactions on Antennas and Propagation,* Vol. 38, No. 9, 1990, pp. 1772–1779.

[36] Las-Heras, F., and T. K. Sarkar, "A Direct Optimization Approach for Source Reconstruction and NF-FF Transformation Using Amplitude-Only Data," *IEEE Transactions on Antennas and Propagation,* Vol. 50, No. 4, 2002, pp. 500–510.

[37] Bucci O.M., et al., "Far-Field Computation from Amplitude Near-Field Data on Two Surfaces: Cylindrical Case," *IEE Proceedings, Part H - Microwaves, Antennas and Propagation,* Vol. 139, No. 2, 1992, pp. 143–148.

[38] Puskely, J. "Application of Iterative Fourier Method in Cylindrical Phaseless Antenna Measurement Technique," *Radioengineering,* 2012, Vol. 21, pp. 422–429.

[39] Yaccarino, R. G., and Y. Rahmat-Samii, "Phaseless Bipolar Planar Near-Field Measurements and Diagnostics of Array Antennas," *IEEE Transactions on Antennas and Propagation,* Vol. 47, No. 3, 1999, pp. 574–583.

[40] Schmidt, C. H., et al., "Phaseless Spherical Near-Field Antenna Measurements for Low and Medium Gain Antennas," *Advances in Radio Science,* 2010,Vol. 8, pp. 3–48.

[41] Costanzo, S., G. Di Massa, and M. D. Migliore, "A Novel Hybrid Approach for Far-Field Characterization from Near-Field Amplitude-Only Measurements on Arbitrary Scanning Surfaces," *IEEE Transactions on Antennas and Propagation,* Vol. 53, No. 6, 2005, pp. 1866–1874.

[42] Schnattinger, G., C. Lopez, and T. F. Eibert, "Fast Near-Field Far-Field Transformation for Phaseless and Irregular Antenna Measurement Data," *Advances in Radio Science,* 2014, Vol. 12, pp. 171–177.

[43] Alvarez, Y., F. Las-Heras, and M. Pino, "The Sources Reconstruction Method for Amplitude-Only Field Measurements," *IEEE Transactions on Antennas and Propagation,* Vol. 58, No. 8, 2010, pp. 2776–2781.

[44] Varela, F. R., et al., "Numerical and Experimental Investigation of Phaseless Spherical Near-Field Antenna Measurements," *IEEE Transactions on Antennas and Propagation,* Vol. 69, No. 12, 2021, pp. 8830–8841.

[45] Guan, J., et al., "High Resolution Millimeter Wave Imaging for Self-Driving Cars," 2019, arXiv:1912.09579.

[46] Shimizu, T., et al., "Millimeter Wave V2X Communications: Use Cases and Design Considerations of Beam Management," *2018 Asia-Pacific Microwave Conference (APMC),* 2018, pp. 183–185.

[47] Li, Z., et al., "Proof-of-Concept of a SDN Based mmWave V2X Network for Safe Automated Driving," *2019 IEEE Global Communications Conference (GLOBECOM),* 2019, pp. 1–6.

[48] Rappaport, T. S., et al., "Millimeter Wave Mobile Communications for 5G Cellular: It Will Work!," *IEEE Access,* 2013, Vol. 1, pp. 335–349.

[49] Nasr, I., et al., "A Highly Integrated 60 GHz 6-Channel Transceiver with Antenna in Package for Smart Sensing and Short-Range Communications," *IEEE J Solid-State Circuits,* Vol. 51, No. 9, September 2016, pp. 2066–2076.

[50] Alvarez-Narciandi, G., et al., "Freehand, Agile, and High Resolution Imaging with Compact mm-Wave Radar," *IEEE Access,* Vol. 7, 2019, pp. 95516–95526.

[51] Alvarez Narciandi, G., J. Laviada, and F. Las-Heras, "Freehand mm-Wave Imaging with a Compact MIMO Radar," *IEEE Transactions on Antennas and Propagation,* Vol. 69, No. 2, 2021, pp. 1224–1229.

[52] Alvarez-Narciandi, G., J. Laviada, and F. Las-Heras, "Towards Turning Smartphones Into mmWave Scanners," *IEEE Access,* Vol. 9, 2021, pp. 45147–45154.

[53] Alvarez-Narciandi, G., et al., "Rapidly Deployable Portable System for Real-Time Antenna Diagnostics and Characterization," *2019 IEEE International Symposium on Antennas and Propagation and USNC-URSI Radio Science Meeting,* 2019, pp. 1689–1690.

[54] Alvarez-Narciandi, G., et al., "Portable Freehand System for Real Time Antenna Diagnosis and Characterization," *IEEE Transactions on Antennas and Propagation,* Vol. 68, No. 7, 2020, pp. 5636–5645.

[55] OptiTrack. Optitrack Motion Capture System; 2021, http://www. optitrack.com.

[56] Texas Instruments. AWR1443 mmWave sensor datasheet,2021, https://www.ti.com/product/AWR1443.

CHAPTER 10

Practical Aspects of Automotive Measurements and Virtual-Drive Testing

B. Altinel, M. E. Asghar, P. Berlt, S. Buddappagari, C. Bornkessel, J. Singh, and M. A. Hein

While previous chapters of this book deal with fundamental aspects of automotive antenna measurements, this chapter focuses on practical aspects related to installed-performance testing approaches for final design verification. Chapter 10 is divided into three sections: Challenges related to the typical limitations of automotive antenna measurement facilities and to the implications of the car body for antenna measurements are treated in Section 10.1. Section 10.2 describes application-oriented post-processing of automotive antenna measurements as needed, for instance, to localize the phase center of an automotive antenna on an absolute geometrical scale, or to derive application-oriented figures-of-merit like total radiated power. Section 10.3 concludes our analysis by discussing the over-the-air vehicle-in-the-loop approach for system validation, starting with a brief introduction and exemplified for two major automotive wireless transmission functionalities: mobile communications in the centimeter-wave range and automotive radar at millimeter-wave frequencies.

10.1 Challenges for Automotive Antenna Measurements

10.1.1 Limitations of Automotive Antenna Measurement Range

With the introduction and advancement of wireless communication and sensor technologies in automotive applications, particularly for automated and connected driving, the demand for antenna measurements and testing under real-world conditions is rapidly growing. Spherical near-field measurements present a widely accepted method in the automotive industry for the characterization of the radiation patterns of automotive antennas, both uncalibrated and calibrated, and performance metrics derived from these. This technique provides precise results over a wide range of frequencies and antenna sizes, where FF conditions cannot be achieved due to space constraints. Some of the key factors differentiating the angular scanning techniques to be considered and balanced under practical and efficiency considerations are listed as follows:

1. Mechanical positioning and electrodynamic system accuracies;
2. Frequency range and spectral sampling density;
3. Measurement speed;
4. Calibration.

The achievable accuracy depends on the measurement technique and calibration procedure applied. Antenna measurement systems that require high-frequency operation or high measurement accuracy have to minimize the angular sampling density to the limit required to perform an adequate NF-to-FF transformation [1–4]. Depending on the size of the AUT, a full spherical coverage with adequate sampling density can become very time-consuming, especially at high frequencies. Various scanning and data post-processing techniques have been devised that can increase the measurement speed while maintaining high accuracy.

Referring to Chapters 2 and 5, spherical measurement systems can be divided into three categories: swinging arm/s (gantry), fixed-probe systems, or robotic arms. Examples of relevance for this chapter are shown in Figure 10.1. The gantry system consists of a single or double arm with the probe antenna mounted at a fixed position [5, 6]. The mechanical movement of the gantry covers the elevation range, and the rotation of a turntable covers the azimuthal range.

One major advantage of such a system is the adaptivity to different frequency ranges and operational bandwidths. Disadvantages of such systems are the mechanical drives and the limited scanning speed resulting from the large moving mass and the finite stiffness of the gantry structure. Reflections off the gantry arm, even if attenuated by absorber covering, can induce artifacts in the radiation pattern. This shortcoming can be mitigated by using a highly directive probe antenna and applying time-gating afterward.

On the contrary, a fixed multiprobe measurement system offers the benefits of hybrid electronic-mechanical drives and fast antenna measurements, though limited to frequency ranges determined by the installed probes [7]. Electronic switching of probes samples the complex-valued fields along one axis, while the AUT is rotated using a turntable. This significantly reduces the measurement time by a factor proportional to the number of antennas in the array. As the probes are fixed, pairs of broadband dual-polarized probes can be installed in two different angular sections. Probes operating at high frequencies require thorough calibration.

(a) (b)

Figure 10.1 Photographs of different spherical NF automotive antenna measurement ranges: (a) dual-arm gantry system (©ATC) and (b) fixed multiprobe measurement arch with two sets of probe arrays (©VISTA). Source: [11].

Three-dimensional scanning techniques are often truncated at or close to the horizon, in order to accommodate the vehicle-under-test while limiting the size and corresponding cost of the chamber. Typically, antennas are measured in the radiating NF region, and the recorded field data are then transformed to FF conditions (e.g., using SWE) [1]. For the transformation to provide accurate results, the entire sphere around the AUT must be covered by measurement points. Truncated measurement ranges, therefore, require advanced interpolation techniques, as the embedding of zeros across the truncated range as a simple basic approach induces artifacts, that may affect the resulting radiation pattern up to elevation angles well above horizon. This effect can be reduced by applying more advanced field reconstruction algorithms. From the mathematical point-of-view, the problem of field reconstruction from truncated measurements can be rephrased as a problem of signal restoration from incomplete or distorted data. In this kind of problem, additional information like the knowledge of the AUT and the minimum measurement sphere is required, to ensure a robust a stable restoration procedure and a unique solution [8, 9].

For automotive applications, the radiation performance of the antennas in their installed state is generally of interest under different environmental conditions. Measurements above reflective ground (semi-anechoic, e.g., [6]) or above absorbing ground [7] offer two complementary approaches. The type of ground can be adjusted according to the electrical conductivity and the dielectric permittivity. In anechoic chambers emulating free-space conditions, the vehicle carrying the AUT is lifted above the ground, covered with absorbers such that ground reflections and scattering off chamber installations are reduced to a tolerable limit. For measurements above reflective—or even conductive—ground, the vehicle is placed on the ground. Such measurements suffer from the strong interaction between floor and measurement system. The main advantage of reflective floor systems is the ease of accommodation of the vehicle-under-test, which is simply parked in the center of the system. In absorber-based systems, instead, more time is needed to prepare the measurement setup, but at the same time, it offers the key advantage to extract the free-space signature of installed antenna that can be further post-processed to emulate realistic ground parameters using numerical tools [10]. Present trends in automotive antenna measurements point toward combined experimental-numerical methods.

Figure 10.2 shows the vertical cuts of the realized gain of two reference antennas measured in the framework of a comparison of two different automotive antenna measurement facilities, namely the Antenna Technology Center located in Itzehoe, Germany (ATC, gantry system, see Figure 10.1(a)), and the virtual road simulation and test area (VISTA, fixed-probe arch, see Figure 10.1(b)), located at the Thuringian Center of Innovation in Mobility at the Technische Universität Ilmenau, Germany. VISTA provides free-space conditions while measurements in ATC are performed over conducting ground plane. The difference between free-space measurements and conducting ground is visible in the results where reflections from the metallic ground produce ripples in the pattern, more pronounced for increasing frequency. Apart from the ripples, the convincingly good comparability between the results obtained for both facilities demonstrates that the type of setup is not relevant as long as these are accurately designed, and measurement and post-processing procedures are well established [11].

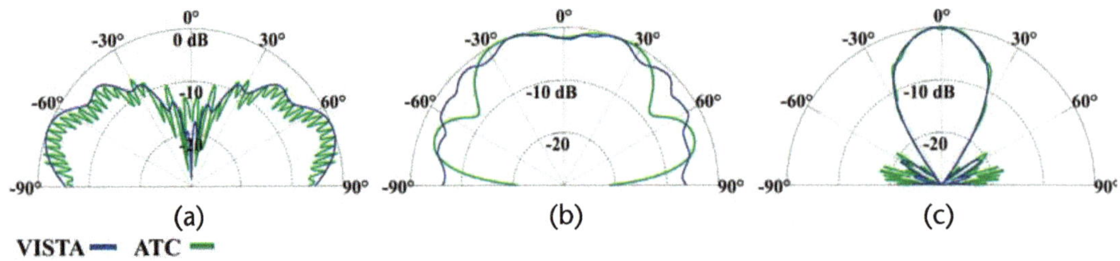

Figure 10.2 Comparison between measured results obtained from the automotive facilities VISTA (blue curves) and ATC (green curves). Vertical cuts of the realized gain (a) for a reference monopole antenna at frequency $f = 2.495$ GHz, and (b, c) for a horn antenna at frequencies $f = 0.7$ GHz and $f = 5.9$ GHz, respectively. ©2019 IEEE.

Automotive NF antenna measurement ranges face certain limitations in terms of conventional NF-to-FF transformation methods, which become critical when the AUT is located outside the minimum measurement sphere (see Chapter 3). The maximum angular sampling distance that meets the Nyquist sampling criterion on the measurement sphere is given as $\Delta\theta \leq \pi/N$, and the maximum electrical size of the minimum sphere at a given frequency f or equivalent free-space wavelength λ can be derived with a safety margin of $10/\pi$ as

$$\frac{d_{SWE}}{\lambda} \leq \frac{1}{\Delta\theta} - \frac{10}{\pi} \approx \frac{1}{\Delta\theta} - 3.2 \tag{10.1}$$

The sphere with the diameter dSWE is referred to as the equivalent measurement sphere. Facilities like VISTA with a fixed multiprobe measurement arch have a fixed angular resolution $\Delta\theta = 1°$ along the elevation axis. This limits the maximum electrical size of an object to be arranged within the minimum sphere. In cases where the AUT (e.g., mounted in the rear spoiler) falls outside of this equivalent sphere, the SWE method will not yield valid results. This problem can be overcome using a TSWE algorithm, where the equivalent measurement sphere is shifted to the location of the antenna-under-test, at the expense of an asymmetric sampling. Practical examples for antennas mounted in the rear of a station wagon are discussed in Section 5.4.5 and [12].

As discussed before, antenna measurements are typically performed in an anechoic (free-space) or semi-anechoic (reflective) environment. Facilities deploying variable boundary conditions for measurements are difficult to devise, although a versatile approach to test and qualify vehicular antennas is desirable. One possible approach is to combine measurements with numerical post-processing of data measured in an echo-free environment [10]. Another approach, entirely based on measurements, has been introduced recently as a modular extension of the automotive test facility VISTA [13]. The setup consists of a mechanically robust plastic frame carrying a silver-plated fabric of thickness 0.1 mm with shielding effectiveness of more than 60 dB presenting the ground plane, as visualized in Figure 10.3. The structure is laid out to be removable and extendable in height. The ground diameter measures 6m, which fits on the turntable and is sufficient to cover the length of the typical cars. The setup allows for both stand-alone antenna measurements and measurements with antennas installed in a car.

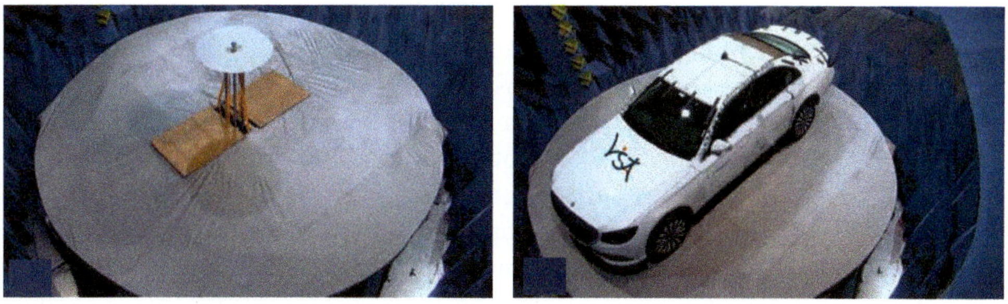

(a) (b)

Figure 10.3 Variable boundary conditions antenna measurement setup in VISTA. The AUT is placed above the metalized fabric stretched across an adjustable plastic frame: (a) reference monopole antenna and (b) installed automotive antenna measurement setup. ©2022 IEEE. (Reprinted with permission.) [13].

For initial proof-of-principle measurements, a monocone antenna with a ground plane was chosen; this type of antenna has also been used for gain calibration. The antenna was lifted to the phase center of the arch, and the woven ground plane was placed underneath at a distance of 1m. This setup is analogous to an antenna installed on the rooftop of a car if we consider a typical height of a car of 1m. To verify the measured results, corresponding full-wave electromagnetic simulations were performed. The model configurations were kept the same as for the measurements.

Figure 10.4(a–c) shows the comparative results of realized gain patterns obtained from simulation and measurement for different frequencies over a bandwidth spanning from 1 to 6 GHz; the gain patterns are normalized to their respective maximum values.

The excellent agreement between the measured and simulated patterns with a correlation of above 90% confirm the test concept. Similar results were also obtained for installed automotive antenna measurements.

10.1.2 Implications of the Car Body for Antenna Design and Measurements in the Installed State

Passenger cars as well as commercial vehicles are no longer just simple modes of transportation but provide wireless access to a variety of safety- and efficiency-relevant services as well as infotainment for the driver and the passengers, so much

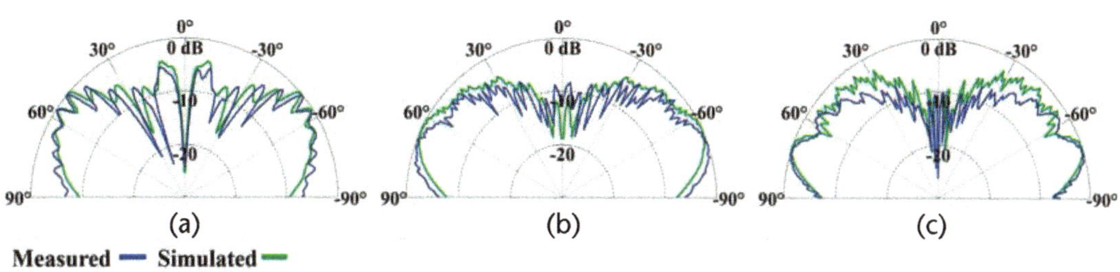

Figure 10.4 Comparison of measured and simulated realized gain patterns for the monopole reference antenna for frequencies (a) $f = 1$ GHz, (b) $f = 3$ GHz, and (c) $f = 5$ GHz. ©2022 IEEE. (Reprinted with permission.) [13].

that nowadays the lack or presence of many of these functionalities is a deciding factor in the purchase of a new vehicle [14, 15]. Accordingly, a wealth of antennas encompassing frequencies from as low as 0.5 MHz for AM radio up to as high as 77 GHz for radar-based collision detection need to be integrated at suitable locations in a passenger car [16]. Not seldom, the total number of antennas installed in high-level passenger cars can reach 30 to 50. With the advent of the 4G (LTE) and 5G mobile communication standards and the demand for MIMO and diversity techniques, the number of antennas is expected to increase further [16]. One may think that since cars are relatively large, finding mounting spaces for all of the current and upcoming antennas would not pose a challenge. However, one factor that plays a key role in the purchase decision of a vehicle is its appearance; potential buyers want a functional car that looks stylish, sleek, and aerodynamic. Antennas like rooftop monopoles may fulfill these requirements but are not robust, can be vandalized, and are exposed to the elements.

Keeping in line with the desired aesthetics of a car, automotive antennas need to be conformal and low-profile, to be embedded into or hidden behind the available plastic/glass parts in a typical passenger car. Such locations are limited in number and are already occupied by a host of antennas nowadays. LTE-MIMO antennas are considered in the rearview mirror in [17]; LTE and V2X antennas have been designed for spoilers [18, 19]; antennas for broadcast services like AM, FM, digital audio broadcast, TV and so on are usually embedded in the rear windscreen or fixed side windows [20–23]; broadcast, mobile communication, and RFID antennas have been designed for side mirrors [24, 25]; and plastic bumpers are favored locations for collision avoidance radars [26, 27]. Shark fin housings hide a variety of antennas including, but not limited to, LTE, AM/FM, V2X, and GPS [28–31]. Therefore, it is apparent that new locations on a car that allow for an integration of antennas need to be identified or made accessible by proper antenna design. As an example, in this direction, antennas were integrated into a cavity constructed on the roof of a car [32]. A further requirement for future automotive antenna design is the integration of multiservice antennas with their front ends in a telematic control unit, while maintaining sufficient decoupling between the multiple wireless functionalities [33].

In addition to finding antenna integration locations, the impact of these locations on the antenna performance also needs to be considered. The electromagnetic interaction between the car chassis and the antenna alters its radiation performance compared to free-space operation. The extent of this interaction depends on the orientation of the antenna at that location as well. For example, if dipole antennas or monopole antennas were oriented parallel to and close above the metal chassis, there would be a significant degradation of the antenna efficiency due to the destructive influence of the image currents in the car body. To avoid this situation, monopole antennas contained in a shark fin housing are oriented perpendicular to the car roof. While the perpendicular orientation of the monopole solves the efficiency problem, the radiation pattern of the monopole shows differences in relation to where it is placed on the roof of the car. The study in [34] showed that shifting a vertically oriented wideband monopole antenna to different positions on a sedan rooftop resulted in nonnegligible, frequency-dependent differences in the shape of its radiation pattern, as the frequency increased from 0.7 to 5.9 GHz, as illustrated in Figure 10.5. Through the example of the monopole antenna, it becomes clear

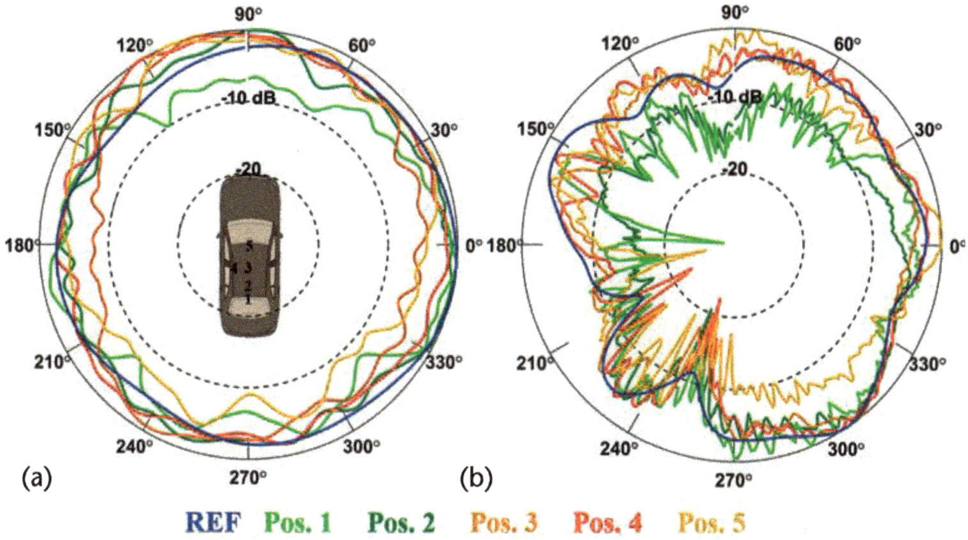

Figure 10.5 Horizontal cuts of the gain patters at (a) 0.7 GHz and (b) 5.9 GHz for five different positions of the vertically arranged rooftop monopole antenna [34].

that an automotive antenna cannot be designed stand-alone and is expected to perform just the same following installation on a car. For an accurate antenna design, the antenna and—at least—a partial/simplified model of the car need to be simulated together to consider the electromagnetic interaction between them.

Previously unused, glossy black-painted plastic covers like in the B-column between the front and rear doors of a passenger car present an attractive new antenna embedding location. There are three principal advantages for this location: (1) The location already exists on typical passenger cars, implying minimal car design changes on the part of the manufacturers, (2) B-column covers are present on the left and right sides of the car; embedding the same antenna on both sides allows for a nearly omnidirectional pattern across the horizontal plane, and (3) durability and robustness over typical rooftop monopoles, and far more invisible integration compared to shark fin housings.

However, making this location viable is not without challenges. The first among them is the injection molding process for embedding the antenna into the plastic cover. The antenna may be exposed to pressures of up to 140 MPa and temperatures as high as 240°C [35]. The second challenge is that following the mounting of the plastic cover onto the car, the embedded antenna lies just 2–3 mm apart from the car chassis. This leads to the aforementioned drawback of poor efficiency for an embedded half-wavelength dipole antenna. Typical patch antennas would work fine in such an environment. However, these, especially for a low-profile design, are too limited in impedance bandwidth to be effectively used for LTE/5G mobile communications applications. Therefore, the need arose for new or improvised antenna designs that suitably accounted for these challenges. The plastic integrated di-patch antenna presented in [35] shows advantages over typical patch antennas. Although called as such, the di-patch antenna originates from a planar full-wavelength dipole antenna that is suitably modified in geometry and the location of its feed, enabling it to radiate efficiently closely above a ground plane. As discussed in detail in [35–37] and as indicated in the inset of Figure 10.6, it comprises two square patches fed differentially at their corners. The two patches lie atop a grounded substrate, with no ohmic contact between the patches and the ground. It is important to note that a half-wavelength dipole would not work closely above a ground plane with such modifications, so the emphasis here is on the phrase "full-wavelength." Unlike a typical patch antenna, a di-patch antenna occupies nearly half the total surface area while not compromising the directivity and realized gain and offering twice the impedance bandwidth. Figure 10.6 also presents measured example radiation patterns with the embedded di-patch antenna mounted onto the front left car door. The measured realized gain and efficiency in both considered cases were approximately 1 dBi and 15%, respectively.

Multiband patch antennas considered in [38] are also well-suited for integration in the B-column plastic cover. These rely on co-planar stacking of parasitic patches and corner feeding of the primary patches. Corner feeding excites two further apart and orthogonal resonant frequencies, while parasitic patches further bifurcate these orthogonal resonances. Similar multiband behavior may also be achieved through the vertical stacking of parasitic patches. Another contender would be high-impedance surface– (HIS-) based half-wavelength dipole antennas [39]. A HIS provides an open boundary condition to an overlying half-wavelength dipole antenna, leading to a constructive superposition of the dipole currents and the image currents on the HIS, enabling the dipole to radiate efficiently.

Global demand for transport is growing fast. Given present trends, passenger and freight traffic will more than double by 2050. Among all modes of transport, railway transportation is the most energy-efficient for freight and passengers. The

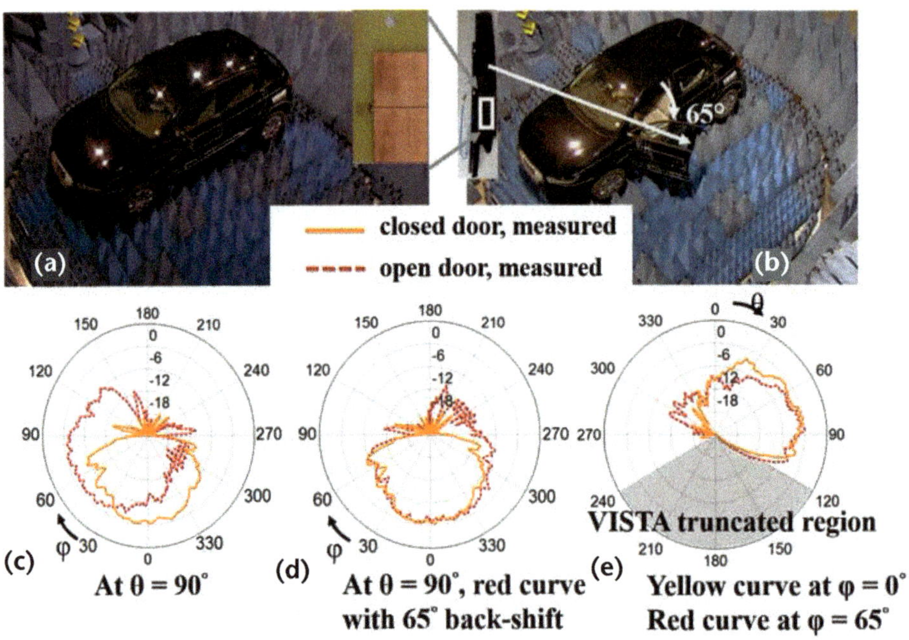

Figure 10.6 Radiation patterns measured with (a) car doors closed, and (b) with the front left door opened; the inset shows the di-patch antenna embedded in the B-column plastic cover that was mounted onto the car, and (c–e) compare the normalized horizontal and vertical cuts of gain patterns measured for the two cases.

requirements from regulation bodies and railway operators demands of an on-board "connected" environment for both customer and operational purposes have initiated a similar trend for antenna multiplicity for rail as for road vehicles [40, 41]. For signaling operations, antennas with omnidirectional patterns are desired, particularly for the co-elevation angular range 60°–90°, as the base stations are generally positioned alongside the track. As the impact of the mounting position on antenna radiation characteristics mounted on the rooftop of trains is even more significant than for cars, due to the larger geometrical size and additional installations, special attention has to be paid to the identification of suitable installation spaces. There can be various types of obstructions on the roof of the train that affect the antenna performance. Examples include air-conditioning units, pantographs, isolators, curved or stepped roof shapes, and power lines.

Another challenge arises when measuring the performance of antennas installed on locomotive roofs. Due to the physically large, heavy, and complex structure of locomotives, it is impractical to accommodate them inside a measurement chamber. Alternatively, scaled model measurements have been used successfully to evaluate the performance of installed antennas. In [42] a realistic, geometrically scaled model of a locomotive mockup for testing at correspondingly scaled frequencies was presented. The test platform reflects essential characteristics and features of typical locomotive roofs, including radiation-relevant superstructures. The scaled model was validated with respect to the original model using electromagnetic full-wave simulations. Subsequently, measurements of the scaled mockup were carried out in VISTA and results were compared with numerical simulations to evaluate the potential of both techniques for the design and analysis of future railway communication locomotive antennas. The results measured at different scaled frequencies and for different mounting positions reveal the strong impact of the mounting position, as well as a significant impact from the different roof geometries and superstructures (i.e., NF effects like parasitic radiation and FF effects like shadowing) on the radiation patterns. The results illustrate that both measurements on the scaled model and numerical simulations present powerful complementary approaches toward analyzing and optimizing locomotive antennas.

10.2 Application-Oriented Post-Processing of Automotive Antenna Measurements

10.2.1 Phase Center Determination

For safety-relevant and high-precision applications like automated driving or satellite navigation, it is desirable to evaluate the phase center and its variations with frequency directly from antenna measurements, in order to localize the antennas three-dimensionally with high spatial resolution. Phase measurements and the derivation of the phase center represent, therefore, important though challenging tasks of antenna measurement and post-processing, especially for automotive antennas in the installed state. The literature describes methods for the localization of the phase center, which often refer to aperture radiators or phase center determination along cuts through the three-dimensional space. Examples are the second-derivative [43, 44], two-point [45], edge diffraction [46], differential phase [47], three-antenna [48], and time-of-arrival methods [49].

One powerful method suitable to calculate the phase center directly from measured phase data is the spherical-phase method (SPM), with which the phase center is obtained as the solution of a linear system of equations. Another variant is an optimization algorithm, in which measured phase values are compared with theoretically expected values, based on the path length between probe antenna and phase center, and the difference is minimized over the entire measurement surface [53].

The SPM uses the translation theorem of antennas [50], which computes the phase angle of an antenna shifted by a vector given by Cartesian coordinates x, y, and z, to calculate the phase center directly from the phase data (compare with Figure 10.7). This method is suited for plane waves (i.e., under FF conditions). The system of equations depends, according to (10.2) and (10.3), on the vector P, which contains the measured phase values at the sampling points (θ_m, ϕ_n); the rows of the matrix E (with its pseudo-inverse denominated E^+) correspond to the normalized wave vector k for each azimuth and elevation angle [52]:

$$(x, y, z, C)^T = \frac{\lambda}{2\pi} E^+ \times P \tag{10.2}$$

$$E = \left(\cos\phi_m \sin\theta_n \sin\phi_m \sin\theta_n \cos\theta_n 1\right)^{(M^*N)\times 4} \tag{10.3}$$

C represents a constant, direction-independent phase offset due to cabling. To run the optimzation algorithm, the geometry of the measuring surface must be known, where any observation point on it is denoted by the distance $\boldsymbol{r}$. For each possible position of the phase center, its distance d from the observation point is used to obtain the calculated spatial phase shift:

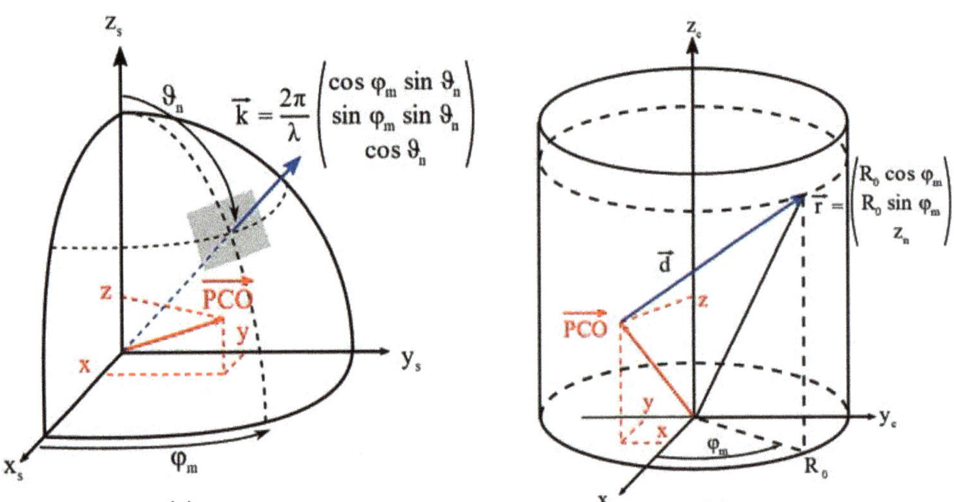

Figure 10.7 Sketch of the setup for (a) localization of the AUT on the basis of spherical FF data, and (b) on the basis of cylindrical sampling of the measurement surface. ©2021 IEEE [53].

$$d = |\mathbf{d}| = \left| \mathbf{r} - \begin{pmatrix} x \\ y \\ z \end{pmatrix} \right| \tag{10.4}$$

$$\phi_{calc}(x, y, z, C) = -\frac{2\pi}{\lambda} \cdot d + C \tag{10.5}$$

With the measured phase value $\phi^{(l)}_{meas}$ for each observation point with index l, the objective function f(x,y,z,C) becomes:

$$f(x, y, z, C) = \frac{1}{L} \sum_{l=1}^{L} \left| \phi^{(l)}_{calc}(x, y, z, C) - \phi^{(l)}_{meas} \right|^2 \tag{10.6}$$

The least-squares minimization of this function yields the coordinates of the phase center as well as the offset C. In case of cylindrical phase sampling, the measurement surface is the outer surface of a cylinder with radius R_0, with its height oriented along the z-axis as illustrated in the right-hand panel of Figure 10.7. According to this setup, with azimuth steps ϕ_m and vertical steps z_n, the expected phase data becomes

$$\phi^{(m,n)}_{calc,cyl} = \frac{2\pi}{\lambda} \times \sqrt{(R_0 \cos\phi_m - x)^2 + (R_0 \sin\phi_m - y)^2 + (z_n - z)^2} + C \tag{10.7}$$

and can be substituted in (10.6). Applying both spherical and cylindrical methods to an antenna integrated in a door frame, the position of an automotive antenna was determined with test signals in the LTE-band 3 (f = 1,870 MHz, λ = 16 cm) with an accuracy below 1 cm (i.e., better than $\lambda/16$) [53].

For this localization method, it is important that the phase data are unambiguous (i.e., phase ambiguities at $\pm\pi$ need to be unwrapped) (see Figure 10.8). If the antenna is offset from the origin of the measurement system by more than one wavelength, this becomes especially relevant. Two-dimensional unwrapping can be realized either by sequential unwrapping along each of the two dimensions (Itoh-algorithm [54]) or by using more complicated algorithms (e.g., based on the transport-of-intensity equation) [55]. Both approaches yield the phase center in a straightforward way. However, determining the phase diagram might be challenging due to difficult-to-access antenna locations in the vehicle or when antennas are integrated with their front end, leaving no access to the RF feed. In the latter case, advanced phase recovery techniques must be used, on the basis of which the phase center can then be derived using the methods described. As described in [56], the phase of integrated LTE antennas can be recovered by indirect holography in the spatial domain [57], using modulated LTE uplink communication signals instead of continuous-wave signals.

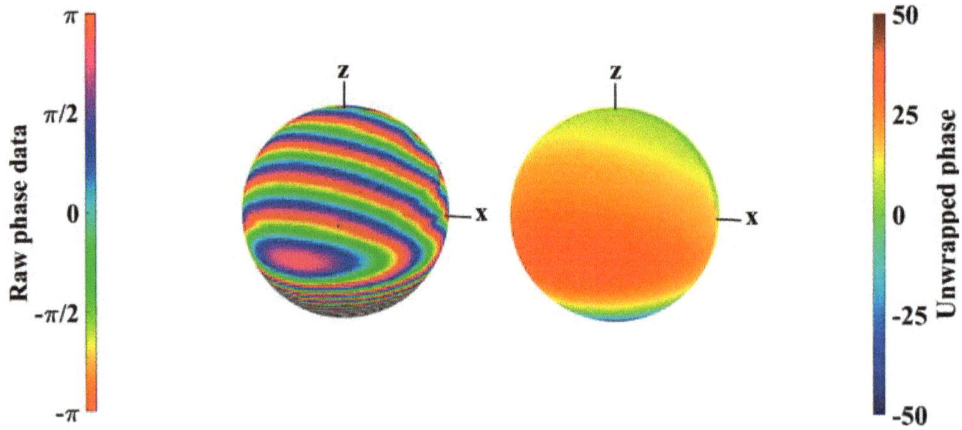

Figure 10.8 Illustration of phase unwrapping for the spherical-phase method. The left-hand panel shows raw phase data in the range $\pm\pi$ with several phase jumps along azimuth and elevation. Unwrapped data can be seen in the right-hand panel where the phase varies smoothly along the surface.

10.2.2 Automotive Antenna Performance Indicators

Over-the-air antenna testing is employed as a viable approach to characterize the expected performance of a wireless link for a given application. Since the car body can affect the radiation pattern of automotive antennas significantly, the testing must be performed in the installed state. Similar to handheld user terminals, the total radiated power (TRP) has been accepted as one important metric to characterize automotive antenna performance. It is a figure-of-merit adopted by WiMax, 3GPP, and CTIA standardization bodies [58–60]. In addition, specific TRP measures have been defined recently, considering limited elevation ranges only, as reflected in many automotive application standards for terrestrial and satellite-based mobile communications. Corresponding antenna performance testing was conducted in the automotive antenna test range VISTA, as published in [61, 62] and further detailed in Chapter 2.

It is important for the correct determination of TRP values that the measured data are available over the complete sampling grid. Measurement range truncation, as discussed in Section 10.1.1, can falsify the TRP values. This is of special concern for cases where the antenna-under-test radiates strongly in the truncated region (e.g., antennas mounted at the front, rear, or side of a car). In [63], the effect of truncation on measured patterns was thoroughly investigated. To extenuate the impact of truncation, different extrapolation techniques were used in post-processing. Extrapolation was performed either in the spectral domain (modal filtering) or in the spatial domain (equivalence principle); these algorithms were studied in detail in [64–66].

The TRP determination was applied to measurement results for the co-planar stacked penta-band patch antenna [38] mentioned in Section 10.1.2. The plastic embedded antenna and the door frame containing the B-pillar were measured in VISTA, as illustrated in Figure 10.9. The measured data were extrapolated using iterative modal filtering (IMF) and equivalent currents (EQC) methods. For comparison, a reference pattern was generated by measuring the AUT twice in two

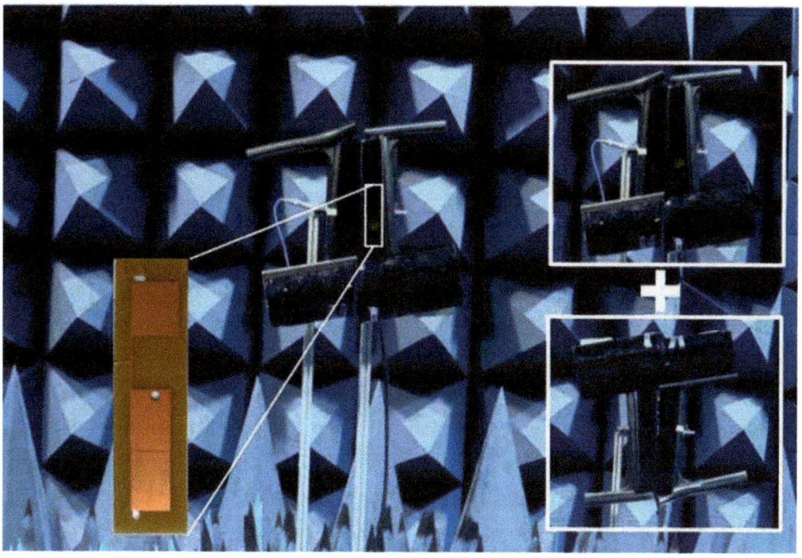

Figure 10.9 Photograph of the setup in VISTA for the measurement of a co-planar stacked penta-band patch antenna (inset on left-hand side). The antenna is embedded inside the plastic B-column cover of an Audi A1. The antenna was measured in two different orientations along the *z*-axis, and the reference pattern was generated by combining the measured upper-hemisphere data, as indicated by the insets on the right-hand side.

different orientations (i.e., straight and 180° rotated), and afterward combining the measured upper-hemisphere patterns.

Figure 10.10 shows the realized gain results of measured, extrapolated, and reference patterns at a frequency of 1.8 GHz. The results illustrate that the extrapolation techniques, in particular modal filtering, can accurately extract the total radiated power values, with a deviation below 1.1 dB compared to the reference.

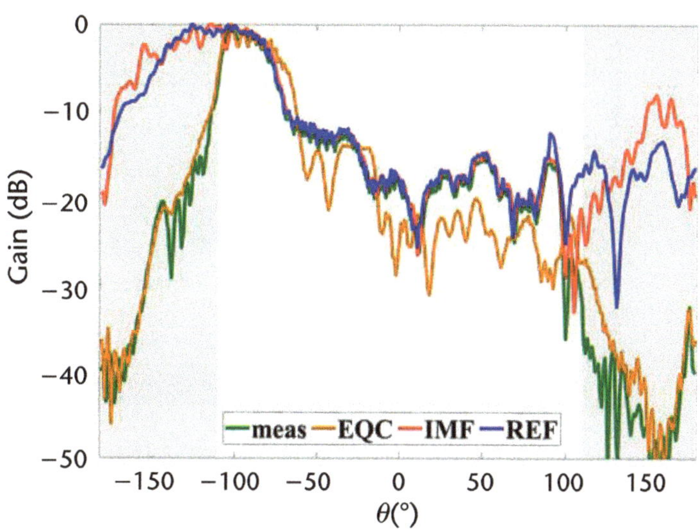

Figure 10.10 Vertical cuts of the realized gain for measured, extrapolated, and reference patterns. The patterns are normalized to their respective peak values. The truncated region is highlighted as grey-shaded areas.

10.3 Over-the-Air Vehicle-in-the-Loop Approach for System Validation in a Virtual Environment

10.3.1 Concept

The transformation toward automated and connected driving (ACD) implies manifold challenges for the testing, verification, and validation of the underlying wireless automotive systems. While the different levels of driving automation [67] have been treated extensively in literature, much less focus was placed on the implications for wireless connectivity until today. Figure 10.11 contrasts the six levels of driving automation with the variety of wireless functionalities required for ACD. As the driving control continues to shift from human drivers to robotic or telecontrolled driving systems for increasing levels of driving automation, the availability, quality, and security of the wireless channel becomes the bottleneck toward fail-operational ACD; we term this need "wireless reliance," which necessitates intensive testing.

Environmental perception through automotive radar has become without a doubt the key sensing technology, very likely to be supported by optical cameras and lidar systems. However, the importance of wireless connectivity has been grossly underestimated until today, as V2X communications in their present layout

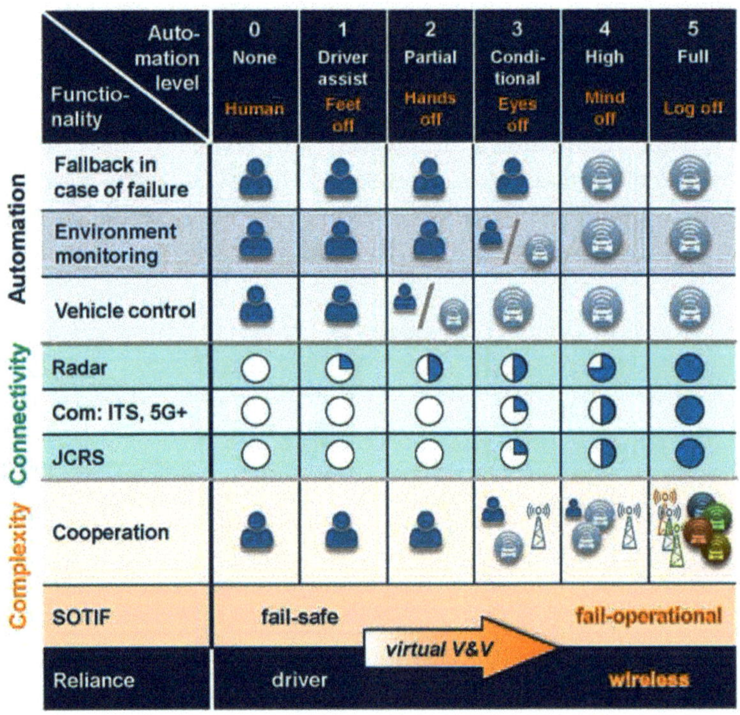

Figure 10.11 Visualization of the vision "wireless reliance." The icons in rows 1–3 and 7 symbolize: control by human driver and control by automated driving system, respectively. The symbols in rows 4–6 indicate the system relevance and functional abundance in four steps from "none" (empty)" to "full" (filled symbols). The tower icons in row 7 represent mobile base stations. Different colors denote different wireless standards; numbers of symbols indicate multiuser, multilink mesh networks. (Please note: Com stands for wireless communication standards; V2X for vehicular-to-everything communications; 5G+ – 5G to new radio and beyond; JCRS to joint communication and radar sensing; and SOTIF to safety of the intended function.)

will definitely not suffice to satisfy the need for exchange of operational data between a multitude of heterogeneous traffic participants in real time [68, 69]. Such intense connectivity is also required to break down the virtually infinitely large decision trees involved in machine learning-based ego-car-oriented driving control algorithms. If cars began to communicate their intentions to each other, the decision trees would collapse to an optimization problem whose solution would no longer require unrealistic computational power and time resources. Therefore, we expect that future releases of the 4G and especially 5G standards will take over key relevance for connected driving.

One plausible next step, which is presently already assuming shape, is the consequent combination of wireless communications and sensing, or joint communication and radar sensing (JCRS) [70]. This development will enable relevant advancements like communication-assisted radar as well as radar-assisted communication. Another step of complication in terms of wireless connectivity will result from the need for driving cooperation, for which—similar to driving automation—various hierarchical levels are under consideration [71].

Altogether, the trends toward sensor fusion, sensor data fusion, and wireless convergence create an urgent need for holistic testing, where the automotive antennas installed on a vehicle must no longer be considered in isolation but should be understood as parts of a complex system. The scientific community has recognized that, in order to verify and validate ACD, distance-based testing cannot meet the safety requirements any more but has to be superseded by scenario-based testing [72]. Figure 10.12 illustrates the complementarity of different testing approaches, which require a trade-off between the degree of realism of a test case and the level of reproducibility and simplification attainable for a given test procedure. Virtual environments become indispensable in this context, in order to save resources, increase efficiency, and cover the innumerable wealth of test cases. Among the different approaches, we have been focusing on the over-the-air vehicle-in-the-loop

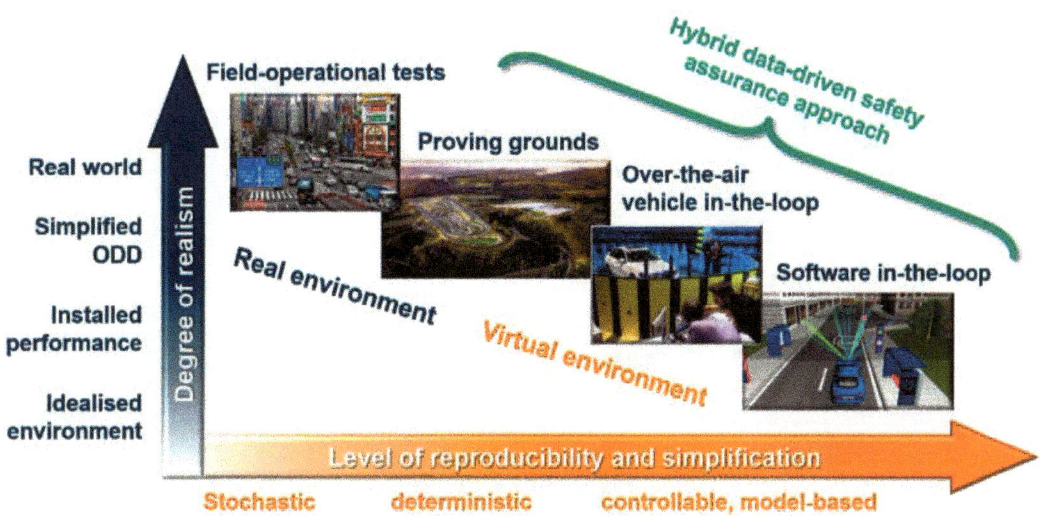

Figure 10.12 Trade-off between the degree-of-realism of testing approaches toward ACD functionalities and their level of reproducibility and simplification. ODD = operational design domain.

(OTA/ViL) method, which enables an efficient combination of powerful numerical simulations with real hardware (antennas and sensors) in their, usually imperfect, installed state.

Figure 10.13 represents the OTA/ViL approach in terms of a V-model, eventually establishing a complete simulation tool chain incorporating the antennas and other wireless sensors in their installed state, the wireless propagation environment, as well as the multifaceted traffic conditions [73, 74].

The consequent merging of antenna measurements and over-the-air testing, as well as the seamless combination of experimental approaches with numerical simulations marks the future of automotive antenna measurement technology. Sections 10.3.2 and 10.3.3 provide simple yet representative examples for two application areas wireless communications and wireless sensing based on automotive radar, respectively.

10.3.2 LTE and V2X Emulation

Wireless communication technologies are key to connected and automated driving. Two complementary transmission standards satisfy the demand for V2X communication: LTE (also referred to as cellular V2X or C-V2X) including the sidelink technology in LTE-V and the dedicated short-range communication (DSRC) with the ad-hoc network-based technology ITS-G5, as standardized by ETSI in Europe. The LTE standard can draw on an existing cellular network with wide aerial coverage. Benefitting from the cellular network, a wider geographical area can be covered, and information such as high-definition maps can be distributed rapidly to many subscribers simultaneously (long-range V2X). Additionally, computational resources can be made available near real-time for applications relying on mobile-edge computing. Short-range V2X can be realized via the LTE side-link or via ITS-G5. Both standards offer the possibility to distribute safety-relevant messages between vehicles, road infrastructure, and further traffic participants in a self-organized manner. This decentralized architecture provides a low latency of packet transmission, which is crucial for safety-related use cases.

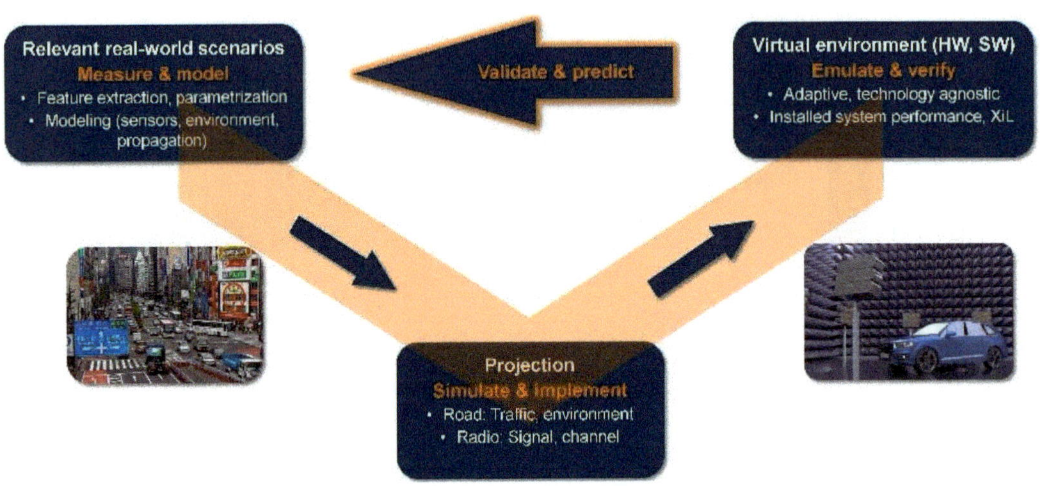

Figure 10.13 V-model representation of the OTA/ViL testing approach.

The LTE-standard uses orthogonal frequency-division multiple access (OFDMA) in the downlink and single-carrier frequency division multiple access (SC-FDMA) in the uplink, allowing for flexible scheduling of radio resources for multiple users. With multiple antennas at the transmitter and the receiver (MIMO) in the downlink case, increased data rates through spatial multiplexing as well as robust data transmission through transmit diversity, LTE provides a complex but reliable transmission system. Key performance indicators for the downlink are the signal-to-interference-and-noise ratio (SINR) and the reference signal received power (RSRP), as used for channel estimation at the user equipment; the modulation method, code rate, and transmission mode are adjusted for optimal signal strength.

The end-to-end properties of ITS-G5 on the link level resemble those of LTE. The performance of ITS-G5 is bound to the received power levels and receiver sensitivity as set in the standards [75]. These properties depend strongly on transmission mode, communication equipment, mobility of the vehicles, and communication environment. Since ITS-G5 is based on the IEEE 802.11p standard, it consists of an ad-hoc based network architecture with packets typically transmitted in a broadcast manner. Therefore, the end-to-end performance is highly correlated with the physical propagation medium, and it is quantified by the packet delivery ratio (PDR) and the latency of the transmitted packets.

10.3.2.1 Emulation of LTE Links in a Virtual Environment

Generally, testing procedures in virtual environments are divided into three categories: multiprobe anechoic chamber [76], reverberation chamber [77], and "wireless cable" [78], all of which attempt to reproduce with high fidelity a specific channel model or an experimentally determined transmission behavior. Here we consider a more practical, "feature-based" approach under the guiding question, "Under which conditions do communication systems fail?" This bottom-up approach is considered highly relevant for the verification and validation of automotive wireless communication systems, while it avoids the complexity related to the top-down approach of wireless propagation channel replay.

As illustrated in Figure 10.14, testing of active devices in a virtual environment requires a communication tester, in order to mimic specific features of a realistic base station such as signaling and communication protocols in the uplink and downlink, and some kind of signal manipulation (channel emulation), in order to reproduce the effects of wireless propagation. Channel emulation is a complex task in terms of signal processing as well as the parameter space and can be realized in various ways (e.g., with commercial emulators or with less expensive software-defined radio modules) [79].

Since LTE supports a wide range of operational modes, and the parameter space for channel emulation offers countless possible combinations, one simple introductory approach consists of identifying and testing critical operational scenarios. Consider one simple example, a static propagation scenario, where the emulated speed of ego-car is zero is considered, and the critical link performance at which the data rate starts to drop significantly is determined for a fixed or an adaptive transmission scheme, as illustrated in Figure 10.15(a). Detailed insights into the setup and the emulated radio channels were provided in [80]. After determination

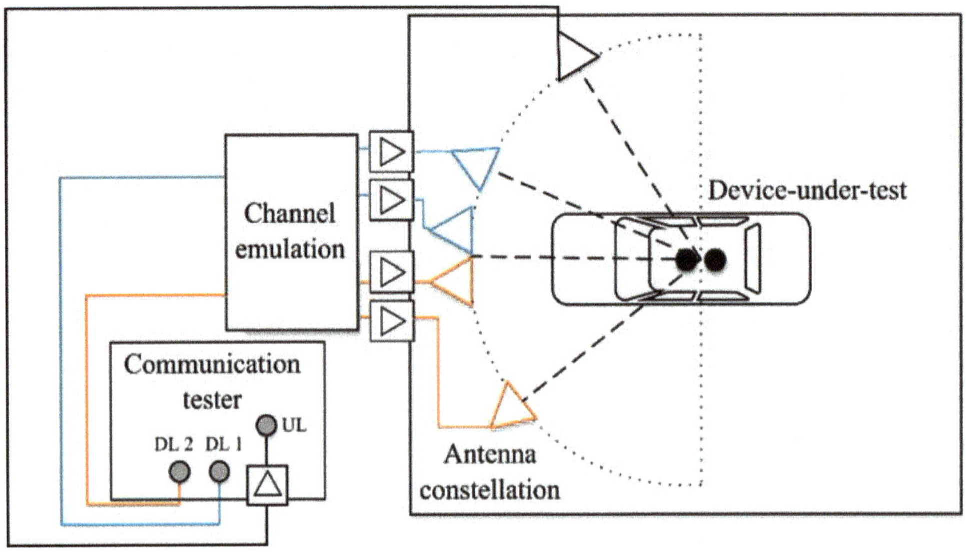

Figure 10.14 Basic setup for over-the-air communication tests in a virtual environment with a communication tester and channel emulator. The outputs of the channel emulator (downlink) and the input of the communication tester (uplink) are connected to amplifiers to achieve signal levels required for stable communication. ©2019 ASTESJ. Source: [80].

of the power threshold, the variation of the vehicle speed was considered. Since the LTE downlink relies essentially on orthogonality between the subcarriers, it is expected that Doppler shifts will stress the wireless link due to reduced orthogonality and thus lead to critical conditions. Figure 10.15(b) shows the relationship between data rate and speed at a constant RSRP = −92 dBm, which was found as the critical power level in this case. In the reference channel mode (green curve), the throughput decreases strongly until the wireless link fails. In the adaptive mode, the performance also decreases strongly but maintains an operational state with a reduced data rate of approximately 7.5 Mbps.

Ongoing efforts, including much more sophisticated link constellations, aim to identify critical LTE link conditions from field-operational tests, extract the most relevant constituting parameters, and reproduce them in virtual environment. With such an approach, the testing of automotive antennas for mobile communications in the installed state proceeds from the bare radiation performance under idealized propagation conditions toward relevant link performance evaluation under close-to-reality conditions.

10.3.2.2 Emulation of ITS-G5 Links in a Virtual Environment

Gathering standard-specific measurement data to comprehend the impact of the electromagnetic propagation channel on the key performance indicator at a given operation frequency is vital for verification and validation of V2X modules [81].

Virtual drive tests for the ITS-G5 standard consist of three main building blocks, as shown in Figure 10.16: road emulation, radio emulation, and a test environment that can combine these two aspects. The road emulation relies typically on road traffic generator software (e.g., SUMO [82]), while the radio environment is generated with a channel emulator that can model a variety of different physical

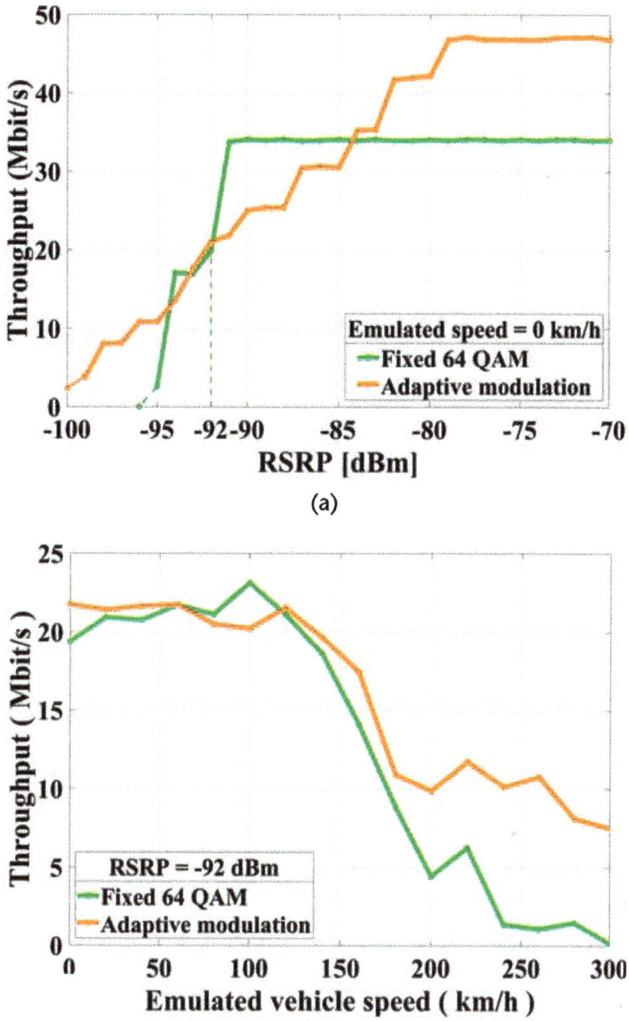

Figure 10.15 (a) Data rate depending on the RSRP for a fixed modulation scheme with 64 QAM and fixed code rate (green curve) and an adaptive transmission scheme (orange curve) depending on the channel quality indicator. The power threshold was determined for the reference channel at about −92 dBm. (b) Data rate depending on the emulated speed of the vehicle. The RSRP is held constant at the power threshold level of −92 dBm. ©2019 ASTESJ. Source: [80].

propagation conditions between on-board-units (OBUs) mounted on the vehicles in a given environment. To set the stage for road and radio emulation and a realistic implementation of the virtual drive tests, reference field-operational tests (FoTs) yield valuable input information, although their quantitative behavior lacks reproducibility. Two different field-operational test approaches with representative outcomes are described in the following. For the first test, the main focus was to obtain information on relevant ranges for the channel condition parameters, such as the distance D between communicating vehicles and the link conditions between them, contrasting line-of-sight (LoS) with non-line-of-sight (NLoS) conditions [83]. This test provided valuable information on the performance of the communication between vehicles, where the link quality, measured in terms of the packet delivery

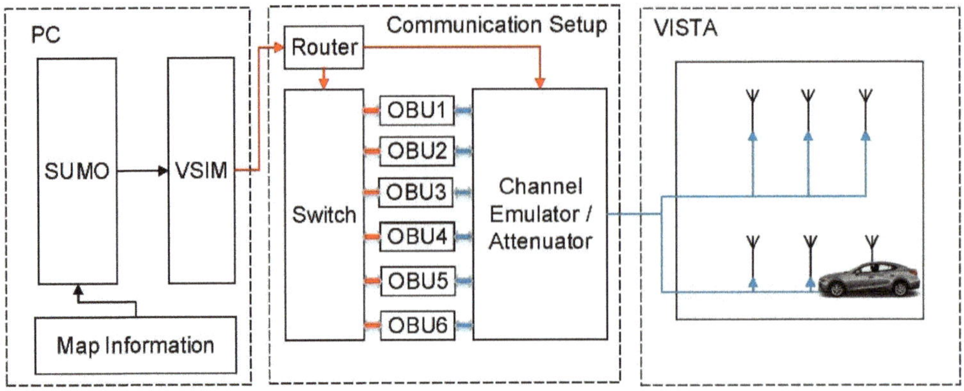

Figure 10.16 Block diagram of the testing framework of a virtual drive test, divided into three stages: PC for road emulation and traffic generation, communication for channel generation, and VISTA as test environment. ©2019 IEEE. (Reprinted with permission.) Source: [94].

ratio PDR dropped at distances $D > 220$m, as shown in Figure 10.17, even under LoS conditions. For the second test, pilot measurements aimed at an assessment of the B-column antennas mentioned before, in terms of their V2X communication performance. For this test, identical antennas were installed on both sides of the car with respect to the driving direction. Configurations of left/left and left/right B-column patch antennas and a rooftop monopole reference antenna were tested simultaneously, with the major outcomes shown in Figure 10.18. The measurements revealed that the B-columns present a promising location for the integration of V2X antennas [83, 84].

10.3.3 Automotive Radar

Verification and validation (V&V) of automotive radar systems with distance-based methods requires exhaustive testing under every environmental condition imaginable. According to, for example, [85], hundreds of millions of test kilometers, and more for any additional functionality, would have to be driven over a period of many years, in order to quantitatively assess achievable reliability and the safety of the intended function. Possible solutions for these challenges are innovative testing methods such as virtual test environments. Virtual test environments offer the highest achievable degree of reproducibility and repeatability, though at the cost of a limited complexity of traffic scenarios, environments, and waveforms, and thus limited realism. For this reason, they demand significant research into their extension, to reduce real-world drive test efforts. Virtual environments can, according to their integration of hardware, be basically divided into software-in-the-loop (SiL), hardware-in-the-loop (HiL), and vehicle-in-the-loop (VIL). The first approach is entirely numerically based, the second employs a stand-alone sensor, and last includes integration of real test vehicles with built-in and fully functional sensors. Modern techniques using ViL enhance the end-of-line V&V of automotive radar systems. They have recommended the use of test beds for closed-loop OTA stimulation using target simulators [86–89]. A dynamic antenna positioning system for radar target simulation was presented in [90].

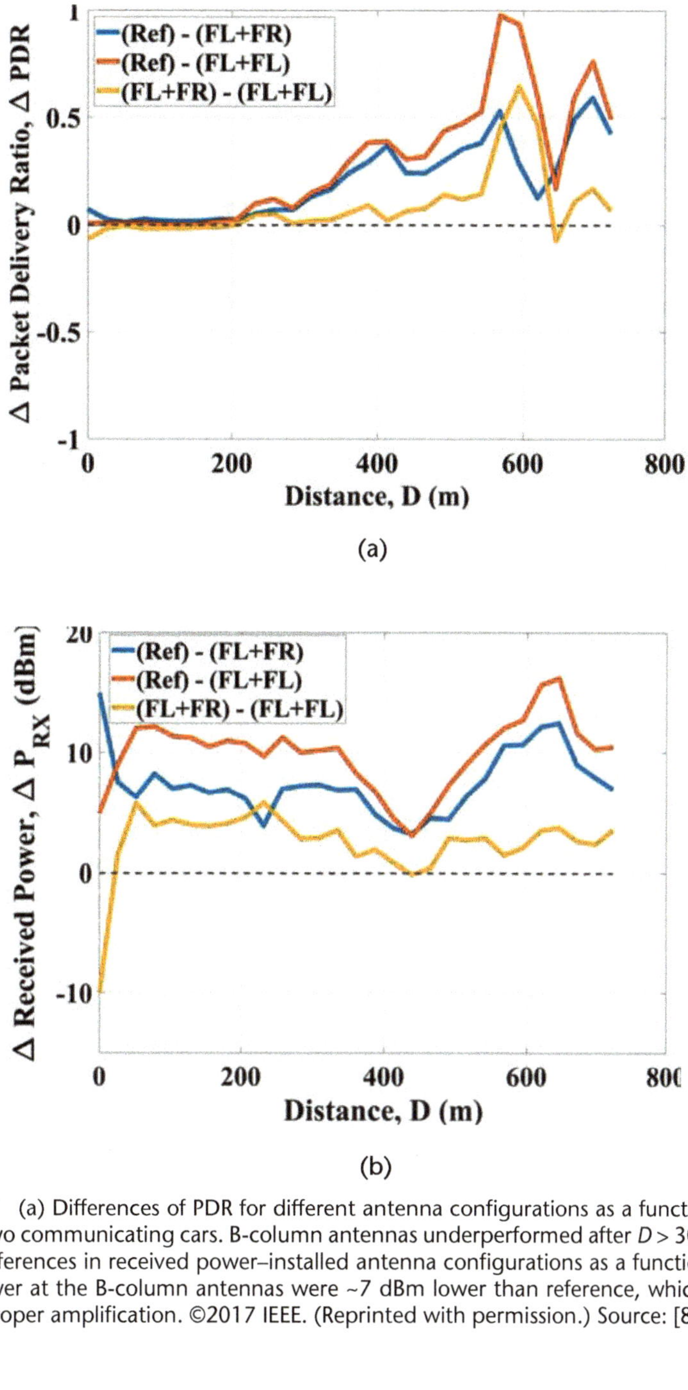

Figure 10.17 (a) Differences of PDR for different antenna configurations as a function of distance *D* between two communicating cars. B-column antennas underperformed after *D* > 300m. (b) Comparison of differences in received power–installed antenna configurations as a function of distance. Received power at the B-column antennas were ~7 dBm lower than reference, which can be mitigated with proper amplification. ©2017 IEEE. (Reprinted with permission.) Source: [84].

In this section, we describe a typical OTA/ViL radar test system architecture along with the most recent stage of completion and evaluation of our test system in VISTA. This is among the first technically reported implementations of a fully operational OTA/ViL test system capable of simulating realistic dynamic targets, simultaneously varying in distance, angular direction, velocity, and signal strength, within the highly relevant field-of-view (FoV) of a typical automotive radar across

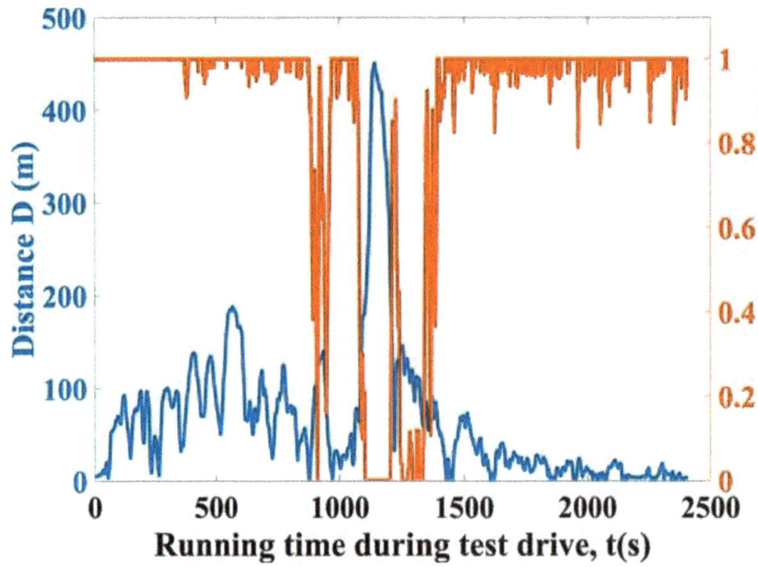

Figure 10.18 Evolution of PDR with distance between two cars along a real-drive test track. Communication failed for $D > 220$m, but rarely at lower distances, which can be attributed to elevation differences between the vehicles and the prominent multipath effects. ©2021 IEEE. (Reprinted with permission.) Source: [83].

the 76–81-GHz range. The system is modular and hence expandable to simulate multiple point targets with few technical limitations.

As sketched in Figure 10.19, the system-level block diagram consists of four main modules: (1) OTA/ViL simulator, (2) radar target simulator (RTS) back end, (3) RTS front end, and (4) the radar-under-test (RUT) installed in a car. The first three modules together determine the virtual environment for the RUT. The physical interfaces between the modules are also indicated in parentheses under the arrow lines. It can be noted that the RUT and the front end have an OTA interface between them, which lets us emulate a realistic wireless signal propagation environment.

The RUT is installed in the center of the radiator grill, behind a plastic shield, and is located 60 cm above ground [91]. The OTA/ViL simulator is the core computer that contains the software and hardware necessary for traffic and radar signal

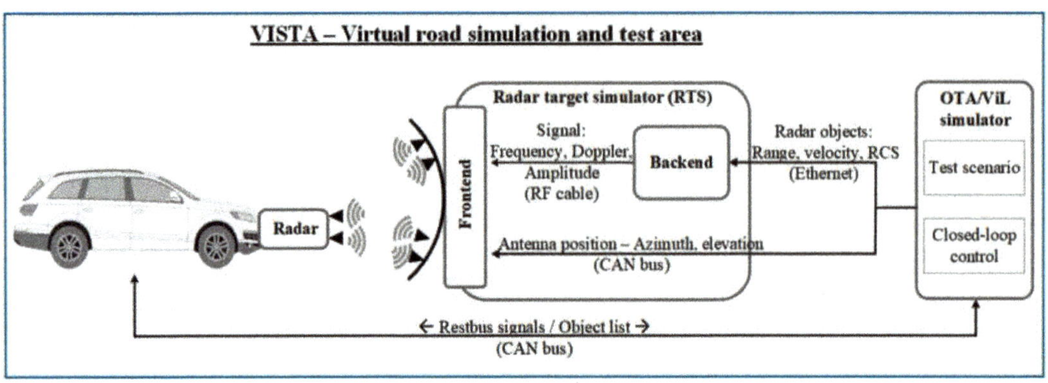

Figure 10.19 System architecture of a typical OTA/ViLtest system for commercial automotive radar systems in a virtual environment.

propagation modeling as well as real-time closed-loop control, and visualization of the test scenarios [88]. A photograph of the functional setup of the system with the RUT installed in a mid-size passenger car, Mercedes Benz E-class sedan, is presented in Figure 10.20. The RTS front end is the antenna positioner that has been custom-designed and constructed using up to seven individual antenna masts, each carrying a MIMO target simulator antenna pair. The masts move along a guide rail of circular geometry. Hence, the distance between the RUT and the positioner ensures FF conditions to the selected RUT.

The positions of the antennas are controlled using electrical motors for movement in azimuth, elevation, and tilt [92]. The individual mechanical degrees-of-freedom are controlled through CAN interfaces by the OTA/ViL computer, so that the point targets radiated by the antennas follow the traffic scenario-under-consideration in real time. The installed RTS back end is a dSPACE automotive radar target simulator that employs digital signal processing, leading to a processing delay presently corresponding to a minimum distance of 5.5m. The minimum simulated range is further limited to 14m because of additional signal propagation delays between RTS and antennas. The dynamic range is greater than 60 dB, and hence our system is capable of simulating the entire range of RCS values of all relevant traffic objects, from pedestrians (–5 dBsm) up to large-sized trucks (40 dBsm) [87]. Further, radar signals of both polarizations can be stimulated, by rotating the horn antennas along their aperture axis. An absorber wall equipped with pyramidal foam absorbers appropriate for the frequency range can be seen in Figure 10.20. This installation reduces the environment clutter and extends the dynamic range of the simulated signal-to-noise ratio (SNR) [86]. In order to close the loop, additional vehicle CAN bus signals specific to the electronic sensor control units (ECUs)

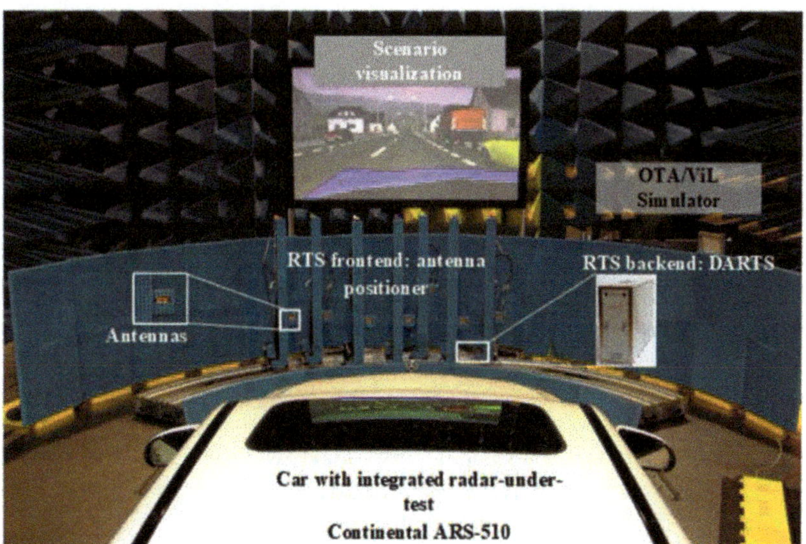

Figure 10.20 Photograph of the OTA/ViL automotive radar test system in VISTA. The positioner with seven masts installed can be seen in front of the Mercedes-Benz test vehicle with a Continental ARS-510 radar sensor installed. The millimeter-wave pyramid absorbers are placed in front of the base and behind the positioner to minimize parasitic reflections from the installed structures in VISTA.

are simulated. The radar output is measured through the same CAN bus as the object list.

In the course of a test run, a simulated point target STi(t) is represented by a set of time-dependent parameters:

$$ST_i(t) = [r, v, \sigma, \phi, \theta](t), i = 1 \ldots N \tag{10.8}$$

where r denotes the range (m), v the relative velocity (km/h) between the ego-car and the radar object, and σ the radar cross-section, RCS (dBsm). The direction parameters are azimuth ϕ and elevation θ, respectively. N represents the number of the MIMO radar targets simulated simultaneously by the corresponding Tx-Rx RTS/antenna pairs. A flow diagram of the measurement process and data analysis is depicted in Figure 10.21.

Initially, a set of especially relevant test scenarios was identified including, for example, critical collision maneuvers along straight roads and crossings. The selected scenario is modeled,and all relevant parameters are computed using a software tool chain for real-time traffic scenario simulation [88]. The appropriate RCS values are read from a look-up table containing values previously measured in VISTA [93]. The scenario-defining parameters are synchronously transferred to the interfaced hardware, to control the stimulation of specific targets.

The ego-car parameters like velocity v_{ego} and steering angle a are simulated and transmitted continuously during the test run to the radar installed in the car through restbus simulation. In our case, the parameter updates are discretized into

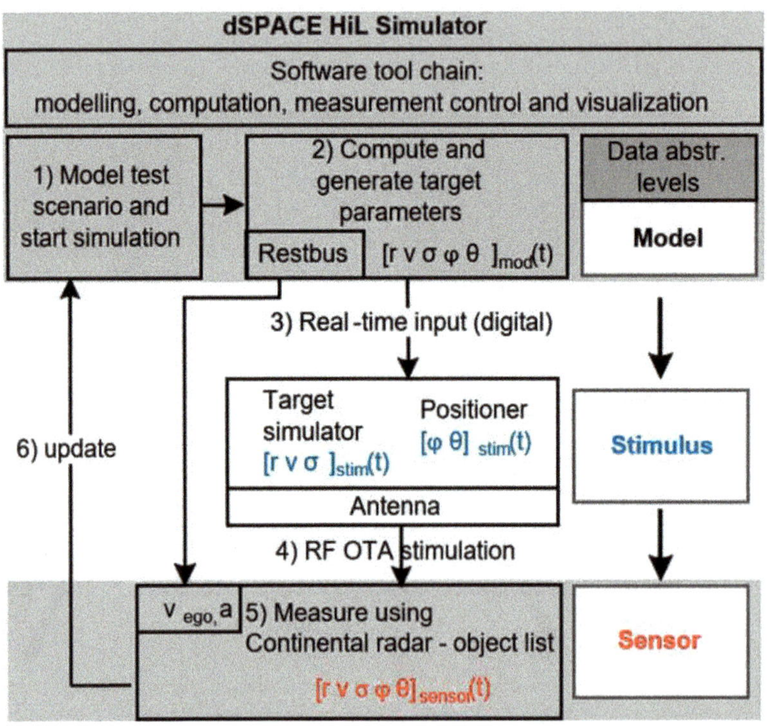

Figure 10.21 Flow chart of the measurement process and the associated analysis in different data abstraction levels. ©2021 IEEE. (Reprinted with permission.) Source: [86].

1-ms time steps. The parameter ranges, Doppler velocity, and the RCS are electronically generated by the respective target simulator back end. The antenna positioner controls the direction of the targets in azimuth and elevation. We model the complete test scenario with a software tool chain, stimulate the sensor in real-time, and measure using an installed radar, altogether forming a closed test loop. We measure and analyze the parameters at three data abstraction levels as indicated in Figure 10.21, for comparison and evaluation of the performance of the test system. These levels are listed as follows:

- *AL1 (model)*: Entirely software-based model input;
- *AL2 (stimulus)*: Hardware-based OTA stimulation;
- *AL3 (sensor)*: Radar signals as perceived by the RUT.

It must be noted that the radar returns objects (tracked detections) that are used at L3. The measurement data is recorded by the OTA/ViL simulator every 1 ms; the RTS and RUT data are interpolated between subsequent updates.

As an example, we select a scenario where the ego-car drives along a straight route (i.e., with a constant lateral displacement) toward a static car on the same lane ahead of it. At the end of the scenario, the car brakes and stops at a sufficient distance in front of the static car. The ego-car and fellow car along with the coordinate system used to parametrize the track are illustrated in Figure 10.22. The origin of the global coordinate system is located at the beginning of the defined road, from where the ego-car starts moving, while the origin of the coordinate system of the ego-car is located in the center of the front-wheel axis. The variation of the scenario-specific target parameters over time are defined as parameter profiles. This scenario offers us the necessary dynamics to validate the performance of the test based on distance, azimuth, and RCS profiles.

Due to the implementation of the test process, the starting time of all scenarios occurs at approximately 13s (i.e., settling interval) as the simulated target enters the FoV of the radar after this period. In order to evaluate the reliability of the scenario representation, we need to compare the parameter profiles at the data levels AL1, AL2, and AL3. In Figure 10.23, we present the distance and the angular profiles, followed by the RCS profiles. The three curves represent level AL1 (model, black solid curve), level AL2 (stimulus, blue dashed), and level AL3 (sensor/RUT, red dotted). In all the target parameter profiles, the curves follow each other closely.

Figure 10.22 Test scenario under consideration and position of the coordinate system. ©2021 IEEE. (Reprinted with permission.) Source: [86].

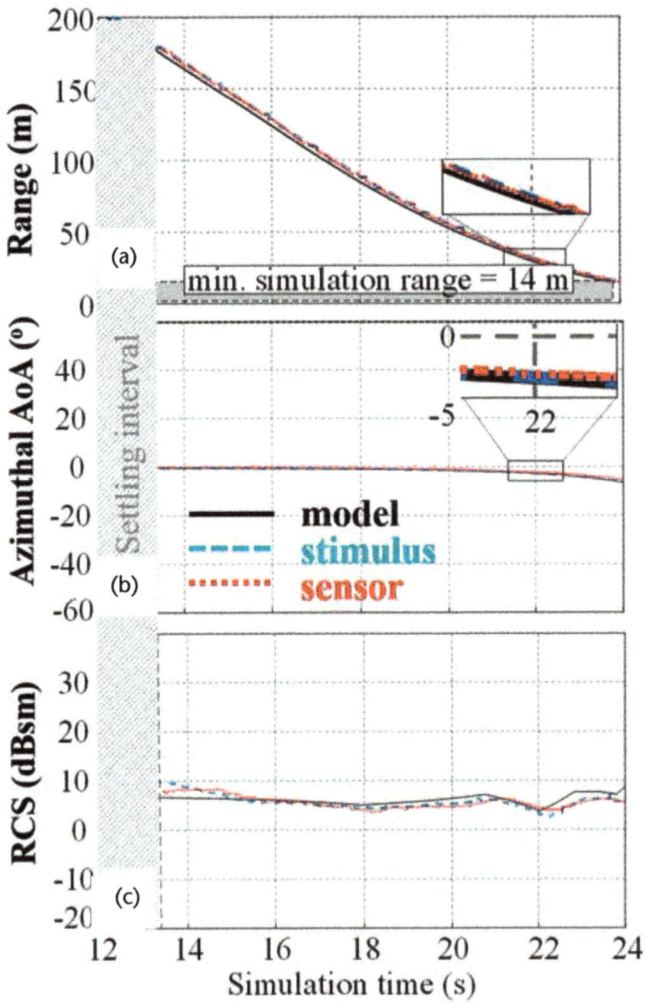

Figure 10.23 Comparison of the (a) range, (b) azimuth, and (c) RCS profiles of the measured test scenario at the three different data abstraction levels AL1–AL3, represented by the different colored curves for comparison of the levels. AoA = angle of arrival. ©2021 IEEE. (Reprinted with permission.) Source: [86].

A very good fit between the three curves is observed, indicating a high level of accuracy. Minor deviations, in RCS profiles, around 3 dB can be noticed; however, they are considered acceptable according to the data sheet of the RUT. Accordingly, quantification of the simulation consistency and analysis of test metrics is a focus of future work.

References

[1] Hansen, J. E. (ed.), *Spherical Near-Field Antenna Measurements* (reprinted edition), IEEE Electromagnetic Waves Series. Stevenage, UK: The Institution of Engineering and Technology, 2008, Vol. 26, pp. 17–26.

[2] Castañer, M. S., et al., "Overview of Novel Post-Processing Techniques to Reduce Uncertainty in Antenna Measurements," Advanced Topics in Measurements," Prof. Zahurul Haq (ed.), *InTech,* 2012.

[3] Hess, D. W., "An Expanded Approach to Spherical Near-Field Uncertainty Analysis," NSI-MI Technologies, *AMTA,* 2002.

[4] Evans, G. E. (ed.), *Antenna Measurement Techniques*, Norwood, MA: Artech House, 1990.

[5] Sanandiya, V., et al., "Using High-Accuracy Swing Arm Gantry in Spherical Near-Field Automotive Measurement Systems," NSI-MI Technologies, *AMTA,* 2020.

[6] Betjes, P. N., et.al., "A Hemi-Spherical Near-Field System for Automotive Antenna Testing," Nearfield Systems Inc., Torrance, CA.

[7] Hein, M. A., et al., "Emulation of Virtual Radio Environments for Realistic End-to-End Testing for Intelligent Traffic Systems," in *2015 IEEE MTT-S International Conference on Microwaves for Intelligent Mobility (ICMIM),* April 2015.

[8] Martini, E., et al., "Spherical Near Field Measurements with Truncated Scan Area," in *EuCAP 2011,* Rome, Italy.

[9] Saccardi, F., et al., "Truncation Error Mitigation in Free-Space Automotive Partial Spherical Near Field Measurements," *2017 Antenna Measurement Techniques Association Symposium (AMTA),* Atlanta, GA, 2017.

[10] Saccardi, F., et al., "Estimation of the Realistic Ground Effect in Free-Space Automotive Measurements," *2018 Antenna Measurement Techniques Association Symposium (AMTA),* Atlanta, GA, 2018.

[11] Asghar, M.E., et.al., "Comparative Analysis of Spherical Near-Field Automotive Antenna Measurement Facilities," *2019 13th European Conference on Antennas and Propagation* (EuCAP), 2019, pp. 1–5.

[12] Wollenschläger, F., et al., "Spherical Wave Expansion Applied to the Measured Radiation Patterns of Automotive Antennas in the Installed State in the GHz Range," in *EuCAP 2019,* April 2019, Krakow, Poland.

[13] Asghar, M. E., et al.,"Automotive Antenna Measurements in the Installed State Under Variable Boundary Conditions," in *EuCAP 2022,* March 2022, Madrid, Spain.

[14] McKinsey & Company, "Competing for the Connected Customer—Perspectives on the Opportunities Created by Car Connectivity and Automation," *Tech. Rep.,* 2015.

[15] GSMA, "2025 Every Car Connected: Forecasting the Growth and Opportunity," GSM Association, *Tech. Rep.,* 2012.

[16] M. Rütschlin, and D. Tallini, "Simulation for antenna design and placement in vehicles," *Antennas, Propagation & RF Technology for Transport and Autonomous Platforms 2017*, 2017, pp. 1–5, doi: 10.1049/ic.2017.0021.

[17] Safin, E., et.al., "Reconfigurable LTE MIMO Automotive Antenna System Based on the Characteristic Mode Analysis," *2015 9th European Conference on Antennas and Propagation (EuCAP),* Lisbon, 2015, pp. 1–3.

[18] Condo, N, E., et al., "Combined LTE and IEEE 802.11p Antenna for Vehicular Applications," *2015 9th European Conference on Antennas and Propagation (EuCAP),* Lisbon, 2015, pp. 1–5.

[19] Hasturkoglu, S., et.al., "A Wideband Automotive Antenna for Actual and Future Mobile Communication 5G/LTE/WLAN with Low Profile," *2017 11th European Conference on Antennas and Propagation (EUCAP),* Paris, 2017, pp. 602–605, doi: 10.23919/EuCAP.2017.7928669.

[20] Lee, J. H., et al., "Transparent Antenna Using a μmetal Mesh on the Quarter Glasses of an Automotive for DMB Service Receiving," *Microw. Opt. Technol. Lett.,* Vol. 60, No. 12, 2018, pp. 3009–3014.

[21] Lee, J. H., et al., "Transparent Dual-Band Monopole Antenna Using a μ-Metal Mesh on the Rear Glass of an Automobile for Frequency Modulation/Digital Media Broadcasting Service Receiving," *Microw. Opt. Technol. Lett.,* Vol. 61, No. 2, 2019, pp. 503–508.

[22] Low, Lester and Langley, J. Richard, and J. C. Batchelor, "Modelling and Performance of Conformal Automotive Antennas," *IET Microwaves, Antennas and Propagation*, 2007, pp. 973–979. ISSN 1751-8725.

[23] Toriyama, H., et al., "Development of Printed-On Glass TV Antenna System for Car," *37th IEEE Vehicular Technology Conference,* Tampa, FL, 1987, pp. 334–342.

[24] Brzeska, M., et al., "New Generation of In-Mirror Integrated Antennas," *2009 3rd European Conference on Antennas and Propagation,* Berlin, 2009, pp. 2704–2707.

[25] Kim, M. S., et.al., "Directivity Design of RFID Tag Antenna Using Side-view Mirror for Vehicle," *2008 Asia-Pacific Microwave Conference,* Macau, 2008, pp. 1–4.

[26] Menzel, W., and A. Moebius, "Antenna Concepts for Millimeter-Wave Automotive Radar Sensors," in *Proceedings of the IEEE,* Vol. 100, No. 7, pp. 2372–2379, July 2012.

[27] Dickmann, J., et al., "Making Bertha See Even More: Radar Contribution," in *IEEE Access,* Vol. 3, pp. 1233–247, 2015.

[28] Marhefka, R. J., et al., "Modeling the Performance of Automotive Bumper Mounted Antennas," *IEEE Antennas and Propagation Society International Symposium (IEEE Cat. No.02CH37313),* San Antonio, TX, 2002, Vol. 3, pp. 316–319.

[29] Liu, Y., et al., "An Integrated Shark-Fin Antenna for MIMOLTE, FM, and GPS Applications," in *IEEE Antennas and Wireless Propagation Letters,* Vol. 18, No. 8, August 2019, pp. 1666–1670.

[30] Demien, C., and Sarkis, R., "Design of Shark Fin Integrated Antenna Systems for Automotive Applications," *2019 Photonics & Electromagnetics Research Symposium—Spring (PIERS-Spring),* Rome, Italy, 2019, pp. 620–627.

[31] Kwon, O., Song, R., and Kim, B., "A Fully Integrated Shark-Fin Antenna for MIMO-LTE, GPS, WLAN, and WAVE Applications," *IEEE Antennas and Wireless Propagation Letters,* Vol. 17, No. 4, April 2018, pp. 600–603.

[32] Artner, G., et al., "Conformal Automotive Roof-Top Antenna Cavity with Increased Coverage to Vulnerable Road Users," in *IEEE Antennas and Wireless Propagation Letters,* Vol. 17, No. 12, pp. 2399–2403, December 2018.

[33] Posselt, A., "Design and Analysis of Multi-Element Antenna Systems and Agile Radiofrequency Frontends for Automotive Applications," Dissertation, Leibniz Universität Hannover, 2019.

[34] Asghar, M. E., et al., "Influence of Antenna Mounting Location on the Radiation Pattern of an Automotive Antenna," *12th European Conference on Antennas and Propagation (EuCAP 2018),* 2018, pp. 1–5.

[35] Singh, J., et al., "Novel Conformal Automotive Di-Patch Antenna Verified through Car Door Frame Measurements," *Antennas and Propagation Conference 2019 (APC-2019),* 2019, pp. 1–5.

[36] Singh, J., et.al., "Low-Profile Wideband Differentially Fed Di-Patch Antenna Closely Above Metallic Ground," in *IEEE Antennas and Wireless Propagation Letters,* Vol. 18, No. 5, May 2019, pp. 976–980.

[37] Singh, J., "Meta Surface Tolerant Conformal Low-Profile Plastic Embedded Antennas for Automotive Applications," Dissertation, TU Ilmenau, submitted (2022).

[38] Singh, J., et.al., "Low-Profile Penta-Band Automotive Patch Antenna Using Horizontal Stacking and Corner Feeding," in *IEEE Access,* Vol. 7, 2019, pp. 74198–74205.

[39] Singh, J., et al., "Metal Chassis Tolerant Conformal High Impedance Surface Based LTE-2600 Automotive Antenna," *2018 IEEE MTT-S International Conference on Microwaves for Intelligent Mobility (ICMIM),* 2018, pp. 1–4.

[40] X2Rail-1, Deliverable 3.3, "Specification of the Communication System and Guideline for Choice of Technology," 2019.

[41] 3GPP, "Study on Future Railway Mobile Communication System (FRMCS)," Technical report 22.889, Release 17.2.0.

[42] Asghar, M. E., et.al., "Simulation and Measurement of Locomotive Antennas in the Installed State Based on a Scaled Mockup," in *Antenna Measurement Techniques Association (AMTA),* Florida, October 2021.

[43] Carter, D., "Phase Centers of Microwave Antennas," *IRE Transactions on Antennas and Propagation,* Vol. 4, No. 4, October 1956, pp. 597–600.

[44] Feng, N., et.al., "New Theoretical Analysis Method for the Phase Center in the Fresnel Region of Complex Structure Antennas," *Journal of Systems Engineering and Electronics,* Vol. 19, No. 5, 2008, pp. 65–871.

[45] Hu, Y. Y., "A Method of Determining Phase Centers and Its Application to Electromagnetic Horns," *Journal of the Franklin Institute,* Vol. 271, No. 1, 1961, pp. 31–39.

[46] Teichman, M., "Determination of Horn Antenna Phase Centers by Edge Diffraction Theory," in *IEEE Transactions on Aerospace and Electronic Systems,* Vol. AES-9, No. 6, November 1973, pp. 875–882.

[47] Raisanen, A., et.al., "Phase Pattern and Phase Center Measurements of Antennas at 105-190 GHz with a Novel Differential Phase Method," *1990 20th European Microwave Conference,* Budapest, Hungary, 1990, pp. 347–352.

[48] Hertel, T. W., "Phase Center Measurements Based on the Three-Antenna Method," *IEEE Antennas and Propagation Society International Symposium,* Digest. Held in conjunction with: USNC/CNC/URSI North American Radio Sci. Meeting (Cat. No.03CH37450), Columbus, OH, 2003, Vol. 3, pp. 816–819.

[49] Liu, H., et.al., "Determination of the Phase Center Position and Delay of a Vivaldi Antenna," *IEICE Electronics Express,* 2013, Vol. 10, No. 21, November 10, 2013, pp. 1–7.

[50] Kunysz, W., "Antenna Phase Center Effects and Measurements in GNSS Ranging Applications," in *2010 14th International Symposium on Antenna Technology and Applied Electromagnetics the American Electromagnetics Conference,* July 2010, pp. 1–4.

[51] Drabowitch, S., et al., *Modern Antennas,* Second Edition, Springer, 2005, p. 55.

[52] Penrose, R., "A Generalized Inverse for Matrices," *Mathematical Proceedings of the Cambridge Philosophical Society,* Vol. 51, No. 3, 1955, pp. 406–413.

[53] Berlt, P., et.al., "Precise Phase Center Localization of Automotive LTE Antennas in the Installed State Through Phaseless LTE Uplink Measurements," *AMTA 2021 Proceedings,* 2021.

[54] Itoh, K., "Analysis of the Phase Unwrapping Problem," *Applied Optics,* Vol. 21, No. 14, p. 2470, 1982, 10.1364/AO.21.002470.

[55] Zhao, Z., et al., "Robust 2D Phase Unwrapping Algorithm Based on the Transport of Intensity Equation," *Measurement Science and Technology,* Vol. 30, 2018, 10.1088/1361-6501/aaec5c.

[56] Berlt, P., et.al., "Accurate 3D Phase Recovery of Automotive Antennas Through LTE Power Measurements on a Cylindrical Surface," *2020 14th European Conference on Antennas and Propagation (EuCAP),* Copenhagen, 2020, pp. 1–5.

[57] Costanzo, S., et.al., "Spatial Domain Indirect Holographic Technique for Antenna Near Field Phaseless Measurements," *Radio Sci.,* Vol. 52, 2017, pp. 432–438.

[58] WiMAX Forum, "Radiated Performance Tests (RPT) for Subscriber and Mobile Stations,"WMF-T25-004-R010v04, April 2012.

[59] 3GPP, "Technical Specification Group Radio Access Network; Measurements of Radio Performances for UMTS Terminals in Speech Mode," Release 9, V9.00, 2009-12.

[60] CTIA: The Wireless Association, "Test Plan for Wireless Device Over-the-Air Performance," Version 3.7.1, February 2018.

[61] Mercier, M., et al., "Evaluation of Automotive Antenna Over the Air Performance," *2021 International Symposium on Antennas and Propagation (ISAP),* 2021, pp. 1–2.

[62] Mercier, M., et al., "Evaluation of Integral Quantities of Over the Air Automotive Antenna Measurements," *IEEE Antennas and Propagation Society International Symposium,* Marina Bay Sans, December 4–10, 2021, Singapore.

[63] Asghar, M. E., et.al., "Experimental Determination of the Total Radiated Power of Automotive Antennas in the Installed State," *EuCAP 2020,* March 2020, Copenhagen, Denmark.

[64] Foged, L. J., et al. "Comparative Investigation of Methods to Reduce Truncation Errors in Partial Spherical Near-Field Antenna Measurements," *EuCAP,* March 2012.

[65] Foged, L. J., et al., "Post-Processing of Measured Field Data Using Equivalent Source Technique," *2011 IEEE Applied Electromagnetics Conference (AEMC),* Kolkata, 2011, pp. 1–4.

[66] Martini, E., et.al., "Reduction of Truncation Errors in Spherical Near Field Measurements," *AMTA,* October 2010.

[67] Society of Automotive Engineers, https://www.sae.org/standards/content/j3016_201806/.

[68] Car-to-Car Communication Consortium, https://www.car-2-car.org/.

[69] Second reference to the subject of [63], i.e., reference towards V2X (ITS-G5 and C-V2X standards.

[70] EuRAD 2020 proceedings (JCRS).

[71] Society of Automotive Engineers, https://www.sae.org/standards/content/j3216_202005/.

[72] Klamann, B., et al., "Defining Pass-/Fail-Criteria for Particular Tests of Automated Driving Functions," *19th IEEE ITSC,* 2019, Auckland, New Zealand.

[73] Schuldt, F., "Ein Beitrag für den methodischen Test von automatisierten Fahrfunktionen mit Hilfe von virtuellen Umgebungen," Dissertation, TU-Braunschweig, 2016.

[74] Weber, H., et al., "A Framework for Definition of Logical Scenarios for Safety Assurance of Automated Driving," *Traffic Injury Prevention,* Vol. 20, 2019, pp. 65–70.

[75] ETSI, TR: 102 638 v1.1.1. In: Intelligent Transport Systems (2009).

[76] Fan, W., et al.,"Recent Advances on OTA Testing for 5G Antenna Systems in Multiprobe Anechoic Chamber Setups," *2017 Sixth Asia-Pacific Conference on Antennas and Propagation (APCAP),* 2017, pp. 1–3.

[77] Kildal, P., et.al., "OTA Testing of 3G-5G Devices with MIMO: From Anechoic Chambers to Reverberation Chambers and ... Back Again?" *2017 IEEE International Symposium on Antennas and Propagation & USNC/URSI National Radio Science Meeting,* 2017, pp. 1697–1698.

[78] Fan, W., et al., "MIMO Device Performance Testing with the Wireless Cable Method," *12th European Conference on Antennas and Propagation (EuCAP 2018),* 2018, pp. 1–5.

[79] Schwind, A., et al., "Over-the-Air MIMO Channel Emulation for Automotive LTE radio systems Using Software Defined Radio," *12th European Conference on Antennas and Propagation (EuCAP 2018),* 2018, pp. 1–5.

[80] Berlt, P., et al., "Over-The-Air Testing of Automotive Antennas and Wireless Links in The Installed State on the Basis of LTE Downlink Communication Parameters," *Advances in Science, Technology and Engineering Systems Journal,* Vol. 4, No. 1, 2019, pp. 282–291.

[81] Hein, M. A., et al., "Emulation of Virtual Radio Environments for Realistic End-To-End Testing for Intelligent Traffic Systems," *2015 IEEE MTTS International Conference on Microwaves for Intelligent Mobility (ICMIM),* IEEE, 2015.

[82] Krajzewicz, D., et al. "SUMO (Simulation of Urban MObility)–An Open-Source Traffic Simulation," *Proceedings of the 4th Middle East Symposium on Simulation and Modelling (MESM20002)*, 2002.

[83] Altinel, B., Bornkessel, C., and Hein, M. A., "Wave-Physical Factors Determining the Link Quality in ITS-G5 Studied with Field-Operational Tests," *2021 15th European Conference on Antennas and Propagation (EuCAP)*, IEEE, 2021.

[84] Singh, J., et al., "V2V Communication Performance of Conformal Plastic-Embedded Side-Mount Antennas Investigated with Field-Operational Tests," *IEEE VTC 2021-Spring.*

[85] Kalra, N., et.al., "Driving to Safety: How Many Miles of Driving Would It Take to Demonstrate Autonomous Vehicle Reliability?" Rep. RR-1478-RC, 2016, Rand Corp., Santa Monica, CA.

[86] Buddappagari, S., et al., "Over-the-Air Vehicle-in-the-Loop Test System for Installed-Performance Evaluation of Automotive Radar Systems in a Virtual Environment," *2020 17th European Radar Conference (EuRAD)*, 2021.

[87] Dallmann, T., et al., "ATRIUM: A Radar Target Simulator for Complex Traffic Scenarios," *ICMIM, Munich,* 2018.

[88] Radar Test Bench, DARTS. ASM and Modeldesk, https://www.dspace.com/en/inc/home/products.cfm.

[89] Rohde & Schwarz, "Better than Real Life: Radar Echoes from a Target Simulator," 2021, https://cdn.rohde-schwarz.com/pws/dl_downloads/dl_common_library/dl_news_from_rs/213/NEWS_213_ARTS9510_english.pdf.

[90] Wayne, D., et al., "RF Target & Decoy Simulator," *AMTA Proceedings*, pp. 489–494, 2011.

[91] Continental ARS 510, https://www.continental-automotive.com/en-gl/Passenger-Cars/Autonomous-Mobility/Enablers/Radars/Long-Range-Radar/ARS510.

[92] Asghar, M. E., et al., "Radar Target Simulator and Antenna Positioner for Real-Time Over-the-Air Stimulation of Automotive Radar Systems," *EuRAD*, 2020.

[93] Buddappagari, S., et al., "Monostatic RCS Measurements of a Passenger Car Mock-Up at 77 GHz Frequency in Virtual Environment," *European Radar Conference (EuRAD)*, 2019.

[94] B. Altinel, et.al., "Interference Tests of ITS-G5 Vehicle-to-Vehicle Communication Networks with Virtual Drive Tests," IEEE, 2019.

About the Editors

Lars J. Foged was born in Denmark in 1966. He received his MS in electrical engineering from the California Institute of Technology in 1990. He is currently the scientific director of the Microwave Vision Group. Since 2004, he has served as secretary to, and is now a vice-chair of, the IEEE Antenna Standards Committee. Previously, he was vice chair of the APS/SC IEEE std 149 working group and now is the chair of the APS/SC IEEE std1720 working group. In 2016/2017 he led the Industry Initiatives Committee (IIC) of IEEE APS. In addition, he has been involved in the evolution of IEC standards on human exposure to electromagnetic fields since 2010.

From 2009 to 2012, Foged was a member of the EurAAP Delegate Assembly and responsible for the Working Group on Antenna Measurements. He was vice-chair of the EuCAP conference in both 2011 and 2022, industrial chair in 2012, 2014, and 2017, and technical program chair in 2016 and 2021. He has been a board member and course organizer in the European School of Antennas (ESoA) since 2006.

Foged is a board member, fellow, and distinguished achievement award recipient of AMTA. He has authored or coauthored more than 300 journal and conference papers on antenna design and measurement topics and received the Best Technical Paper Award at the 2012 AMTA symposium and the Best Measurement Paper Award at the EuCAP 2021 conference.

Manuel Sierra Castañer was born in 1970 in Zaragoza, Spain. He obtained a master's degree in telecommunication engineering in 1994 and a PhD in 2000, both from the Technical University of Madrid (UPM). He worked for the cellular company Airtel from 1995 to 1997, and from 1997 to 1998 for the University Alfonso X. Since 1998, he has served at the Technical University of Madrid, becoming a full professor in 2017. In addition, he has been a visiting researcher at Tokyo Tech and EPFL at different times.

Sierra's current research interests are in planar antennas and antenna measurement techniques. Among other awards, he has receieved the IEEE APS 2007 Schelkunoff Prize Paper Award for "Dual-Polarization Dual-Coverage Reflectarray for Space Applications." Currently, he is a senior member of the IEEE and a fellow of the AMTA Society for his contributions to antenna measurement theory.

He has been a member of the EurAAP board of directors from 2016 to 2021, becoming a vice-chair during the last three years. He is also a member of the board of the European School of Antennas, organizing courses in Madrid, Paris, Shanghai, and Beijing and teaching courses at Aalto University.

Sierra has been involved in the organization of European Conferences on Antennas and Propagation, serving as the general chair of EuCAP2022 for the 2022 edition in Madrid.

In addition, he has served as the dean of the UPM Telecommunication Engineering School, the oldest school in Spain in this area, since May 2021. He has also led the UPM Radiation Group (antenna area) since 2019.

About the Contributors

Berk Altinel received his MS from Technische Universität Ilmenau (Germany) in 2018. He has worked in the RF and Microwave Research Group in Technische Universität Ilmenau since 2019. His current activities include the performance evaluation of V2X communication technologies, with a focus on the physical layer phenomenon (communication link) and emulating link conditions in virtual environment.

Yuri Álvarez López received his MS and his PhD from the University of Oviedo (Spain) in 2006 and 2009, respectively. He joined the signal theory and communications research group of the University of Oviedo in 2006, where he has been a professor since 2018. His research interests include antenna measurement techniques, phaseless methods for antenna characterization, and inverse scattering and imaging for nondestructive testing applications.

Guillermo Álvarez Narciandi received his PhD in 2020 from the University of Oviedo, Gijón, Spain, where he is currently a postdoctoral researcher. Since 2022, he has also served as a research fellow at the Centre for Wireless Innovation (CWI), Queen's University Belfast (United Kingdom). His research interests include radar systems and imaging techniques, antenna diagnosis and characterization systems, localization and attitude estimation systems, and RFID technology.

Muhammad Ehtisham Asghar received his BE in electronics from the National University of Science and Technology (NUST), Islamabad, Pakistan, in 2012, and his MS in communications and signal processing from Technische Universität Ilmenau, Germany in 2017. He joined the RF and Microwave Research Group TU Ilmenau as a research assistant and a doctoral student in 2017. His research contributions include the characterization and testing of automotive and locomotive antennas and radar test systems in a virtual electromagnetic environment.

Philipp Berlt received his MS in electrical engineering with a specialization in information and communication technology from the Technische Universität Ilmenau in 2015. He joined the RF and Microwave Research Group at Technische Universität Ilmenau in 2016 and is working as a research assistant and doctoral student in the field of virtual test environments for automotive radio systems. His research includes antenna measurement techniques for automotive antennas as well as measurement and test procedures for LTE-based vehicular communication.

Thierry Blin started his career as a planar phased-array antenna designer in 1999 (at THALES) and led several studies about some innovative antenna measurement techniques. He joined Microwave Vision Group (MVG) in 2006 where he managed some construction projects on large antenna measurement facilities, including a few automotive test ranges. Currently, he is the director of operations at MVG.

Christian Bornkessel (Dipl-Ing 1990, Dr-Ing 1993) worked as the head of the Test Center at IMST GmbH in Kamp-Lintfort, Germany. Since 2014, he has been with the Technische Universität Ilmenau (RF and Microwave Research Laboratory). His current activities involve radio-based car communication (V2X) as well as several EMC topics with a focus on human exposure to RF and LF electromagnetic fields. He is a member of the German Commission on Radiological Protection and the German ITG technical committee HF1 "antennas."

María García Fernández received her PhD in telecommunication engineering in 2019 from the University of Oviedo (Spain), where she currently works as a postdoctoral researcher. In 2022, she also joined the Centre for Wireless Innovation, Queen's University Belfast (United Kingdom). Since 2015 she has been promoting a new research line focused on innovative microwave systems on board drones for two major applications: subsurface radar imaging and antenna characterization at operational conditions.

Andrea Giacomini was born in Rome, Italy in 1977. He received his MS (cum laude) in electrical engineering from the University of Rome Tor Vergata, Rome, in 2002. Since finishing his master's thesis, he has worked at SATIMO and later Microwave Vision Italy, first as an antenna engineer and then as the antenna department manager. His professional interests and research activities relate to antenna design and measurement techniques. In particular, he is involved in the development of reference antennas, wideband radiators, measurement probes for PNF/SNF applications, CATR feeds, and passive microwave components, such as orthomode junctions and beamforming networks. He is also actively involved in several ESA programs focused on novel antenna concepts for space applications. He has authored and coauthored more than 50 journal and conference papers on antenna design and measurement topics and holds two patents on microwave orthomode junctions.

Sreehari Buddappagri Jayapal Gowdu received his BE from VTU, Karnataka, India in electronics and communication engineering in 2011 and his MS in communications and signal processing from Technische Universität Ilmenau in Germany in 2016. Since 2017, he has worked as a research assistant at the RF and Microwave Research Lab in TU Ilmenau and is pursuing his doctorate. His current research interests and contributions are in the field of over-the-air testing of automotive radar in virtual environments, radar cross-section measurements, signal post-processing, sensor and environment modeling, and EM wave propagation.

Nicolas Gross received his engineering degree from Ecole Nationale Supérieure de Techniques Avancées Bretagne in 2005, after which he joined SATIMO where he worked for several years as an antenna engineer. In 2007, he took charge of the development of multiprobe antenna measurement systems. In August 2009, he became director of the system and engineering department, leading the software, system, and product development in the field of antenna measurement. In addition to his industrial responsibilities, he has participated in several studies and has to his credit several publications and patents in these fields.

Matthias A. Hein (Dipl-Ing 1987, Dr rer nat 1992, habil 1998) started his academic career at the University of Wuppertal, Germany, with superconductor research for mobile and satellite communications. Following a British EPSRC senior research fellowship at the University of Birmingham, he joined the TU Ilmenau in 2002 as head of the RF and Microwave Research Laboratory. His research focus is on intelligent automotive wireless sensor and communication systems, over-the-air testing, and virtual test drives with a focus on antenna and microwave engineering. He is the director of the Thuringian Center of Innovation in Mobility at TU Ilmenau.

Jaime Laviada Martínez became an associate professor of the Universidad de Oviedo (Spain) in 2019. His research interests include numerical techniques applied to EM imaging, antenna measurements, method of moments, and antenna pattern synthesis. He received his MS in telecommunication engineering and his PhD from the Universidad de Oviedo, Spain, in 2005 and 2010, respectively.

Fernando Las-Heras joined the University of Oviedo in 2001 where he now serves as a full professor and heads the research group Signal Theory and Communications (TSC-UNIOVI). He was the vice-dean for telecommunication engineering and held the Telefónica chair on RF technologies. He received his MS and his PhD in telecommunication engineering from the Universidad Politécnica de Madrid, where he was associate professor until 2001. His research interests include inverse problems with applications to antenna measurement and EM imaging and metamaterials and technologies ranging from microwaves to terahertz bands.

Mathieu Mercier received his MS in electronics and microwave in 2000 in France. He then worked as a RF engineer in Sweden for two years, and then joined Satimo in 2003 in Paris as project manager for antenna measurement systems. He moved to the MVG Hong Kong office in 2005 and is now technical director for the Asia Pacific region.

Fernando Rodríguez Varela was born in Ourense, Spain, in 1994. He received his master's degree in telecommunication engineering and his PhD from the Technical University of Madrid (UPM), Madrid, Spain, in 2016 and 2021, respectively. He was a visiting scholar at the Technical University of Denmark, Lyngby, Denmark during 2019. From 2019 to 2021, he worked as an external consultant for Huber+Suhner AG, Herisau, Switzerland. Since 2021 he has served at Universidad a Distancia de Madrid, Madrid, Spain, as a professor. He is the author or coauthor of more than 20 journal and conference articles. His current research interests include antenna measurement techniques and millimeter-wave antenna design.

Kim Rutkowski received her BSE and her MS from the University of Central Florida, studying under Christos Christodoulou, with a focus on electromagnetic fields and optics. She worked at Ericsson and Panasonic as a senior RF antenna engineer, researching and designing antennas for mobile devices. She is currently employed by MVG as a senior OTA specialist, actively contributing to standardization groups such as 3GPP, CTIA, and 5GAA.

Francesco Saccardi received his MS in telecommunication engineering from the University of Siena, Italy. In 2010 he joined MVG where he works in the R&D department. His activities are mainly focused on the research and development of innovative antenna measurement techniques for several applications. He has authored and coauthored more than 80 journal and conference papers on antenna measurement and related post-processing techniques. He has taught in several seminars and workshops focused on general near-field antenna measurements, automotive measurements, and advanced post-processing techniques. Since 2020 he has been a senior member of AMTA.

Maria Alberica Saporetti received her MS in astronautic engineering cum laude from Sapienza Rome University (Italy) in 2008. From 2010 to 2012 she worked as project engineer at Deimos Engenharia, Lisbon (Portugal). From 2012 to 2021 she worked at Microwave Vision Italy, part of MVG. Since October 2021 she has worked at the Italian Space Agency as a technologist in the telecommunication and navigation unit. She has been the coordinator of the EurAAP Working Group on Antenna Measurements since 2018. In June 2021, she was elected director of the EurAAP board. She is also part of the EurAAP Working Group on Women on Antennas and Propagation and was the technical program committee cochair for the EuCAP 2022 symposium.

Alessandro Scannavini joined Microwave Vision Group (MVG) in 2007. He serves as a researcher in the field of antenna measurement techniques, mainly on the topic of over-the-air measurements. Since 2009, Scannavini is the MVG delegate in 3GPP and CTIA standardization bodies for wireless communications. He was with the Motorola Design Center in both Turin, Italy, and Chicago from 2000 to 2006, where he was involved in designing antenna and RF components for 2G and 3G smartphones. He holds a master's degree in electronic engineering (2000) from La Sapienza, Rome.

Lucia Scialacqua received her BS and her MS in telecommunication engineering from the University of Siena, Italy. She has worked at Microwave Vision Italy since 2007. Her research interests include antenna design and validation, antenna measurement post-processing, and software management. She has coauthored more than 90 publications, including journal and conference papers.

Jasmeet Singh received his BS in electronics and communication engineering from Guru Gobind Singh Indraprastha University, New Delhi, India, in 2012, and his MS in communications and signal processing from Technische Universität Ilmenau, Germany in 2016. He worked as a research assistant with the RF and Microwave Research Group at the TU Ilmenau for the following five years. His research interests include antenna and microwave engineering, with a strong focus on low-profile and compact automotive antenna design. Currently, he is employed at Bosch, Hildesheim as an antenna design engineer for automotive applications.

Frank Wollenschläger (Dipl-Ing 2008) was a research assistant at the RF and Microwave Research Laboratory at the Technische Universität Ilmenau, where his work was related to microwave antenna design as well as automotive antenna and system (V2X) characterization. Since 2019 he has been a member of the microwave instruments group at Airbus Defence and Space, where he works on satellite radar antenna design and verification. His main fields of interest are antenna theory and antenna measurement techniques and systems.

Index

A

Absolute gain, 9, 16
Absorber floor systems
  about, 98
  gain calibration, 99–101
  gain calibration accuracy, 99, 101–3
  measurement examples, 103–6
  scaled vehicle measurement over, 103
Absorbers
  about, 73–74
  absorption from, 74
  Chebyshev, 76
  ferrite tiles, 75–76
  frequency and, 76
  layout, 76, 77
  multiprobe systems, 124
  pyramidal, 73, 74
  wedge, 80
    See also Chambers
Accuracy
  achievable, 222
  efficiency substitution method and, 100
  gain calibration, 99, 101–3
  measurement, 55, 82, 89, 111–12, 122, 125
  multiprobe systems and, 120
  PEC-based systems and, 49
  phase, 19
  scaled measurement, 129
Acquired amplitude, 203, 214
Acquisition subsystem, 208–9
Active antenna system (AAS), 138
Advanced driver-assistance systems (ADAS), 208
Advanced modulated scattering technique (A-MST), 121
Airborne-based measurement techniques
  about, 195–96
  acquisition schemes, 200
  antenna characterization, 199–202
  application example, 202–8
  challenges, 199–200
  overview of, 197–99
  subsystems, 197–98
    See also Unmanned aerial vehicles (UAVs)
Analysis methods
  about, 81
  full-wave models, 81
  image theory-based technique, 83–85
  ray-tracer solvers, 81–83
  scaled model method, 85–90
    See also Chambers
Anechoic chambers, 71–73
Angular sampling, 23
Angular scanning techniques, 221
Antenna measurement(s)
  about, 7–8
  application-oriented post-processing of, 229–33
  calibration antenna, 12–14
  car body implications for, 225–29
  challenges for, 221–29
  gain and OTA, 8–12
  practical aspects of, 221–46
  range limitations, 221–25
  scenario illustration, 8
  system validation, 234–46
  testing parameters, 14–17
    See also specific types of measurements
Antenna pattern comparison, 91–93
Antenna performance indicators, 232–33
Aperture efficiency, 20
Application-post-processing
  automotive antenna performance indicators, 232–33

Application-post-processing (*continued*)
phase center determination, 229–31
See also Post-processing
Automated and connected driving (ACD), 234–35
Automotive antenna measurements. *See* Antenna measurement(s)
Automotive radar, 240–46
Automotive testing
about, 97–98
absorber and conductive floor systems, 98–106
common configurations, 97–131
FF ranges, 106
implementation and calibration of, 97–131
multiprobe systems, 98, 120–29
scaled measurements, 129–31
single-probe (SP) system, 97–98, 108–19

B

B-column covers, 227–28
Biconic antennas, 112

C

Calibrated OTA setup, 12
Calibration antenna, 12–14
Car body implications, 225–29
Chambers
absorber layout, 76
absorbers, 73–76
analysis methods, 81–90
anechoic, 71–73
with conductive floors, 80
design criteria, 73–76
full-wave method analysis, 82
image-based method schematic, 84
measurement ranges and, 71–73
minimum height and width, 78
minimum length, 78–79
performance of, 71
ray-tracer method analysis, 82–83
rectangular, 76–79
scaled model method measurements and, 85–90
tapered, 79
Chebyshev absorbers, 76
Combinational method, 138
Computational electromagnetic (CEM), 63, 162–63
Conductive floors
chambers with, 80
ground, 123
as PEC, 80
single-probe (SP) system, 109
SWE with, 59–60
Conductive floor systems
about, 98
example of, 98
gain calibration, 99–101
gain calibration accuracy, 101–3
measurement examples, 103–6
power radiation in backward hemisphere, 101
scaled vehicle measurement over, 103
Conductivity, 186
Conjugate gradient (CG), 51
CST Studio Suite, 181
Cylindrical wave expansion (CWE), 107–8

D

Differential phase method, 229
Digital audio broadcasting (DAB), 202, 206
Direct FF ranges, 106–8
Direct gain calibration, 99–100
Direct MIMO OTA testing, 146, 147–50
Direct OTA measurements, 138–39, 141, 145–46
Down-sampling in translated SNF measurements, 53–58
Drones. *See* Unmanned aerial vehicles (UAVs)
Dual-polarized probes
about, 110
angles, 115
broadband, 222
directivity radiation pattern, 116
ENL, 117–19
gain radiation pattern, 117–18
measurement configurations, 116
QH400 and QH1400, 114–19
See also Probes
Dynamic range, 26

E

Echo suppression, 38
Edge diffraction method, 229
Effective isotropic sensitivity (EIS), 2, 14–16, 134, 137–38
Efficiency substitution method, 100–101
Electrical condition, 18
Equivalent box, 165
Equivalent current (EQC) technique
about, 60, 156
antenna measurement equipment and, 161–62
benefits of, 62–63
calculation from, 160
computational time, 166, 167
demonstration of, 156–57
diagnostics and filtering, 156
equivalent box and, 165
example NF calculation from, 160
example of, 157–58
extrapolation of truncated areas, 158–59
features, 61–63
filtering principle and, 157–58
formulation, 60–61
in INSIGHT, 174
link between measurements and simulations, 160–63
measurement examples and, 63
NF source modeling exchange data and, 162–63
performance of, 167
post-processing by, 156–63
processing NF source modeling, 162
reconstruction, 170
in representing radiator, 160–61
workflow calculations, 62
working procedure and results, 157
Equivalent electric currents, reconstructed, 205
Equivalent impedance, 186
Equivalent isotropic radiated power (EIRP)
about, 2, 14–15
defined, 16
measurement techniques, 137, 145
sampling grids and, 142, 143
Equivalent material wave number, 186–87
Equivalent noise level (ENL)
about, 88–90
comparison of measured frequencies and, 114
dual-polarized probes, 117–19
probe measurements and, 112–13
truncation errors and, 163–64
Equivalent sampling area (ESS), 38, 126–27
Equivalent sampling cylinder (ESC), 38
Euler's angle, 211
Extrapolation of truncated areas, 158–59

F

Far field. *See* FF
Fast Fourier Transform (FFT), 35
Ferrite tiles, 75–76
FF
about, 1–2
calculation from expansion currents, 159
desired measurement condition, 2
direct, 18
direct ranges, 106–8
distance evaluation, 20
distances for frequencies, 19
evaluation with SWC, 34
indirect, 18
measurements, 23, 106
outdoor system, 106
pattern margin, 207
patterns from magnetic currents, 215
range dimensions, 77
See also NF; NF/FF transformations
FFT/matrix inversion approach, 59
Field-operational tests (FoTs), 239
Field probe measurements, 90–91
Filtering
principle, 157–58
spatial techniques, 168–70
SWE-based, 168
Finite-difference time domain (FDTD), 81
First-order probe compensation (FOPC) technique
about, 36
application example, 52–53
patterns obtained with, 117
reducing large differences with, 113
use of, 51, 58, 112

First-order probes, 36
Free-space impedance, 186
Free-space path loss attenuation, 10
Free-space retrieval (PEC-based systems)
about, 168
example with scaled measurements, 170–74
schematization of, 169
spatial filtering techniques, 168–70
workflow, 168–69
Free-space truncation, 44–47, 187
Friis transmission formula, 9
Full probe compensation, 51–53
Full-wave models, 81

G

Gain
absolute, 9, 16
determination, 16
pattern comparison, 113
pattern cuts, 13, 14
performance derived from, 12
realized, 9, 163, 223
reference, 46
of unknown DUT, 10–11
Gain calibration
accuracy, 101–3
direct method, 99–100
efficiency substitution method, 100–101
equation, 100–101
error comparison, 103
types of, 99
uncertainty, 100
Gain-substitution technique, 9, 10
Gantry arm system, 110
Generalized SWE method, 50–51
Georeferring uncertainty, 215–16
Global positioning system (GPS), 195
Ground emulation
about, 183–85
of measured vehicle antennas, 183–90
parameters, 185

H

Handheld systems
about, 195–96
acquisition subsystem, 208–9
application example, 212–13
description, 208–10
illustrated, 212
offset correction, 210–11
overview of, 208
particular characteristics of, 210–11
peculiarities, 210
processing subsystem, 210
scheme of subsystems, 209
splitting plates into 3-D cells and, 210
summary, 216
tracking subsystem, 209–10
Hankel function, 32
Hardware-in-the-loop (HiL), 240
Hemispherical acquisition domains, 199, 203, 204
Hertzian dipole antenna, 35, 36, 58
High-frequency measurements, multiprobe systems, 125–29
High-impedance surface (HIS), 228
Hood antenna, 88–89

I

Image theory (IT), 185
Image theory (IT)-based technique
about, 83–85
arbitrary grounds and, 188
illustrated, 187
multiprobe systems and, 188
mutual coupling and, 189
validation of, 190
Infinite plane boundary condition (IPBC), 180
INSIGHT software, 169, 171, 174
Integral OTA power, 17
Intelligent transportation system (ITS), 126, 129
Iterative modal filtering (IMF) technique
about, 45, 165
computational time, 166, 167
data extrapolation using, 232
error mitigation and, 49
pattern obtained with, 47
performance of, 166, 167
technique diagram, 46
Itoh-algorithm, 231
ITS-G5

links, emulation in virtual environment, 238–40
performance of, 237
V2X and, 236

J

Joint communication and radar sensing (JCRS), 235

L

l1-minimization, 51
Least-squares minimization, 231
LEHA-UPM, 91–94
Limited ground plane evaluating method, 140–41
Local measurement approach
about, 43, 54
application of, 126
gain and TRP errors from, 55
illustrated, 54
implementation, 54–55
Low-frequency measurements, multiprobe systems, 124–25
LTE
links, emulation in virtual environment, 237–38
operational mode support, 237
standard, 236
V2X and, 236
LTE-MIMO antennas, 226

M

Magnetic currents, 215
Measurement errors, 21, 100, 103, 125, 139, 141–42
Method of moments (MoM), 61, 81
MIMO OTA testing
about, 146
challenges, 149
direct, 146, 147–50
example of, 148
implementation, 148–49
introduction, 146
of LTE-enabled devices, 148
system performance and, 147
test environment, 146
two-stage, 150–51
See also Over-the-air (OTA) testing
Modal filtering, 38
Monocone antennas, 13, 46–47, 119, 125
Multiplexers, 26
Multiprobe anechoic chamber (MPAC), 146
Multiprobe systems
about, 98, 120–21
absorbers, 124
design challenges, 120, 121–22
example illustrations of, 120–21
high-frequency measurements, 125–29
hybrid electronic-mechanical drives, 222
implementations of, 123–24
IT technique and, 188
low-frequency measurements, 124–25
primary objective of, 120
probe array elements, 121
probes, 121–22
sampling in, 122–23
Mutual coupling, 189
MV-Echo, 168, 171, 172–74
MVG SH600 dual-ridge horn, 39–41

N

Near field. *See* NF
Near-horizon PRP (NHPRP), 104–6
Nested skeletonization scheme (NSS), 61
NF
about, 18
acquired, amplitude of, 203
hemispherical acquisition domain, 204
measurements, 22, 23
partial acquisition, 159
processing, 30
ranges, 78
samples computation, 45–46
simulated, on hemispherical acquisition domain, 204
See also FF; Spherical NF (SNF) measurement
NF/FF transformations
about, 3, 29–30
algorithms, 18, 22, 25
based on spherical wave expansion, 30–60
equivalent current (EQC) technique, 60–63

NF/FF transformations (*continued*)
new methods for, 63–65
optimal sampling interpolation and, 64–65
parallax and, 139–40
performing in different scenarios, 49–50
with phase recovery, 136–37
schematic depiction, 29, 30
sparse recovery of SWC and, 64
spherical, 26
theory, 24
virtual drive testing (VDT), 185
Numerical modeling, 177
Nyquist criteria, 122
Nyquist-Shannon sampling theorem, 37

O

OEW probes, 111–14
Offset correction, 210–11
On-board units (OBUs), 239
On-the-fly acquisition scheme, 110
Open-ended waveguides (OEWs), 111
Optimal sampling interpolation, 64–65
Organization, this book, 2–4
OTA measurements
about, 4, 133
combinational method, 138
desense and, 138
direct, 138–39, 141, 145–46
examples of, 142–46
measured EIRP, 145
methods, 136–41
MIMO testing and, 146–51
NF to FF transform using phase recovery, 136–37
parallax and, 139–40
performance metrics, 144
radiated power, 143, 144
sampling and, 141–42
sensitivity, 143, 145
separate, with conducted RF, 137–38
system calibration and, 12, 135–36
testing parameters and, 134–35
test setup and, 133–34
TRP/TIS testing of devices on limited ground plane, 140–41
two-step, 138
See also Over-the-air (OTA) testing
OTA system calibration, 12, 135–36
OTA/ViL approach
about, 235–36
automotive radar, 240–46
concept, 234–36
LTE and V2X emulation, 236
measurement process flow chart, 244
operational test system, 241–42
performance comparison, 246
performance evaluation, 245
simulator, 242–43
test scenario, 245
test system photo, 243
V-model representation of, 236
Over-the-air (OTA) testing
about, 2, 8
complexity and, 134
MIMO, 146–51
parameters, 134–35
sampling and, 141–42
setup, 133–34, 238
of wireless devices, 14
See also OTA measurements
Over-the-air vehicle-in-the loop. *See* OTA/ViL approach

P

Parallax, 139–40
Partial radiated power (PRP), 104
Patch antenna, 41–43, 227
PEC-based measurements
schematic illustration, 48
spherical wave spectrum of, 169
TSWE technique applied to, 59
PEC-based systems
accuracy and, 49
emulation, 171
free-space retrieval in, 168–74
PEC mode, 123
Perfect electric conductor (PEC)
about, 44
boundary condition, 44, 47–48, 172
conductive floors as, 80

interface, 49, 50
removal, spatial filtering techniques for, 168–70
truncation, 47–49
Phase center determination, 229–31
Phase retrieval techniques, 213–15
Phase unwrapping, 232
Plane wave expansion (PWE), 62
Plane wave generator (PWG), 85
Positioners, 27
Post-processing
about, 155–56
advanced techniques, 155–74
application-oriented, 229–33
truncation errors and, 163–67
virtual drive testing (VDT), 183
Probe compensation (PC), 111
Probes
dual-polarized, 114–19, 222
higher frequency, 222
MP system, 121–22
OEW, 111–14
wideband, 110–14
Processing subsystem, 210
Proof-of-principle measurements, 225
Prototype measurements, 156
pSRM algorithm, 201–2
Pyramidal absorbers, 73, 74

Q

QH400 and QH1400 probes, 114–19
Quiet zone (QZ)
about, 71
evaluation with measurements, 90–93
field probe measurements over, 90–91
illumination, 83
probing, 85
scattering on, 79
at UMP CATR, 92

R

Radar cross section (RCS), 244–46
Radar target simulator (RTS), 242–43
Radar-under-test (RUT), 242–43
Radiation characteristics, 229
Radiation patterns, 228
Radiation performance, 223
Radio communication tester (RCT), 14
Radiofrequency system, 25–26
Rao-Wilton-Glisson (RWG), 61
Ray-tracer solvers, 81–83
Realized gain, 9, 163, 223
Reconstructed equivalent electric currents, 205
Reconstructed magnetic currents, 215
Rectangular chambers, 76–79
Reference signal received power (RSRP), 237–39
Reflection coefficients, 185–86, 187
Relative permeability, 186
Relative permittivity, 186

S

Sampling
angular, 23
in MP systems, 122–23
NF/FF transformations, 63
OTA measurements, 141–42
SNF measurement, 37–38
Sampling grids, 142, 143
Satsim, 82–83
Scaled measurements
accuracy, 129
conductivity of materials and, 130
examples of, 130
free-space retrieval and, 170–74
radiation pattern and, 129–30
relationship among quantities in, 131
Scaled model method
advantages of, 85
analysis objective, 85–86
automotive setups, 87
ENL and, 88–90
example of, 85
measurements with, 85–90
scaling factor and, 86
Second-derivative method, 229
Sensitivity performance, 17
SH600 probe, 112
SH2000 antenna, 101–2, 119
Shark fin antenna, 162, 182

Single-probe (SP) system
about, 4, 97–98, 108–9
absorber-covered floors and, 109
conductive floors, 109
examples of, 109
probe bandwidth and, 110
wideband and dual-polarized probes and, 114–19
wideband probes and, 110–14
Sources reconstruction method (SRM), 201–2
Sparse recovery of SWC, 64
Spatial channel models (SCMEs), 147
Spatial filtering, 158
Spherical measurement systems, 222
Spherical NF (SNF) measurement
about, 30
alternative sampling schemes for, 63
of 5-m vehicle, 53–54
formulation, 31
minimum sampling rate, 38
nonprobe-compensated, 122
sampling criteria, 37–38
system illustration, 31
translated, down-sampling in, 53–58
truncated, 43–49
of x-band standard gain horn, 52
Spherical-phase method (SPM), 230
Spherical wave coefficients (SWCs)
about, 33–34
computation of, 34, 50
extraction of, 35
FF evaluation from, 34
minimum number of, 37
nonzero, 35
of patch radiated field, 42
probe-corrected, 58
recomputation of, 42
representation and examples of, 39–43
sparse recovery of, 64
translation of, 58
Spherical wave expansion (SWE)
advanced techniques, 49–60
angular behavior, 33
with conductive floor at arbitrary height, 59–60
down-sampling in translated SNF measurements, 53–58
full probe compensation, 51–53
generalized method, 50–51
image theory-based technique and, 83–84
as key tool, 34
NF/FF transformation based on, 30–43
theory, 43
truncated SNF measurements and, 43–49
Spherical wave functions, 33
Spherical wave spectrum, 36, 48, 169
Spherical wave summation (SWS), 34
System calibration, 10–12, 135–36

T

Tapered chambers, 79
Telematic control unit (TCU), 8
Testing parameters
EIRP, 14–16, 137, 142–43, 145
EIS, 2, 14–16, 134, 137–38
TIS, 15–17, 140–41, 145
TRP, 2, 15–17, 54–55, 140–42, 232
TRS, 15–17, 137–44, 145
3-D radiation pattern amplitude, 181
Three-antenna method, 229
Three-dimensional scanning techniques, 223
Time-of-arrival method, 229
Total isotropic sensitivity (TIS), 2, 15–17, 145
Total radiated power (TRP)
about, 2, 15–17
correct determination of, 232
defined, 15
errors, 54, 55
as performance metric, 232
of scaled vehicle, 142
Total radiated sensitivity (TRS), 15–17, 137–44, 145
Tracking subsystem, 209–10
Translated SWE (TSWE)
about, 55
application to PEC-based measurements, 59
capabilities of, 56–57
concept illustration, 55
downside of, 56
ESSs and, 126–27
local coordinate system and, 57
in NFFC transformation, 123
SNF offset measurements processed with, 56

undersampled measurement based on, 57
Translation-propagation approach, 58
Translation theorem of antennas, 230
Transmission angle, 186
Transport-of-intensity equation, 231
Transverse electric (TE), 33
Transverse magnetic (TM), 33
Truncated areas, extrapolation of, 158–59
Truncated SNF measurements
about, 43–44
emulated, 163
error, 45
free-space truncation, 44–47
PEC truncation, 47–49
Truncation errors
about, 163
case study, 163–65
ENL and, 163–64
example of, 45
investigation and mitigation of, 163–67
iterative modal filtering (IMF) and, 165–67
mitigation of, 165–67
Two-dimensional unwrapping, 231
Two-point method, 229
Two-stage MIMO OTA testing, 150–51
Two-step measurement, 138

U

UAV-based measurements
amplitude and phase information, 201
application example, 202–8
architectures, 197
challenges, 199–200, 213–16
complexity, 213–15
hemispherical acquisition using, 203
reconstructed currents, 216
Unmanned aerial vehicles (UAVs)
acquisition schemes, 200
in automotive measurements, 198
AUT placement and, 202
hemispherical acquisition using, 199
photos, 198, 199
in radiofrequency applications, 197
subsystems, 197–99
technology development, 196
See also Airborne-based measurement techniques
Upper hemisphere efficiency (UHE), 11, 100–102, 113
Upper-hemisphere radiated power (UHRP), 104–5, 113–14
Urbanmacro (Uma), 147
Urbanmicro (Umi), 147

V

V2X communications, 190–92, 236–40
Validation, 234–46
Vehicle antenna test methodology (VATM), 133
Vehicle-to-vehicle (V2V) communications
antennas, 126, 183, 184
coupling and coordinate systems, 191
emulated, 191–92
evaluation illustration, 192
V2X coupling evaluation, 190–92
Verification and validation (V&V), 240
Virtual drive testing (VDT)
about, 177
block diagram, 240
ground emulation, 183–90
levels of, 179
link between measurement and simulation, 180–81
NF/FF transformations, 185
post-processing, 183
scopes, 178–79
simulation of complex scenarios, 181–83
two-stage method and, 179
for V2V, 184
V2V and V2X coupling evaluation and, 190–92
workflow, 182–83
Virtual Road Simulation and Test Area (VISTA) MP system, 125–29, 223–25, 229, 232–33

W

Wedge absorbers, 80
Wideband probes
about, 110
drawbacks, 110–11
gantry arm and measurement layout with, 111

Wideband probes (*continued*)
use of, 111–14
See also Probes
Wireless reliance, 234

Z

Zero-padding (ZP), 43, 44, 46–47

Recent Titles in the Artech House Antennas and Propagation Library

Christos Christodoulou, Series Editor
Vince Rodriguez, Series Editor

Adaptive Array Measurements in Communications, M. A. Halim

Antenna-on-Chip: Design, Challenges, and Opportunities, Hammad M. Cheema, Fatima Khalid, and Atif Shamim

Antenna Design for Cognitive Radio, Youssef Tawk, Joseph Costantine, and Christos Christodoulou

Antenna Design for CubeSats, Reyhan Baktur

Antenna Design with Fiber Optics, A. Kumar

Antenna Engineering Using Physical Optics: Practical CAD Techniques and Software, Leo Diaz and Thomas Milligan

Antenna Measurement Techniques, Gary E. Evans

Antennas and Propagation for Body-Centric Wireless Communications, Second Edition, Peter S. Hall and Yang Hao, editors

Antennas and Site Engineering for Mobile Radio Networks, Bruno Delorme

Antennas for Small Mobile Terminals, Kyohei Fujimoto and Koichi Ito, editors

Analysis, Design and Measurement of Small and Low-profile Antennas, Kazuhiro Hirasawa

Analysis of Radome-Enclosed Antennas, Second Edition, Dennis J. Kozakoff

Applications of Geographic Information Systems for Wireless Network Planning, Francisco Saez de Adana, Josefa Gómez Pérez, Abdelhamid Tayebi Tayebi, and Juan Casado Ballesteros

The Art and Science of Ultrawideband Antennas, Second Edition, Hans G. Schantz

AWAS for Windows Version 2.0: Analysis of Wire Antennas and Scatterers, Antonije R. Djordjević, et al.

Broadband Microstrip Antennas, Girsh Kumar and K. P. Ray

Broadband Patch Antennas, Jean-François Zürcher and Fred E. Gardiol

Circularly Polarized Dielectric Resonator Antennas, Raghvendra Kumar Chaudhary, Rajkishor Kumar, and Rakesh Chowdhury

Design and Applications of Active Integrated Antennas, Mohammad S. Sharawi and Oualid Hammi

Design and Measurement of Antennas and Propagation in Mobile Cellular Systems, Hiroyuki Arai

Designer Notes for Microwave Antennas, Richard C. Johnson

Dielectric Resonator Antenna Handbook, Aldo Petosa

Dielectric Resonator Antennas, Biswajeet Mukherjee and Monika Chauhan

Electromagnetics and Antenna Technology, Alan J. Fenn

Electromagnetics, Microwave Circuit and Antenna Design for Communications Engineering, Peter Russer

Engineering Applications of the Modulated Scatterer Technique, Jean-Charles Bolomey and Fred E. Gardiol

Fixed & Mobile Terminal Antennas, Akhileswar Kumar

Foundations of Antenna Engineering: A Unified Approach for Line-of -Sight and Multipath, Per-Simon Kildal

Four-Arm Spiral Antennas, Joseph A. Mosko

Frequency-Agile Antennas for Wireless Communications, Aldo Petosa

Fresnel Zones in Wireless Links, Zone Plate Lenses and Antennas, Hristo D. Hristov

Handbook of Antennas for EMC, Second Edition, Thereza Macnamara with John McAuley

Handbook of Reflector Antennas and Feed Systems, Volume I: Theory and Design of Reflectors, Satish Sharma, Sudhakar Rao, and Lotfollah Shafai, editors

Handbook of Reflector Antennas and Feed Systems, Volume II: Feed Systems, Lotfollah Shafai, Satish Sharma, and Sudhakar Rao, editors

Handbook of Reflector Antennas and Feed Systems, Volume III: Applications of Reflectors, Sudhakar Rao, Lotfollah Shafai, and Satish Sharma, editors

Interference Suppression Techniques for Microwave Antennas and Transmitters, Ernest Freeman

Introduction to Antenna Analysis Using EM Simulators, Hiroaki Kogure, Yoshie Kogure, and James C. Rautio

Introduction to Antennas and RF Propagation Analysis, Dean James Friesen

LONRS: Low-Noise Receiving Systems Performance and Analysis Toolkit, Charles T. Stelzried, Macgregor S. Reid, and Arthur J. Freiley

Measurement of Mobile Antenna Systems, Second Edition, Hiroyuki Arai

Microstrip Lines and Slotlines, Third Edition, Ramesh Garg, Inder J. Bahl and Maurizio Bozzi

Microstrip Antenna Design, K.C Gupta and Abdelaziz Benalla

Microwave and Millimeter-Wave Remote Sensing for Security Applications, Jeffrey A. Nanzer

Microwave Cavity Antennas, Akhileswar Kumar and Hristo D. Hristov

Millimeter Wave Microstrip and Printed Circuit Antennas, Prakash Bhartia

Mobile Antenna Systems Handbook, Third Edition, Kyohei Fujimoto, editor

Modern Automotive Antenna Measurements, Lars J. Foged and Manuel Sierra Castañer, editors

Moment Methods in Antennas and Scattering, Robert C. Hansen

Multiband Integrated Antennas for 4G Terminals, David A. Sánchez-Hernández, editor

Near-Field Antenna Measurements, Dan Slater

Noise Temperature Theory and Applications for Deep Space Communications Antenna Systems, Tom Y. Otoshi

Phased Array Antenna Handbook, Third Edition, Robert J. Mailloux

Phased Array Antennas, Arthur A. Oliner

Phased Array Antennas with Optimized Element Patterns, Sergei P. Skobelev

Plasma Antennas, Second Edition, Theodore Anderson

Polarization in Electromagnetic Systems, Second Edition, Warren Stutzman

Practical Antenna Design for Wireless Products. Henry Lau

Practical Microstrip and Printed Antenna Design, Anil Pandey

Practical Phased Array Antenna Systems, Eli Brookner

Practical Simulation of Radar Antennas and Radomes, Herbert L. Hirsch

Printed MIMO Antenna Engineering, Mohammad S. Sharawi

Radiowave Propagation and Antennas for Personal Communications, Third Edition, Kazimierz Siwiak

Reconfigurable Antenna Design and Analysis, Mohammod Ali

Reflectarray Antennas: Analysis, Design, Fabrication and Measurement, Jafar Shaker, Mohammad Reza Chaharmir, and Jonathan Ethier

Reflector Lens Antennas: Analysis and Design Using Personal Computers, Carlyle J. Sletten, editor

RF Coaxial Slot Radiators: Modeling, Measurements, and Applications, Kok Yeow You

Shipboard Antennas, Second Edition, Preston Law

Significant Phased Array Papers, Robert C. Hansen

Smart Antenna Engineering, Ahmed El Zooghby

Solid Dielectric Horn Antennas, Carlos Salema, Carlos Fernandes, and Rama Kant Jha

Sparse Phased Array Antennas: Theory and Applications, Ashutosh Kedar

Ultrawideband Antennas for Microwave Imaging Systems, Tayeb A. Denidni and Gijo Augustin

Ultrawideband Short-Pulse Radio Systems, V. I. Koshelev, Yu. I. Buyanov, and V. P. Belichenko

Waveguide Components for Antenna Feed Systems: Theory and CAD, Jaroslaw Uher, Jens Bornemann, and Uwe Rosenberg

Wearable Antennas and Electronics, Asimina Kiourti and John L. Volakis, editors

For further information on these and other Artech House titles, including previously considered out-of-print books now available through our In-Print-Forever® (IPF®) program, contact:

Artech House
685 Canton Street
Norwood, MA 02062
Phone: 781-769-9750
Fax: 781-769-6334
e-mail: artech@artechhouse.com

Artech House
16 Sussex Street
London SW1V HRW UK
Phone: +44 (0)20 7596-8750
Fax: +44 (0)20 7630 0166
e-mail: artech-uk@artechhouse.com

Find us on the World Wide Web at: www.artechhouse.com